'This book is a great contribution to the cyber canon and offers a comprehensive reference for both students and policymakers. The authors cover down on the many dynamic facets of cyber conflict, providing a strong foundation for anyone interested in this critical aspect of international relations.'

—**General Michael V. Hayden,** *Former Director of the Central Intelligence Agency and National Security Agency*

'This second edition is an absolute must-read for anyone looking to learn more about the critically important threat of cyber warfare. This updated and enhanced text helps prepare readers for how to think about the historical, technical, and strategic context of conflict in this critically important domain.'

—**Samantha F. Ravich,** *Former Commissioner, U.S. Cyberspace Solarium Commission and Chair, Center on Cyber and Technology Innovation*

'An authoritative tutorial on the arcane complexities of cyber warfare. This edition updates a previous version and makes the book more contemporary. It is a must-read for those who are serious about mastering this unique medium of combat, in all its dimensions.'

—**James R. Clapper,** *Former Director of National Intelligence*

Praise for the first edition:

'This book is an important summary and tour-de-force of historical and contemporary Cyber issues. It is a go-to comprehensive reference for both students and policymakers interested in perhaps the most transcendent strategic topic of our time. The authors of this book have captured the core essence and components of this story.'

—**Admiral William O. Studeman USN (Ret.),** *Former Director NSA and Deputy Director CIA*

'This remarkable effort represents the true first textbook of cyber conflict. The volume avoids the hype and bluster of the typical comprehensive accounts to accurately survey the issues, facts, and problems inherent in the cyber domain. A sure must-adopt for anyone teaching a cyber-security course.'

—**Brandon Valeriano,** *Donald Bren Chair of Armed Politics, Marine Corps University, USA*

Understanding Cyber Warfare

This textbook offers an accessible introduction to the historical, technical and strategic context of global cyber conflict. The second edition has been revised and updated throughout, with three new chapters.

Cyber warfare involves issues of doctrine, strategy, policy, international relations (IR) and operational practice associated with computer network attack, computer network exploitation and computer network defense. However, it is conducted within complex sociopolitical settings alongside related forms of digital contestation. This book provides students with a comprehensive perspective on the technical, strategic and policy issues associated with cyber conflict, as well as an introduction to key state and non-state actors.

Specifically, the book provides a comprehensive overview of the following several key issue areas:

- The historical context of the emergence and evolution of cyber warfare, including the basic characteristics and methods of computer network attack, exploitation and defense;
- An interdisciplinary set of theoretical perspectives on conflict in the digital age from the point of view of the fields of IR, security studies, psychology and science, technology and society (STS) studies;
- Current national perspectives, policies, doctrines and strategies relevant to cyber warfare;
- An examination of key challenges in international law, norm development and deterrence; and
- The role of emerging information technologies like artificial intelligence and quantum computing in shaping the dynamics of global cyber conflict.

This textbook will be essential reading for students of cybersecurity/cyber conflict and information warfare, and highly recommended for students of intelligence studies, security and strategic studies, defense policy and IR in general.

Christopher Whyte is an assistant professor in the program on Homeland Security and Emergency Preparedness at the L. Douglas Wilder School of Government and Public Affairs, Virginia Commonwealth University. He is co-editor of *Information Warfare in the Age of Cyber Conflict* (2020) and co-author of *Information at War* (2022).

Brian M. Mazanec is an adjunct professor in the Defense and Strategic Studies program at Missouri State University and a senior executive with the U.S. government. He is author of *The Evolution of Cyber War* (2015), co-author of *Deterring Cyber Warfare* (2014) and co-editor of *Information Warfare in the Age of Cyber Conflict* (2020).

Understanding Cyber Warfare

Politics, Policy and Strategy

Second Edition

Christopher Whyte and Brian M. Mazanec

Routledge
Taylor & Francis Group
LONDON AND NEW YORK

Cover image: © Getty Images – Diverstudio

Second edition published 2023
by Routledge
4 Park Square, Milton Park, Abingdon, Oxon OX14 4RN

and by Routledge
605 Third Avenue, New York, NY 10158

Routledge is an imprint of the Taylor & Francis Group, an informa business

First edition published by Routledge 2019

British Library Cataloguing-in-Publication Data
A catalogue record for this book is available from the British Library

Library of Congress Cataloging-in-Publication Data
Names: Whyte, Christopher, 1988- author. | Mazanec, Brian M., author.
Title: Understanding cyber warfare : politics, policy and strategy / Christopher Whyte and Brian M. Mazanec.
Description: Second edition. | Abingdon, Oxon ; New York : Routledge, 2023. | Includes bibliographical references and index.
Identifiers: LCCN 2022042735 (print) | LCCN 2022042736 (ebook) | ISBN 9781032159393 (hardback) | ISBN 9781032159317 (paperback) | ISBN 9781003246398 (ebook)
Subjects: LCSH: Cyberspace operations (Military science)
Classification: LCC U163 .W49 2023 (print) | LCC U163 (ebook) | DDC 355.4—dc23/eng/20220907
LC record available at https://lccn.loc.gov/2022042735
LC ebook record available at https://lccn.loc.gov/2022042736

ISBN: 978-1-032-15939-3 (hbk)
ISBN: 978-1-032-15931-7 (pbk)
ISBN: 978-1-003-24639-8 (ebk)

DOI: 10.4324/9781003246398

Typeset in Bembo
by Apex CoVantage, LLC

Contents

Figures

Tables

Text boxes

Abbreviations

AI	Artificial intelligence
APT	Advanced Persistent Threat
ARPA	Advanced Research Projects Agency
AS	Autonomous system
C2	Command and control
CBM	Confidence-building measure
CCDCOE	Cooperative Cyber Defence Centre of Excellence
CDC	Cyber Defense Command
CDMA	Cyber Defence Management Authority
CERT	Computer Emergency Response Team
CI	Critical Infrastructure
CNA.	Computer network attack
CNCI	Comprehensive National Cybersecurity Initiative
CND	Computer network defense
CNE	Computer network exploitation
CSIRT	Computer Security Incident Response Team
CSP	Cloud service provider
CYBERCOM	United States Cyber Command
DCIRT	Defence Computer Incident Response Team
DCSA	Defence Communications Services Agency
DDOS	Distributed denial of service
DHS	Department of Homeland Security
DISA	Defense Information Systems Agency
DNS	Domain Name System
DoD	Department of Defense
DPRK	Democratic People's Republic of Korea (North Korea)
ER97	Eligible Receiver 1997
FBI	Federal Bureau of Investigation
FPO	Federal Protective Service
FSB	Federal Security Service
GCHQ	Government Communications Headquarters
GEOINT	Geospatial intelligence
GGE	Group of government experts
GRU	Main Intelligence Directorate
GSD	Korean People's Army General Staff Department
HTTP	HyperText Transfer Protocol
HUMINT	Human intelligence
IAB	Internet Architecture Board
ICANN	Internet Corporation for Assigned Names and Numbers

ICTs	Information and communications technologies
IDS	Intrusion Detection System
IESG	Internet Engineering Steering Group
IETF	Internet Engineering Task Force
IMINT	Imagery intelligence
IoT	Internet of Things
IPv4	Internet Protocol (version 4)
IPv6	Internet Protocol (version 6)
IR	International relations
ISOC	Internet Society
ISP	Internet Service Provider
ISR	Intelligence, surveillance and reconnaissance
IT	Information Technology
ITU	International Telecommunication Union
JFCC-NW	Joint Functional Component Command - Network Warfare
JTF-CND	Joint Task Force Computer Network Defense
JTF-CNO	Joint Task Force Computer Network Operations
JTF-GNO	Joint Task Force Global Network Operations
LOAC	Law of Armed Conflict
MAC	Message authentication code
MASINT	Measurement and signature intelligence
MOD	Ministry of Defense
MSS	Ministry of State Security
NASA	National Aeronautics and Space Administration
NATO	North Atlantic Treaty Organization
NIST	National Institute of Standards and Technology
NRC	National Research Council
NSA	National Security Agency
OCO	Offensive Cyber Operations
OSCE	Organization for Security and Cooperation in Europe
OSD	Office of the Secretary of Defense
OSINT	Open source intelligence
PDD-63	Presidential Decision Directive 63
PECU	Police Electronic Crime Unit
PLA	People's Liberation Army
PRC	People's Republic of China
RA	Response action
RAT	Remote Access Trojan
RGB	Reconnaissance General Bureau
SIGINT	Signals intelligence
TCP/IP	Transport Control Protocol/Internet Protocol
TOR	The Onion Browser
UNIDIR	United Nations Disarmament and International Security Committee
USSTRATCOM	United States Strategic Command
WSIS	World Summit on the Information Society
WWW	World Wide Web

Acknowledgments

Christopher Whyte

As with the first edition of *Understanding Cyber Warfare*, there are a great many people who contributed to the completion of the second version of this project. I am thankful for such excellent and thoughtful colleagues at the Wilder School at Virginia Commonwealth University, particularly my fellow faculty within the Homeland Security and Emergency Preparedness program. Their diverse, inter-disciplinary perspectives on the national security enterprise are a constant source of new thought for me, something particularly valuable given the crosscutting nature of life in the digital age. And I would like to note a special thanks to Miguel Alberto Gomez, a friend and coauthor with whom regular conversations about our professional and other parts of life regularly help me think more clearly.

Most importantly, I would like to thank my wonderful wife, Susan, and my beautiful daughter, Rose. They are my constant inspiration and respite. For me, the world is wonderful because of them and I hope my writings in some way contribute, for their sake, to a world that is slightly better than otherwise might be the case.

I would also like to thank my co-author, Brian M. Mazanec, and our editor at Routledge, Andrew Humphrys. Collaborating with Brian has never stopped being an engaging experience, and it is a genuine pleasure to have been able to work on this second edition. And Andrew's guidance and patience with our work progress here have been nothing short of exceptional, for which both Brian and I are extraordinarily grateful.

Finally, I would like to thank Willie and Rupert, the two best Man's Best Friend(s) a person could ask for. You two are my constant companions as I go about my research, writing and more. My job would go a whole lot less smoothly if I weren't able to turn away to take a mental break with both of you.

Brian M. Mazanec

First and most importantly, I would like to thank my four incredible children: Charlotte, Reagan, Peter and Ben. As with the first edition, without your boundless energy, this updated edition would have been finished sooner but without the joy and balance you bring into my life. This book is for you and in some small way hopefully contributes to a safer and more secure world for you to thrive in. I would also like to thank my wonderful wife, Abby, for her support of this undertaking and all of my academic endeavors.

I remain grateful for the wisdom, diligence and persistence of my co-author, Christopher Whyte. It has truly been a pleasure to continue to work with him on this project, for which he did so much of the heavy lifting.

Finally, I would also like to thank our editor at Routledge, Andrew Humphrys, for his incredible patience and support of this project and help in guiding it to fruition.

Any errors herein are solely those of the authors. Additionally, the views expressed herein are those of the authors and are not representative of any organization.

Preface to the second edition

Almost a decade ago, the authors of this book were introduced to one another by a mutual mentor. After just a few phone calls and a couple of meetings for coffee in Fairfax, Virginia, we set about the idea of building a resource for students of cyber conflict studies that, so far as we could see, simply did not yet exist. Cyber conflict is an expanding and crosscutting area of concern in world affairs, as relevant to those working in advanced technology or the military as it is to diplomats, journalists and consumers of everything from electricity to soap. At the time we set about the task of imagining what a book that effectively framed cyber conflict might look like, there was precious little in the way of cohesive discussion of the diverse issues we cover in this book. Education on cybersecurity in international affairs made use of scattered articles, policy reports and media commentary to build engagement with everything from the history of the Internet to the contours of global cybercrime. What was missing was the core text one expects to see in higher education as a guiding resource. We wanted to develop that textbook.

In 2018, we published the first edition of *Understanding Cyber Warfare.* It was well received and made its way into the hands of thousands. Other books like ours have also appeared, a welcome set of additions to a field much in need of greater accessibility for the average reader. These books and ours all suffer from the same malady, however, which is the need to account for quickly shifting conditions and evolving knowledge about insecurity and conflict in the age of the Internet. Because of the ubiquity and rapidity of digital security issues, texts like the first edition of *UCW* are out of date almost from the start, at least insofar as each year brings new threat types, new uses of malicious code and new costs in the fight against digital insecurity into the public eye. The principles and lessons in our first edition represent a strong foundational resource but even the space of just a few years has made it necessary to provide updates on the knowledge foundations of global cyber conflict.

A second edition

This second edition of *UCW* represents an effort to address major evolutions in thought and circumstance along four lines. First and most broadly, we seek to reflect the ever-diversifying and ever-intensifying global cyber conflict. In addition to new focus on the contours and implications of attacks that have transpired since the first edition was published, we have substantially revamped one of the core chapters of this book that previously engaged in qualitative overview of a handful of major incidents. Now, in Chapter 7, we present empirical information on broader conflict trends using a set of emergent data resources, including the 2.0 version of the Dyadic Cyber Incidents and Disputes (DCID) dataset. We also substantially enhance our conversation about cyber operations as just one element of the broader global information environment. In doing so, we engage more deftly with, on the one hand, recent shifts in Western defense community thinking about the nature of digital insecurities for democratic societies and, on the other hand, the significance of competing conceptualizations of digital conflict that dominate in Russia, China, India and elsewhere.

Our renewed focus on the information environment, consisting of a new chapter entitled "Shape: subvert, deceive, poison" and numerous smaller updates, corrects for an overuse of the cyberspace domain concept as an organizing principle that is all too common in Western writing about digital issues. In these new sections, we engage arguments about strategic culture and the social construction of technology to contextualize the development of the domain concept and place it amidst alternative perspectives that have policy and strategic significance in the 21st century.

The second set of updates we provide in this second edition reflects topical advances in research and practice across several core national security thematic areas, which we now outline more clearly in our discussion of cyberspace in the context of the IR field in Chapter 3. Specifically, in this edition, we now survey and contextualize a large array of developments in strategic thought on cyber deterrence, coercion, repression and subversion. Over the past several years, an immense body of scholarship has advanced theory and provided evidence regarding the potential for successful deterrence of cyber aggression. In the United States, evolving thought on the shortcomings of pre-existing deterrence postures has produced new strategy based on persistent engagement and proactive action taken beyond domestic networks to signal intent to adversaries. In Europe and Asia, much practitioner and researcher discourse revolves around the complexities of cross-domain interstate engagements and potential escalation where cyber capabilities are used alongside information operations, clandestine force and more. These evolutions are significant for students of cyber conflict and are now anchors of discussion in *UCW*.

While those who teach about cyber conflict often characterize the phenomenon by addressing the techno-institutional context of the Internet's rise, it is increasingly clear that other forthcoming revolutionary information technologies will shape and – perhaps – change the nature of digital warfare. A third set of updates provided in this edition of *UCW* involves detailed discussions of several such technologies, particularly artificial intelligence (AI) and machine learning, quantum computing, blockchain ledger technology and 5G wireless standards. Far from simply adding these discussions as an addendum, however, we view the imperative of addressing new ICT as an opportunity to talk about the future of cyber conflict in a more detailed fashion than that exists in other publications (at the time of writing). As part of the book's reorganization, we have built two concluding chapters that split these prospective developments into two categories – technologies that (1) may change the logics of cyber conflict established in the foregoing sections of the book, such as AI, and those that (2) may transform the nature of digital conflict by way of altering or supplanting global Internet architecture, such as blockchain ledger technology or quantum computing. In our thinking, this approach presents learners with the best possible framework for applying new information about technological advances to what we know about the contemporary landscape of cyber conflict.

Finally, though much of *UCW* inevitably focuses on the actions of state security forces, national governments and transnational stakeholders in defining the shape of global cyber conflict, this second edition updates our focus on non-state actors along several lines. First, we have rebalanced focus on terrorist uses of the web, relying on new case examples to compress existing content that is overly academic in its descriptiveness. Second, the chapters in this edition now better contextualize the social context of Internet usage among the fringe elements of global society, guiding students from the experiences of early hacktivists to both the use of the web by protest movements around the world and the more recent emergence of extreme, countercultural movements tied to political violence in numerous Western countries. We talk herein about the rise and role of cyber mercenaries, both as a unique feature of non-state contestation in international affairs and as a unique facet of 21st century conflict between state actors.

In addressing these major evolutions in both practice and thought, we believe that *UCW* represents the highest quality pedagogical content available on issues of global cyber conflict. Indeed, we saw the task of revising *UCW* for this edition as a necessary one, a point reinforced by developments and incidents that continue to change the landscape of global cyber conflict even as this edition prepares for publication.

Cyberspace and the Russo-Ukrainian war: a country under siege

The months immediately preceding the preparation of this second edition have been defined – not just in the context of cybersecurity but also in truly global terms – by the invasion of Ukraine by the Russian Federation under the direction of Vladimir Putin. Beginning in late February, Russian forces have advanced into Ukrainian territory in several phases of destructive attacks, first targeting key territory around the capital city Kyiv and thereafter placing emphasis on securing land in the country's east. In particular, after losing the battle for Kyiv, Russian forces have attempted to secure and expand territory that they have contested less formally since 2014, specifically around Luhansk and Donetsk alongside urban areas north of the occupied Crimean Peninsula.

As we send this manuscript to press, the conflict remains substantially unresolved. Despite Kyiv's successes and the groundswell of support for Ukrainian sovereignty among Western partners, the war of attrition continues. As such, the lessons and impacts of the conflict remain undetermined, even though the airwaves and the pages of Beltway publications have been filled with little but attempts to forecast coming developments since the early days of the invasion. The implications for cybersecurity and global cyber conflict as a result of the conflict likewise remain undetermined at this time. Thus, we do not commit a major section in the chapters to come to what is most assuredly an unprecedented evolution of cyber conflict conditions, namely the presence of so much digital combat and contestation during a real shooting war between two capable state actors. Nevertheless, there are some initial points that can be made that bear upon the information imparted in *UCW*.

No cyber blitzkrieg

For students and scholars of international relations (IR), what's perhaps most interesting about the digital dimensions of the ongoing conflict in Ukraine in its early days has been the fact that events seem to bear out much of what cybersecurity scholars have said for years about the utility of cyber instruments for enhancing state power. As various experts put it, in the weeks after the invasion, there was no Russian "cyber blitzkrieg." As later chapters in this book suggest, this is likely because cyber-instruments just aren't good tools for controlling escalation or affecting the battlefield. The war and Russia's reaction to global sanctions may dictate a new phase of heightened digital insecurity in international affairs. But the strategic utility for using cyber tactics in Ukraine in support of an initial physical invasion itself just wasn't there. Cyber tools produce only temporary victories and so aren't all that good for direct coercion. And an expected quick victory on Russia's part took sophisticated cyber tools off the table immediately by the logic of "don't break what you're about to buy." And so while a great amount of interference and disruption targeting civil society and industry actors accompanied the opening of hostilities, Russian cyber attacks largely weren't tied to military objectives.

Sabotage, disruption and siege

Of course, the lack of a "blitzkrieg" in cyberspace is a narrow description of the situation in Eastern Europe in the first half of 2022. The reality is that Ukraine is a country under siege in digital terms on top of the military invasion. Over the past eight years, since the events of the Euromaidan protests and the onslaught of Russian-backed separatist civil conflict in the east, Ukrainians have often been subjected to siege-like conditions. Particularly in the context of major ransomware attacks like WannaCry and NotPetya or sophisticated malware threats to industrial control systems, Ukrainians have increasingly been forced to deal with reliability and availability issues stemming from digital effects. Network interference preventing the use of train ticketing systems or ATMs has become almost commonplace, particularly during the winter periods when Russian threat actors have typically launched exploratory disruption attacks. Even more seriously, direct attacks on critical infrastructure have frustrated the development of Ukraine's economy and of Kyiv's capacity to ensure its own security as Russian threat actors have increasingly used the country as a laboratory for deploying ever more devious weapons of mass digital disruption.

The early phases of the Russo-Ukrainian war in 2022 have continued and escalated this dynamic. The opening weeks of the war saw major wiper malware attacks that destroyed data in use by countless companies, utilities, media outlets and state service providers. Denial of service attacks almost too numerous to reasonably catalog have persistently hit sites and services across Ukraine and communications systems – notably commercial satellite providers like Viasat that help service the Ukrainian military – have been targeted for disruption. And major energy grid attacks mirroring those encountered in 2015 and 2016 utilized the BlackEnergy malware have been attempted. The cyber reality of the war for Ukraine has been, in short, a lesson – if not quite, perhaps, a masterclass – in besieging an entire national society as an adjunct to military campaigns.

Two sides, dozens of players

It would be disingenuous of us to suggest, however, that the cyber conflict surrounding Ukraine is one-sided affair. Hackers associated with the Ukrainian government or, more broadly, the Ukrainian cause have attacked all manner of targets in Russia and Belarus in the best traditions of both military cyber forces and hacktivists. Hacks aimed at media outlets for the purpose of introducing information on the war to the Russian public have attempted to weaken the domestic position of Putin's government, as have attacks on state websites, oligarchs and critical services like banks. Both sides are, quite simply, heavily invested in the cyber fight playing out alongside the military contest.

This being said, "two sides" does not only mean two actors. As Figure 0.1 illustrates, this cyber conflict involves dozens of actors, from threat actors directly associated with the

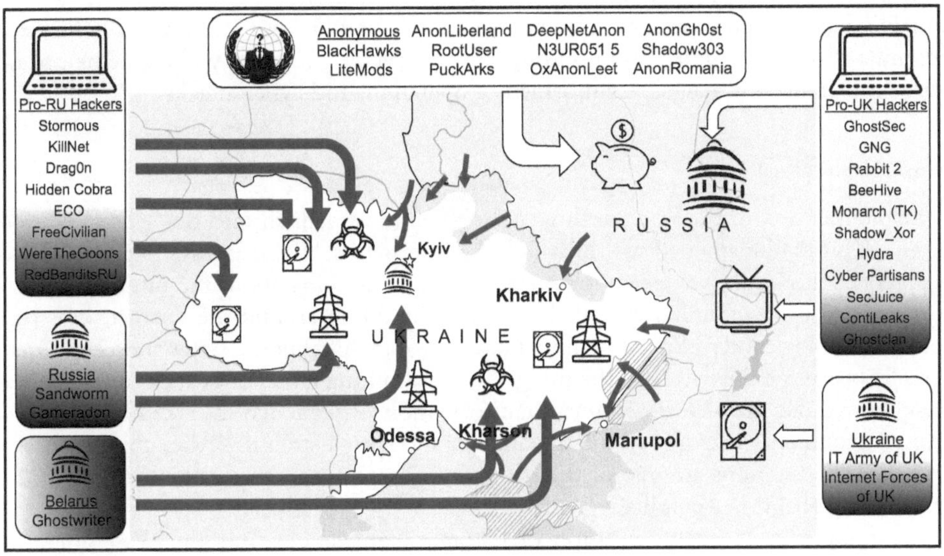

Figure 0.1 Two sides, dozens of players

governments of Ukraine, Russia and Belarus to ad hoc formations of volunteer hackers. Ukraine has been bolstered by an army of such hackers reported at more than 300,000 strong, while Russia has leveraged its deep connections with organized cybercriminal enterprise to present the façade of a similar popular front supporting Moscow in cyberspace. Hacking outfits tied to Anonymous have also played a major role in attempting to "crack" the Russian state by attacking media outlets and stealing data on all manner of organizations linked to the Kremlin. And American officials have even confirmed the deployment of "hunt forward" teams in aid of Ukrainian cyber defensive efforts, something that is likely being supported by the governments of Kyiv's most vocal Western advocates, from Poland to Slovakia and the Czech Republic.

The primacy of the information war

As one of the themes of this second edition is the ever-degrading distinction between cyber operations and adjacent forms of digital threat, it seems only reasonable to highlight the degree to which the cyber situation in Eastern Europe has been dominated by the need to win the information war. On the one hand, early conflict – both before and after the invasion – was dominated by Russia's attempts to dissuade Western unity against the efforts of Volodymyr Zelenskyy's government to build widespread support for Kyiv's struggle. President Zelenskyy's bid, on this front, clearly succeeded, in part due to proactive release of U.S. intelligence to warn of Russian aggression and preempt false narratives justifying Russian aggression. On the other hand, Russia's media strategies have clearly been influenced by the need to keep the domestic population from expressing outrage about the war and becoming a source of unrest too potent to ignore. Both of these dynamics have motivated a great amount of cyber behavior, from Anonymous's naming and shaming operations to moves made by numerous technology firms to support continued access to the web and to social media tools even within the warzone. Going forward, the information dynamics of this conflict and whatever comes after will almost certainly dictate even greater developments, not least of which is the approach Russia and potential partners of the Kremlin – from Iran and Syria to China – will take when planning interference campaigns, disinformation operations and cyber attacks.

Russia's splinternet move

A final dimension of the war worth us noting before we launch into this second edition of *UCW* is the fundamental way in which the Internet itself is changing. In the final two chapters of this book, we take a unique approach to thinking about the future of cybersecurity and cyber conflict, essentially framing the future around the distinct ideas that some technological developments might alter the logic of cyber conflict and others may actually overwrite the nature of the thing. And yet, shifting logics of cybersecurity is quite possible without major technological transformation. Recent actions taken by the Russian government in response to the Ukraine conflict, pro-Ukrainian cyber counter-attack and Western sanctions are one of the strongest examples of such a potential shift in the history of global cyber conflict.

In response to Moscow's invasion of Ukraine, companies across the West shuttered physical locations and shut off access to services for Russians. In retaliation, the Russian government itself took sweeping action to nationalize the assets of withdrawing companies and ban access to certain Western social media services and information services. Since then, the Russian government has appeared to commit to broad changes in what web access within Russia will look like going forward. In March 2022, a document promulgated by Russia's Ministry of Digital Development (MDD) demanded extreme changes in how state-linked enterprises must function online. In addition to beefing up security, all online services of state-owned or state-affiliated organizations were forced to shift to Russian-based hosting options and remove certain code elements from site pages. Most significantly, those same services were forced to switch to using Domain Name System (DNS) servers within Russia. These changes were enforced in the weeks that followed.

This move by Russia is perhaps the most significant shift toward the reality of a global "splinternet" – wherein some of the Internet remains the open, decentralized space idealized by Western societies, while other bits become closed-off spaces in which authoritarian regimes dictate the reality of information access for their citizens – in at least a decade. The DNS demand made by the MDD is particularly telling of Moscow's long-term intention to disconnect from Western systems and move to emulate the Chinese model of web control (discussed in later chapters), replete with strong constraints on free access to data and services. As we discuss in Chapter 2, devices attempting to navigate the Internet use DNS as a digital address book, sending a request for technical information about the location of desired spaces and services to a specialized server. By locating this process entirely within Russia, Putin's government is attempting to both prevent traffic from leaving the country and stake control over what information is presented to users as they navigate the web.

The long and short of such moves is that political power leveraged toward specific technological goals has the real potential to produce a world of nested logics. In other words, the future may bring a real world that is functionally divided between open web spaces we are all presently accustomed to and an expanding series of walled-off gardens where national governments – most notably authoritarian regimes – dominate the information sphere and create realities of societal engagement for their citizens that diverge radically from today's baseline.

Looking forward

This second edition represents an effort to provide students of international relations and cyber conflict studies with tools and knowledge that are, amidst the context of rapidly shifting conditions in the real world, accessible and durable to test of time. As readers move into the chapters of this book, we hope that they come to view cyberspace and digital issues the way that we do, as a set of technological issues that, while complex, are nevertheless shaped by very human considerations. Information technologies hack society and produce unprecedented insecurities, but these technologies are in turn shaped by the sociopolitical, economic and strategic dynamics of today. Only by understanding both can we think clearly about the security implications of cyberspace, of web technologies and of technological revolutions to come.

1 Introduction

The year was 1983 and President Ronald Reagan was relaxing on a weekend retreat in much the way that national leaders often do – by listening to some music or watching television in the background, while getting on with the grueling 24/7 task of reading reports, making policy plans with staff and generally running government. This particular weekend, Reagan was watching a new movie, one that had just come out the previous Friday. In it, a young hacker accesses a military nuclear response system in the course of being generally mischievous. The only issue is that the hacker, played by Matthew Broderick, thinks he's just found some sort of game. What follows is a harrowing series of events wherein the Americans and the Soviet Union move closer to nuclear war, all due to the hacker's actions in the "game." The next week, Reagan brought the movie – *War Games* – up in a meeting of his national security staff. After describing the plot in detail, he asked the room: "Could something like this really happen?" A week later, after some light inquiries had been made, his senior staff came back to him with a worrying answer: "The problem is far worse than we think."[1]

Reagan's interest in threats like the one outlined in *War Games* was the first time that someone at the head of government really gave serious thought to the notion that computer security might impact upon real national security imperatives. Others, however, had given the matter a great deal of thought in the preceding decades. In 1983, the Internet still existed in a nascent form and the predecessor of that global network of networks – the Department of Defense's ARPANET – was barely 15 years old. Nevertheless, entire industries were already growing up around the Internet, particularly those that had come of age in the 1960s and 1970s as a result of the rapid development of computers in the years following World War II. Both the private sector and government institutions were engaged in the wholesale adoption of new technologies based around something (the Internet) that had been built to ensure greater efficiencies in communications. Herein lay the basis of the answer that Reagan was given regarding the challenge of what would come to be called cybersecurity: "The problem is far worse than we think."

The Internet and the computers that it connects are present everywhere in the world today, at every level and in every facet of modern society. Networked information systems have opened new doors for progress across every sector of business and government. The result of all of this has been a proliferation of security challenges that is largely premised on something related to how the Internet came into existence – it was not conceived of with security in mind. Those who collaborated to develop the web in its earliest forms aimed to resolve obvious problems with inefficiency in inter-computer communication in a world without network connections. Then, before security could be effectively

DOI: 10.4324/9781003246398-1

addressed, commercial and governmental forces fueled an explosion of development of systems based around the new paradigm.

Because of this dynamic, the threat that concerned Reagan was very much a realistic one in the mid-1980s. Indeed, the decade or so that followed would see the manifestation of several threats that roughly took the form of what *War Games* described. In 1986, East German hackers would breach the MILNET (the Department of Defense intranet) looking for military secrets to sell to the Soviet Union. In 1998, an Israeli hacker and two teenagers in California got into Air Force, Navy, Pentagon, NASA and other government systems. And for several years in the late 1990s, the U.S. defense community was the target of an enormous, persistent espionage campaign in which Russian-based hackers systematically searched for sensitive national security information.[2]

Beyond the kind of direct attacks on government control systems fretted about by Reagan, however, the past three decades have *also* seen the manifestation of an incredibly broad range of national security threats enabled in new ways by action taken online. Digital weapons have disrupted the ability of countries to enrich uranium for the production of nuclear bombs. Malware has caused the disruption of electrical systems in Ukraine, causing blackouts that affected hundreds of thousands over a period of hours. Cyber attacks have been targeted in support of propaganda campaigns targeting a growing list of countries' political processes. And hackers have been seen to engage in broad-scoped campaigns to steal commercial and governmental secrets in what can only be described as the most significant series of economic espionage efforts since at least the early 20th century.

Cyber conflict is almost ubiquitous in the world today. It not only occurs frequently and with increasing intensity; it is diverse in how it manifests. The upshot of this is quite simply that understanding cyber conflict is not simply the task of comprehending computer or network security challenges. Understanding cyber conflict demands discussion of the topic in the context of world politics. This book undertakes exactly this task, blending content regarding the form and context of cyber operations with the theoretical and empirical perspectives of the international relations (IR) field of study. It is our hope that in doing so, these issues – which both appear to be everywhere in modern society and are, at the same time, frustratingly inaccessible at times – will become more readily understandable to those interested in grasping how computers and the Internet are changing the world.

Aims of the book

This book is, again, not about cybersecurity. Rather, it is about cyber conflict and the range of historical, empirical, theoretical and policy issues that entails. The distinction between these different mission statements is meaningful for a number of reasons. Foremost among these is the much greater scope of the topics we must cover herein by focusing on cyber conflict – which has today is often used to describe an immensely broad range of conflict forms that are being augmented by cyber actions – rather than simply on traditional issues of cybersecurity.

Cybersecurity is an inherently technical field. However, as any computer scientist will tell you, cybersecurity is also inherently defined by the interaction of technology with sociopolitical institutions and systems, from the human users of information technologies to the institutions that require their use. Cybersecurity is, in the broadest possible terms, about the security of socio-technical systems emerging from or impacted by the information revolution. If the purview of our discussion were merely those issue areas at the heart of the cybersecurity field, this definition would see us extend our gaze only to, say, the study of organizational hygiene (e.g. the mandate of better password practices among a company's employees) and the implications of network technologies for risk management procedures. As it is, our focus on cyber conflict involves systematically merging a study of cybersecurity issues with assessments of human interactions from the lowest levels to the highest. It involves understanding how new information systems augment and alter political systems at the level of institutions, countries and the global system itself.

Our aims in writing this book are fairly simple. As professors and researchers of cyber issues, we have been struck time and again by the lack of cohesive resources available for those attempting to impart knowledge about cyber conflict to students from a social science perspective. As we noted earlier, cybersecurity is not merely a technical subject; indeed, the study of cyber conflict must naturally be a mixed course in comprehending technical foundations, significant historical context, international and national security theory, policymaking and doctrine. And yet what resources exist from which to learn about cyber conflict virtually demand assignment alongside supporting readings. Instructors are called upon to construct syllabi around a hodgepodge of content that aims to support an interdisciplinary curriculum on cyber conflict. Often, such syllabi are excellent. But even those good examples force students to consume a narratively inconsistent set of works such that learning is hampered.

We certainly do not and cannot claim to cover every element of cyber conflict in the course of this book. Far from it. This topic is broad enough and varied enough that a compendium of extant knowledge on cyber conflict, even just from a political or social science perspective, would reach some thousands of pages in length. Rather, what we offer here is a great deal of context and in-depth introduction to key issues on the topic. We offer great historical context at times, scholarly discussion of relevant theories elsewhere and robust empirical discussion of the actual conduct of cyber conflict as seen from the 30,000 feet vantage point often favored by scholars of world politics. More importantly, we aim to provide a consistent narrative voice such that this book serves as an excellent introduction to the field of cyber conflict studies for all levels of students.

What is cybersecurity?

Asking someone to define what "cybersecurity" means seems like a naturally deceptive act. In reality, there are few "cyber" terms (i.e. "cyberwar," "cyberterrorism," "cyberbullying," "cyberspace," etc.) for which there is universal agreement on what is exactly meant. Cyberspace, the nature of which we talk about in Chapter 2, is a particularly slippery concept. Cybersecurity, likewise, would be tricky to define if you were to ask a diverse enough audience. From a purely technological perspective, cybersecurity is constituted of all processes, procedures, design considerations and actions concerned with the security of information systems. This means that anything bound up in protecting computers systems and the networks that connect computers from attack, disruption and infiltration constitutes cybersecurity. In reality, this definition is, roughly speaking, the one likely to receive broadest approval by practitioners and scholars across the fields concerned with cybersecurity.

But it also naturally speaks to the technological drivers of the cybersecurity field at the expense of any focus on the global context of the information revolution of the past 50 years or so. Is security in the digital age just about the function of information systems and computers that now underwrite most major functions of 21st century global society?

Or is it also characterized by the ways in which the information revolution has affected non-technical features of international relations, from the function of state militaries to the way in which humans approach problem solving? This book's sections and chapters grapple with these questions as uniquely relevant to questions on the nature of cyber conflict. As such, we adopt a more general format of definition for cybersecurity from the start, arguing that cybersecurity *as it pertains to cyber conflict* is simply constituted of all processes, procedures, design considerations and actions concerned with the security of socio-technical systems.

What is the scope of cyber warfare and conflict?

Throughout this book, we will refer to different security layers of the digital world. Cybersecurity issues are, in many ways, best understood as layered around basic mathematical, technical and social principles. From basic issues of cryptographic security to the design of computer and network systems, cybersecurity is constituted of an incrementally expansive sphere of security considerations. How might developers best approach issues of information privacy? How should man-made programming languages and hardware be adapted so as to minimize the subversion of computer systems? And how should organizations regulate their user base so as to achieve an optimal defensive posture?

At the same time, such security considerations themselves constitute only the foundation of cybersecurity considerations from the perspective of the international security researcher. In many ways, it is useful to think of such considerations, the shape of which is described in detail in Chapter 2, as the most incremental level of analysis of cybersecurity issues. Much in the same way that IR scholars teach students to think of human capabilities and psychology as the most fundamental possible category of explanations for why different things happen in world affairs, we emphasize that fundamental insecurities and threat realities bound up in cryptography and information technology design constitute the base category of explanations for why different cybersecurity issues are of more or less importance for national and international security.

Beyond socio-technical issues of logic and design at that level, students of cyber conflict must seek to assess the manner in which different national institutions and non-state actors adopt information and communications technologies (ICTs). National experiences with cybersecurity have produced radically different approaches to doctrine, to policy and to practice across intelligence, military and law enforcement units the world across. Likewise, scholarship on non-state actors' use of ICT is increasingly emphasizing understanding of actor-specific characteristics – i.e. the different incentives that terrorists, criminals and "patriotic" hackers might have to develop cyber capabilities and use them in conflict – as more critical than comprehension of technical factors for the construction of knowledge on different actors.

And beyond focus on specific socio-technical systems and actors, students of cyber conflict must consider the broader shape of the digital world. Cyberspace is, in many ways, an artificial construct overlaid on top of the international system. It is, however, constituted of a basic terrain – a global infrastructure characterized by certain underlying features and organizing mechanisms. These macro features, as much as anything else, also drive the development of new trends in cyber conflict.

So what is the scope of cyber conflict? Much as might be said of conventional warfare, cyber conflict is constituted of everything from basic issues of computer network security to both the shape of those who operate in the digital world and the actions of every kind of political actor in attempting to govern its features. The task before students of cyber conflict studies is in teasing apart the different elements of a given issue and attempting to discern where – across different technical, social, political or economic features – explanatory power lies.

In this book, we will be addressing questions like "does the Law of Armed Conflict apply to cyber conflict?" and "is it possible for actors to use cyber 'weapons' to coerce others into changing their behavior?" Answering these questions will inevitably involve differential assessment of the degree to which developments at different levels of analysis drive conflict dynamics. In other words, finding answers will mean adjudicating on whether or not conflict emerges more from technical complexities, institutional dynamics, strategic realities or some other force acting on those who have chosen to operate in cyberspace. As such, the book addresses issues of cyber conflict from a range of perspectives.

Themes of the book

The ongoing information revolution – centered on computers and the Internet – has changed the world. This is, of course, a bold and overly grandiose statement. Indeed, it is a statement that might be challenged by many.[3] For instance, the question of whether or not ICT constitute a revolution in military affairs wherein a new paradigm of security applies today that did not several decades ago is particularly hotly contested among political scientists. And yet, that debate is largely centered on the question of warfighting as a distinct political activity. We, the authors of this volume, do not argue that the information revolution has necessarily (and controversially) changed the way in which future wars will be fought to such a degree that conventional military strategy will never be the same again. But we do assert that the information revolution has altered fundamental features of international affairs. Information technologies constitute the wiring of global infrastructure and enable almost all human interaction in the technologically developed world (meaning most of the planet) today. As such, cyber conflict issues – regardless of whether one thinks that wars fought entirely online are possible or probable – are inevitably and enduringly relevant in a way that other topical conflict programs are not. This is not only because ICTs constitute the wiring of international affairs but also because information and computer security challenges manifest in such a way as to link social, political and economic spheres that might previously have rarely touched. One need only consider the number of times in recent years that issues of cryptographic encryption or network privacy addressed in the context of cybercrime or terrorism have prompted national conversations about civil liberties to know what is meant in this regard.

The content in this book is intended as an introduction to the topic of cyber conflict and to key questions bound up in the history, practice and study of warfare in the digital domain. As such, there are inevitably certain themes the recur throughout as we grapple with similar concepts or issues across different levels of analysis and in different settings (i.e. academic vs. historiographical). Beyond the crosscutting nature of all things cyber, several of these are worth mentioning up front.

Enduring barriers between the digital and the real

A defining characteristic of security in the information age is the degree to which kinetic and digital actions are disconnected from one another in meaningful ways. With cyber conflict, it seems reasonable to assert that most experts would point to attribution problems – i.e. problems in accurately ascertaining responsibility for digital actions – as the foremost drivers of uncertainty and aggression in the digital domain. Even where it is possible to forensically describe cyber attacks, attributing cyber aggression to real-world actors – from individuals to terrorist organizations and state entities – is additionally difficult, particularly that evidence on who is responsible for cyber actions rarely reflects a record of the sociopolitical machinations at work behind such operations. Non-state hackers may actually work for state intelligence organizations or military institutions, while threat agents traceable to foreign military Internet Protocol (IP) addresses may actually be attempting to lay false blame at the feet of a third party. These issues are pervasive and have a number of implications for the incidence and likelihood of cyber conflict.

The physicality of the digital domain

Despite the reality that enduring barriers between the digital and real worlds exist to complicate cyber conflict dynamics, it is also worth noting up front that there is an underlying physicality to everything we talk about in this book. Many scholars and practitioners problematize and develop perspectives on cyber conflict that assume an essentially unlimited potential for digital actions that can affect national security processes. The truth of the matter, however, is that the digital world runs on physical infrastructure and logical programming implemented in reference to that infrastructure. We discuss this further, particularly in Chapter 2, but it is worth remembering from the start that physical circumstances have dictated and will continue to dictate the manner in which states, non-state actors and private industry use the Internet for security purposes.

Exogenous development

Leading on from the broader point about physicality, the topography of the digital world is decentralized and its development is driven by a large number of factors. Security dynamics are regularly determined by technological circumstances rather than by institutions that dictate technological conditions. Take, as an example, the logical processes set up to ensure that information from one person sent across the Internet to another makes it to the right place. At present, there are about 130 root servers controlled by 12 different private companies.[4] Five of these companies are in Northern Europe; the rest are in North America. Root servers play an important role in Internet functionality, essentially controlling the process of registering website addresses to specific IP addresses (the identification digits given to each computer on the web). As a result of the current physical location of these servers (each of which operates with a great number of redundancies distributed around the world, admittedly), most Internet traffic around the world travels through the United States. That is to say that packets of data sent from one networked device to another invariably visits the United States before reaching its final destination. This technological dynamic has been uniquely impactful in the development of cybersecurity programs in the United States, as access to such through-traffic provides a number of opportunities to enhance the national security (and, particularly, intelligence) functions of government.[5]

But this will not always necessarily be the case. The number of root servers in the world has largely remained static because of the packet-level requirements of the IPv4 standard currently used to generate IP addresses. But the world has essentially run out of IPv4 addressed and is now in the early stages of transition to a new, longer standard called IPv6. Here, data packets can be bigger than was previously desirable. As a result, new standards of network reliability and performance will likely make new space for the addition of more high-level routing servers to accommodate more of the world's Internet traffic. Those servers need not be in the United States or Europe. Indeed, it is highly likely that many will be in countries with increasingly complex technology sectors and growing economic prowess, like China or Israel. Thus, the shape of the digital world will, in a reasonably fundamental way, shift over time and exogenously affect the security capabilities of some countries. This is but one example of the way in which the information revolution has fundamentally altered the wiring of the national and international security realms for states from the outside in.

Embeddedness leading to proliferation

The story we open with here about Ronald Reagan serves a few purposes. An obvious one was to suggest that security challenges related to the Internet emerged in the way they did during the 1970s, 1980s, and 1990s because of the way in which governments and the private sector co-opted and innovated new products around network technologies. Simply put, the natural process of modern societies recognizing the immense promise of a new information technology and then moving to benefit from it has created a host of negative externalities that drive insecurity. Security engineers are encouraged to produce mediocre security systems so that customers are not hampered by the need for lengthier login wait times or significantly more powerful processors. Technology developers are not uniformly subject to regulation aimed at standardizing product security across countries. Thematically, this amounts to a dynamic that we've seen innumerable times in human history. Where new information technologies become extensively and rapidly embedded in societal functions, security challenges proliferate.

Inherent insecurity

What much of this amounts to is simply that the networked world is inherently insecure. This is the case on several fronts, but at the highest levels, again, this is so primarily because the history of network technologies is the history of people favoring convenience over security. The Internet was not conceived and thought of primarily with security management in mind. Indeed, the Internet is a beast of communications efficiencies. The basic difference between ICT and previous kinds of telecommunications technologies – which is discussed later, notably in Chapter 2 – essentially dictates a form of information insecurity insofar as the process of sharing and accessing information across networks is a highly disaggregated and uncoordinated process. On top of this, the international community remains in a state of marked disagreement over the right approach to governing the digital world. At present, a broad range of non-profit entities govern the Internet's "address book" (more on that to come), technology standards oversight functions and more, with help from a small subset of state-supported security units. The result is a ponderous global ecosystem that supports the function of a digital world overlaid on top of the real one and constituted of systems inherently possessed of vulnerabilities. Moreover, the commercial dynamics of technology development have, over time, worked to create negative externalities in the system via the encouragement of practices that favor business convenience over the best security. This complex reality both enables cyber conflict in various forms and complicates questions of cybersecurity cooperation at both national and international levels.

How do we know what we know about cyber conflict and digital insecurity?

Before we proceed, an elementary question for any would-be student of cyber conflict is "how do we know what we know" when we're talking about issues of digital insecurity? Whether our focus is on cyber operations undertaken by states to spy on strategic competitors, interference campaigns aiming to manipulate social media platforms or more simple computer security challenges, it doesn't take a clairvoyant to notice that the actors, actions and intentions involved are invariably a lot less visible than other phenomena of interest to social scientists. Things like international diplomatic engagements, economic activity or militarized disputes are affairs with substantial visibility to global governments and publics alike. By contrast, so much of global cyber conflict is the stuff of clandestine operations; of capabilities developed and deployed in private settings without the extensive infrastructure or paper trail that makes most security activities – such as building major armaments or funding allies – hard to miss. Indeed, with manifestations of cyber attack – though not all – success ideally means maintaining secrecy and avoiding attribution, even long after the event takes place.

So, given this, how can we speak of the empirical dimensions of global cyber conflict, crime and politics with any amount of confidence? Aside from the fact that much learning about cybersecurity as an international relations issue involves discussion of concepts and the established historical record, we have confidence in the lessons constructed and conveyed in this book for two primary reasons. The first of these is quite simply that we are standing on the shoulders of giants, albeit fewer than might be found in other areas of study. By that, we mean that a great groundswell of scholarly and analytic focus on cyber conflict in the first two decades of the 21st century have seen the production of extensive resources cataloguing different facets of the digital dimensions of world affairs. In particular, many of the resources and work we rely on here to convey meaning to students of IR and cyber conflict have emerged from a simple logic – while most cybersecurity activity occurs in settings that are difficult to observe for those in the public sphere, those incidents and campaigns that are meaningful and relevant to the conduct of national security and statecraft invariably spillover into the public eye over time. Where government or major corporate operations, national security processes or societal functionality are seriously threatened, there are invariably ripple effects that are difficult to hide over following the months and years. Since these are the events that are most relevant to the study of cyber conflict in an IR setting, this has translated to a meaningful production of data resources from which scholars – and the authors of this book in chapters to come – have built nuanced depictions of cyber operations and other digital activities as new mainstays of political behavior in the 21st century.

The other reason that a growing cyber conflict studies field can credibly talk about such inherently clandestine activities as those discussed in this book is simply that the digitally infused world of the 21st century is immensely complex. While parsing apart information and finding meaning amidst the noise can be challenging, an unprecedented deluge of data about diverse – and always further diversifying – Internet-aided activities gives researchers the basis from which to develop a rich understanding of the world as it changes around us. True, the basic characteristics of cyberspace that include difficulty in observing others' actions and attributing digital events to real-world intention make generalization tricky. But complex conditions also make it possible for researchers to narrowly examine specific persons, communities and events in rich detail. In that way, much of the research that constitutes what we call here the "cyber conflict studies" field has given us in-depth touchstones from which to pursue yet further narrow analyses and, thereafter, begin to generalize about cyber behavior.

Plan of the book

This book is designed so as to lead students from general precepts and principles through typological discussion of cyber conflict to chapters focused on empirics and theory. The next two chapters take aim at cybersecurity foundations. Chapter 2 discusses technological foundations of the security dynamics we are grappling with cyber conflict, while Chapter 3 undertakes the task of contextualizing the information revolution from the perspective of IR's main theoretical perspectives. Chapters 4 through 6 – entitled "Exploit," "Attack" and "Shape," respectively – then take different approaches to grappling with the actual conduct of cyber warfare. Chapter 4 takes a historical approach to introducing students to the context of cyber conflict's emergence from the practices of intelligence communities in North America and Western Europe. Chapter 5 leads on from there and overviews military perspectives – particularly those of the Department of Defense in the United States – on cyberspace as a domain of warfighting, as well as the special characteristics of operating online. Then, it briefly (since cyber defense is naturally discussed in later chapters in the context of particular forms of cyber threat) outlines what is involved with cyber defense, before Chapter 6 introduces an alternative view of cyber conflict than is often adopted in the West – that of a phenomena not particularly useful for warfighting or signaling, but exceedingly useful for actors who want to manipulate the rules of the game that is strategic competition in international affairs.

The latter half of the book turns to describe and explain the actual conduct of cyber warfare. Chapter 7 provides a (necessarily incomplete) history of cyber conflict incidents and an empirical picture of the landscape thereof. Chapter 8 discusses how different national experiences with cybersecurity have shaped approaches to cyber conflict practices. Chapters 9 and 10 then debate the nature of cyber warfare as useful for major warfighting, coercive signaling and "grey zone" actions taken below the threshold of declared war. Chapter 11 then moves to discuss how the information revolution has affected non-state actors. Chapter 12 covers international cooperation, norms and deterrence before we conclude. Our two concluding chapters, 13 and 14, take a unique approach to thinking about the future of cyber conflict, namely by discussing first how new technologies might evolve the logics of cyber engagement outline in this book and then how new technologies might even change the nature of digital conflict.

Notes

1 For perhaps the best discussion of Reagan's experiences in grappling with computer security as a national security issue for the first time, see Kaplan, Fred. *Dark Territory: The Secret History of Cyber War*. Simon and Schuster, 2016.

2 Though we cover these episodes in great detail in later chapters, a good narrative resource that unpacks America's early experiences with cyber conflict is Healey, Jason and Karl Grindal, eds. *A Fierce Domain: Conflict in Cyberspace, 1986 to 2012*. Washington, DC: Cyber Conflict Studies Association, 2013.

3 For an overview of the arguments involved in this debate in the IR field, see *inter alia* Kello, Lucas. "The meaning of the cyber revolution: Perils to theory and statecraft." *International Security* 38, no. 2 (2013): 7–40; Valeriano, Brandon and Ryan C. Maness. *Cyber War versus Cyber Realities: Cyber Conflict*

in the International System. Oxford University Press, USA, 2015; Gorwa, Robert and Max Smeets. "Cyber conflict in political science: A review of methods and literature." (2019); Choucri, Nazli and David D. Clark. *International Relations in the Cyber Age: The Co-Evolution Dilemma*. MIT Press, 2019; and Maschmeyer, Lennart. "The subversive trilemma: Why cyber operations fall short of expectations." *International Security* 46, no. 2 (2021): 51–90.

4 For an overview of the routing functions of the Internet in historical context, see Mockapetris, Paul and Kevin J. Dunlap. "Development of the domain name system." *ACM* 18, no. 4, (1988).

5 Though, again, we discuss this unique issue in later chapters, Kaplan's *Dark Territory* provides more extensive narrative description of the manner in which this logical dynamic of Internet function has shaped national security processes in the United States.

Further readings

Choucri, Nazli. *Cyberpolitics in International Relations*. Cambridge, MA: MIT Press, 2012.

Guiora, Amos N. *Cybersecurity: Geopolitics, Law, and Policy*. Boca Raton, FL: CRC Press, 2017.

Healey, Jason, and Karl Grindal, eds. *A Fierce Domain: Conflict in Cyberspace, 1986 to 2012*. Washington, DC: Cyber Conflict Studies Association, 2013.

Reveron, Derek S., ed. *Cyberspace and National Security: Threats, Opportunities, and Power in a Virtual World*. Washington, DC: Georgetown University Press, 2012.

Singer, Peter W., and Allan Friedman. *Cybersecurity: What Everyone Needs to Know*. Oxford: Oxford University Press, 2014.

Van Puyvelde, Damien, and Aaron F. Brantly. *Cybersecurity: Politics, Governance and Conflict in Cyberspace*. Cambridge, UK and Medford, OR: John Wiley & Sons, 2019.

Manjikian, Mary. *Introduction to Cyber Politics and Policy*. CQ Press, 2019.

2 The technological foundations of insecurity in the digital age

In listening to presidents and pundits alike in recent years, one could be forgiven for thinking that all national security issues feed into broader problems of a cyber nature. And, indeed, there is some element of truth in that sentiment. If forced to elect a single theme for this book on cyber conflict, "cross-cuttingness" wouldn't be a bad one. In the 21st century, most areas of politics, social engagement, economics and security are shaped in some sense by digital insecurity. This is because, in a very foundational sense, the Internet – it's physical infrastructure, logical routing architecture and informational substrates – today acts as a plane upon which all other elements of society function.

Thought about another way, one often hears commentators speaking of diverse cyber issues use paired terms – "tightly coupled" and "loosely coupled" – that are regularly used by engineers to describe the design of a given system. In the simplest sense, tightly coupled systems are those where a disruption at one point is quickly felt at other points, such as when a pipe bursts in a house and water pressure tanks. Loosely coupled systems are those where this is far less the case. With national critical infrastructure sectors, for instance, one might think of possible disruptions and quickly see that agriculture sectors are far less tightly coupled than is the banking sector.[1] A localized drought affecting cattle farming in one part of the country will only lead to rising beef prices gradually over a course of weeks, if at all. By contrast, a disruption of the financial services industry – a major shift in interest rates, for instance, or the burst of a market bubble – will quickly overtake and shape an entire national economy. The thing is, no matter what kind of system one considers – political, infrastructural, economic, social – it is hard to avoid the reality that the information revolution has broadly acted more to tightly enjoin parts of state national security apparatuses than loosen them. Not only do information technologies essentially constitute the "wiring" of most societal functions today, but security challenges in one vein invariably prompt logistical, legal or moral challenges elsewhere, as common security foundations are meaningful beyond how technologies are applied more narrowly.

But what are these common security foundations? Put better, why are the systems and technologies that have emerged from the information revolution vulnerable to malicious actions? Answering these questions allows for focus on specific questions that are more commonly the domain of political scientists and security studies scholars. Is it possible to design perfectly secure information systems and procedures? Is there a security dilemma in cyberspace? Can cyber aggression be deterred? And to what degree is international cooperation on different cybersecurity issues, from technology development standards and Internet governance to the weaponization of code, possible or desirable? As such, the

DOI: 10.4324/9781003246398-2

task of explaining the foundational insecurities of the digital world is of central significance to students of world politics in the information age.

This chapter takes up the challenge of concisely and simply outlining the roots of the myriad security challenges that characterize life in the digital age. Chapters to come – particularly Chapters 4 through 7 – go into greater detail on the history and dynamics of cyber conflict, but here the idea is to acquaint the reader with foundational concepts in an effort to lay the groundwork for further conversations about the nature of cyber threats.

Broadly, doing this involves selling the idea that cyberspace is inherently insecure. This is the case for a number of reasons, not least because the Internet was not designed with security management in mind and has proliferated to a point where we have a veritable "tragedy of the commons" with regard to incentivizing positive cybersecurity behavior (more on that in Chapter 3). But it is hard to overlook the significance of the ever-present human factor in all of this. How people design technology, empower technology to change over time and interact with both design processes and resultant technologies all matter a great deal. The implication of this is quite simple. Cybersecurity is inherently about the interaction of human systems with important technological ones. More than just being the practice of protecting information systems, "cybersecurity" – i.e. security that is of a cyber nature – describes the security of socio-technical dynamics related to the impact of the information revolution on pre-existing mechanisms of human interaction the world over.

The sections that follow attempt to lay out the fundamentals of security and insecurity in the digital age in a format familiar to most students of international politics. The following sections treat different elements of the cyber insecurity puzzle – information security challenges, technology design and usage issues, and infrastructure setup – as different levels of analysis. Much as is the case with other realms of study in political science, dissecting the digital world in such a way is intended to allow students to better deliberate on the value of different factors for explaining insecurity. All cybersecurity issues with political, economic, social or strategic dimensions – meaning all of them – are defined by the interaction of security dynamics at different levels. Information security problems have driven political conversation about civil rights' protections. Political happenstance has, consistently in recent history, led to faulty assumptions about the motivation behind cyber attacks. The following sections take steps to outline different areas of concern and the logic of insecurity at different levels so that students might be better prepared to think adaptively about the substance of cybersecurity issues in the chapters to come.

The technological foundations of cyber insecurity

The first step in understanding the insecurities of a world wired by novel and evolving information technologies is to understand how the networks that connect all manner of digital systems – from personal computers to infrastructural control systems – work. Though not everything networked is a part of the Internet, the history and nature of the Internet are nevertheless the only logical starting point for any effort to outline the broad shape of the digital world. From there, it becomes possible to springboard, first, to questions about the nature and extent of cyberspace and from there, second, to more fundamental principles of information security and technology design that impact upon the full gamut of cybersecurity and conflict issues.

Networking the world: how the Internet works

The history of what is now called the **Internet** is about half a century old and centers, as has been the case with many breakthrough innovations over the past few hundred years, on a government-funded research and development organization. In this case, the setting is the network of laboratories at government facilities and universities across the United States linked to the programs of the Advanced Research Projects Agency (ARPA, now called DARPA after the word "Defense" was added at a later date).[2] Chapter 4 describes in greater detail the backdrop of computerization that drove stakeholders within the U.S. Department of Defense (DoD) community to invest in the networkization of increasingly sophisticated computer systems. In short, however, the motivation for funding what would essentially become a new kind of information technology emerged almost entirely from the transatlantic scientific community.

Specifically, researchers at laboratories funded by the DoD were regularly faced with the need to send sizable amounts of information to their sister labs and to fellow government collaborators.[3] Likewise, the advent of early "supercomputers," which were situated at only a handful of facilities around the world (largely in the United States), meant that researchers were vying for both physical and remote access to new processing capabilities for a variety of purposes. Whereas telephone, snail mail and fax were obvious existing mechanisms via which information could be sent and telephone connections could support some remote usage of computing power, the fact of the matter was that using these methods of communication represented a serious inefficiency. If computers being developed and used in these labs could speak directly with one another, wouldn't life for these researchers and their teams be both simpler and cheaper? Likewise, if methods for linking computer systems could be developed and implemented, wouldn't the perpetual concern about vulnerable control systems held by the military consistently through the 1950s and 1960s disappear?[4]

The answer in both instances is, of course, yes. The development challenge was that existing methods of linking labs to ensure data connection between computers was a reasonably expensive proposition. The reason for this was twofold. First, the laboratories and their computers would – as technological practices at the time dictated – have to be physically connected with one another. After all, the only method of sending the relevant information in physical form was over the telephone network. Thus, there was a clear infrastructural cost to hooking up numerous facilities and personnel across government. Second, there was a serious risk of failure from an information communications perspective insofar as data would be sent from one computer – essentially functioning like a computer switchboard – to another and another down the line of physical connection until it reached the intended recipient machine. This was **circuit-switching**, where "switching" referred to physical connections that were manually adjusted to allow for new interconnections between endpoints. This was the basic technological principle behind the telephone and telegraph networks. But the issue with this approach was that computer failures – or the inability of a recipient program to read the data and resend it due to a glitch, for instance – would lead to a break in the circuit. Information would not (or not entirely) make it through.

The solution lay in a shift to a new technological approach to sharing information between endpoints.[5] Called **packet-switching**, this approach – which was not originally developed in the United States – would allow information to be broken up into constituent parts and sent through a network of redundant inter-computer linkages to an intended endpoint. Much like transporter systems in *Star Trek*, the point was that information would be broken down to an appropriately small level and then reassembled in the right way at the final destination. The transmission of these packets of data worked via the logic of multiple failsafe routes to the final source. By networking computers together – and then by connecting networks of computers with other networks of computers – data packets could simply try every possible inter-machine connection node in order to continue down the pathway to an intended destination. If an issue arose with one machine failing to accept the transmission of a data packet, it would be sent via the next fastest route and then joined with other packets to be reassembled into meaningful information at the end of the line.[6]

With what would become the Internet, information transmission can be best thought of by envisioning a tree of networks – a hierarchical series of interconnected networks of computers that constitute the digital world. Information transmission across networks and between networks of networks is enabled through a series of **protocols**, which are specifically delineated sets of rules about how and when information should be transmitted between computer systems. For the purposes of a discussion on the history of the Internet, there are two specific protocols worth mentioning.

The first of these is the set of protocols – incorporated in the design of all networked information systems – that allows information to be passed between networks regardless of differences in the design of underlying physical technologies. An important concept here is the **handshake protocol**, which was suggested and designed by Vint Cerf and Robert Khan in the 1970s.[7] The idea held by these gentlemen was simple – machines connected to others should be programmed to use a standard set of rules for contact so that there is no need for requiring the *physical* compatibility of underlying technology. The idea was that, if one wanted to ensure greater utility and higher adoption potential for Internet technologies, there needed to be some way to compensate for the inevitable reality that computers built by different companies in different countries across different generations over time would not share most physical design characteristics. This led initially to what is generically called a three-way handshake protocol, wherein machines were directed to request the right to send information to another, would receive acknowledgment of authorization to do so and would then acknowledge that acknowledgment before attempting to send over data. Included in the protocol was specific information about what kind of data packets were preferred by the receiving machine and how best to accommodate the recipient's access to the nascent Internet.

Less generically, information is transferred from computer to computer across networks via reference to a **protocol stack** that moves information from applications on a device, packages it, addresses it and transforms it into different kinds of physical signals to be sent across the Internet to another end point.[8] Specifically, the two most critical elements of this stack, developed to accommodate the function of the growing Internet, are so often described in the same breath that they bear mention together. The **Transport Control Protocol** and the **Internet Protocol** (together referred to as **TCP/IP**) constitute the rules of how computers would break apart different kinds of data coming from one computer and then transmit that data across networks to intended recipients. On the one hand, TCP determines how data will enter and leave a computer. To match various pieces of software that use different kinds of incoming network information to support user functions, TCP directs data through different **ports**. Data constituting email messages might leave and be accepted through one port, while streaming music, video or website data might enter through others. The Internet Protocol then enables the direction of such diverse data between linked networks (more on this later).

Another protocol worthy of specific mention is called the **HyperText Transfer Protocol** (HTTP). HTTP addresses the human element in information transmission between networked machines and between networks of network machines insofar as it allows for the identification of formatted "documents" online. These documents (i.e. websites) can, thanks to HTTP, be recognized as discrete formatted resources able to be displayed in a standard format.[9] Moreover, HTTP allows for such documents to be imbued and enriched with applications to enhance user experience, from "hyperlinks" that allow users to jump quickly between webpages to more complex tools (e.g. widgets) hosted on websites.

Knowledge of what these protocols do in broad terms should offer an idea of how computers ultimately interact with each other across the digital world. But it doesn't yet tell us how the Internet works; mapping the Internet requires understanding how information makes it from your computer connected to Network A to another computer on Network Z. To understand how this happens requires greater focus on the IP as a mechanism employed in order to ensure effective movement of information across the digital world.

The IP works in direct reference to the **Domain Name System** (DNS). In essence, the DNS is the address book of the Internet.[10] If we think again of the Internet as a family tree of sorts, with different kinds of networks linked to other networks via physical linkages and protocol-enabled agreements on acceptable forms of communications, the DNS is the function that acknowledges the specific identification information of each layer of the Internet and translates it into something usable for humans. In other words, as the IP assigns a number to different networks and specific computers (i.e. your personal computer will have an IP address), outgoing connections will inevitably target a specific external numeric location when requesting information. The DNS is a registry of locations – specifically a registry of IP locations tied to web addresses (from high-level ones ending in. com,. uk,. edu and so on to lower level ones like. ac.uk or. co.ch) – that your computer queries when it is trying to find out where to send information.

Box 2.1: Sections of the Internet

In all of this, it's worth teasing out that there are different component sections of the Internet. What most people think of as the Internet – public-facing websites, news sources, etc. – is in actual fact something called the **surface web**. This part of the Internet is based on reasonable knowledge of the content of web "documents" (i.e. websites) and on an ability to access them without precondition. Search engine providers map the Internet through the use of web crawlers – sophisticated code devices that follow the pathways of the Internet from one webpage to the next, cataloguing the contents of the sites they find and following embedded links yet further out into the networked world. The result is, in essence, what most would think of as the Internet.

But not all sites are able to be catalogued in such a way. Consider the paywall. Paywalls are access prevention mechanisms that allow a site to deny access to anybody without, say, pre-existing login credentials or the ability to purchase such credentials. An example would be a newspaper like the *Wall Street Journal.* Access to *WSJ* is restricted beyond the main webpage to those who have not purchased a subscription pass to read online content. Since web crawlers are not naturally possessed of such credentials (though they may be enabled by inter-organization arrangements), they do not, in reality, scour all available content in what we might call "the networked world." Thus, there exists a part of the Internet beyond the surface web where it is not possible to access content via conventional search engine means. This is called the **Deep Web** and is, in reality, far larger than the surface web.

And content exists in the networked world beyond even what is not indexed by the world's extant web crawling services. Some content exists on overlay networks that send information along the same physical connections that the Internet is premised on, but have unique criteria for access. Specifically, overlay networks usually require specific credentials *and* the use of specialized software to access information. These kinds of networks (**darknets**) are often referred to as the **Dark Web** (though that is a misnomer because these networks are not connected to one another via followable links as websites in the surface web or Deep Web are). Entry to these parts of the cyber world requires the use of both keys – pre-obtained credentials – and specific software to enter. Mapping the fullest extent of this bit of the digital world is tricky, if not impossible, because the primary function of enabling software (such as **The Onion Router** (TOR)) is to mask the IP address of web users.

The result is a *hierarchical* network of networks wherein your packets of information keep going upward until they find the address of the person you are trying to send a message to. The functionality of the Internet essentially rests with a series of **root servers** that regulate the registration of high-level domains and redirection of information to those domains. Traffic is routed at the highest level through such servers to a massive series of **autonomous systems** (AS). Taken together, all ASs constitute what is called the **Internet backbone** – the entirety of the physical infrastructure of the Internet, including fiber optic cables, telephone networks and satellites. This backbone infrastructure is owned and operated by **Internet Service Providers** (ISP). Sometimes your information will leave an AS and be routed through higher-level **servers** to other locations; where your machine is attempting to communicate with a local partner – say, where you are sending an email to someone located within the network of the ISP you are currently using – this may not happen.

This description of how the Internet works should result in a few realizations. Three are particularly noteworthy. First, most of what we tend to think of as cyberspace is actually privately owned. ISPs own (or rent from other backbone owners) the infrastructure of the Internet, and even the highest-level DNS functions are operated privately (though with many redundancies and failsafe features in operation at this time). This point is worth bearing in mind in later chapters when we discuss the shape of cyber policy in the United States and the relationship between the government and actors in private industry.

Second, this should give readers an idea of exactly "where" the Internet is. Of course, the Internet is in the tubes and airwaves connecting various pieces of the world via network connectivity. But in reality, the Internet is more significantly concentrated at important points where connectivity between the layers of the thing is most critical – at the level of functional control of different autonomous systems and at the root servers.[11] When it comes to thinking about the function of the Internet in the context of potential conflict between powerful countries in later chapters, it's worth bearing in mind the fact that the massive amount of redundancy vested in such a decentralized system for transmitted information around the world *does not* mean that there are no points of unique vulnerability. While the Internet would not be destroyed by a strike on critical high-level infrastructure, such as the submarine cables that carry much Internet traffic across oceans, it is certainly possible that much service could be disrupted and that some parts of the Internet could be cut off from others.

Third, and related to the first two points, it is perhaps most important to realize that the Internet is not an inherently secure system. It was not set up and augmented at various developmental inflection points with security management in mind. Rather, it was developed with greater efficiency in communications between networked computer systems in mind. Though security considerations were brought up even during the initial development of ARPANET, most famously by Willis Ware and his colleagues, they were ultimately sidelined by the argument that it was more important to get the technology right before saddling it with regulatory requirements.[12] Due to the rapid adoption of the technology across entire industries in the West in the 1970s, 1980s and 1990s, this *ex post facto* correction was not feasible.

As a result, the existence of a global network that undergirds global commerce and society is based on rather a lot of trust. And the history of the Internet is replete with instances of actors violating that trust – a trust which generally takes the form of ISPs not making false representations with regard to information pathways – and causing a great many headaches for users. This is a significant point, particularly given that the Internet will not look the same forever. Indeed, by the 2030s, it is entirely feasible to think that many of the high-level functions of the Internet will have migrated to other parts of the globe. At present, the Internet has 13 root servers, a number maintained largely because of the efficiencies to be had in squaring the requirements of the IP standards – currently IPv4, which dictates a particular length of address for **client** computers (i.e. the lowest-level computers employed to use the web, such as personal computers) – with standards on the most efficient size of data packets. Most of these are in North America, with a few being in Northern Europe.[13] However, under IPv4 the world is running out of IP addresses. Thus, the world is transitioning to IPv6. This is a complex process and adoption of the more lengthy address format is going to take time. However, it changes the calculation regarding packet composition such that more high-level DNS servers could make sense out into the future. By the 2040s, much web traffic could run through emergent leaders in digital industry – Israel, China, Russia, Germany or Brazil. Given this, the issue of trust as the basis of Internet functionality becomes increasingly more problematic. Can an expanded set of countries be trusted to maintain the shape and functions of what we've come to know as the digital world?

Box 2.2: Is the Internet American?

The Internet is largely owned and operated by private enterprise, both in the United States and around the world. But does any one country have more control over or access to the logical functions of the Internet than others? Put another way, is there such a thing as a "national-level" Internet (i.e. an Israeli Internet, a U.S. Internet, a Russian Internet, etc.)?

There are two ways to think about this – in the context of (1) the physical infrastructure and informational content of the Internet and (2) the logical functionality of the Internet. In terms of the physical setup of the Internet, the modern Internet is underwritten by a broad system of submarine cables connecting the world. Though the Internet was birthed in the United States, these cables and the on-land network infrastructure that they plug into are reasonably evenly distributed around the world, with substantial development in the 21st century focused on South American (particularly Brazilian), South and East Asian and Middle Eastern/Indian Sub-continent connections. From this point of view, it would be unreasonable to say that the Internet is intrinsically "American" beyond the historical context of its genesis. Likewise, it would be reasonably accurate to say that there are such things as "national Internets." After all, the most common locations for exchange points – i.e. for those nodes where data traffic is transferred from one autonomous system to the next – are on national borders. Thus, in the event of a conflict, it is possible to target by either physical or cyber attack the Internet of a specific country (or regional sub-unit, like a province or in some cases a municipality), and, in the event of political interference with the Internet, the scenario of a **Balkanized Internet** divided along national lines is not fantastical.[14]

In terms of the information environment enabled by this underlying Internet infrastructure, it is also somewhat reasonable to assume that much information is stored along national lines. Websites run by German citizens are more likely than not hosted physically inside Germany somewhere. The wrinkle here is that major consumer and enterprise services for storing content and deploying applications online have increasingly moved to **the cloud**. The cloud is quite simply a configuration of information services that allows businesses to offer immense resources and storage to customers without the need for significant local infrastructure. In other words, **cloud service providers** (CSP) use the web to provide customers with access to processing power, data storage and development environments remotely. Instead of having to buy servers and the like to set up a business, entrepreneurs can simply pay a CSP for access to pre-existing resources that are accessible online.[15] Resultantly, the trick here is that consumer content is often not stored close to the consumer – oftentimes not even in the same country. Massive server farms operated by multinational corporations in cold climates (to save on cooling costs) constitute an incredibly large portion of the information layer of cyberspace, all of which adds an additional dimension to the question of "national" information ecosystems.

Whereas it might be reasonable to argue that there exist national Internets based on how Internet infrastructure is internationally distributed, the same cannot be said when one considers the logical functionality of the Internet. Data traffic flows upward as people send messages and otherwise transmit information around the world. Specifically, data flows upward toward root servers that are able to route packets between high-level domains. Here, the reality of U.S. first-movership in Internet affairs is that most root servers are physically located within the United States and, to a lesser extent, Northern Europe. Thus, while the physical infrastructure of traffic exchange is well distributed around the world, most Internet traffic inevitably flows through the United States.

From this point of view, one might reasonably argue that the Internet is American. Though the Internet is not federally owned, the United States has an unusual degree of privileged access to Internet communications, something that has been the source of unique security developments (discussed in later chapters) over the past few decades.

Naturally, this is not the entire picture when it comes to understanding insecurity in the world in the digital age.[16] Hopefully, this lends some insight as to why the governance of cyberspaces (a topic more fully covered in later chapters) is so fraught in international affairs and where the main perspectives on that governance lie – from the **multilateralism** of states that favor state control of the web to the **multi-stakeholderism** of those that want to include civil society. But there are few answers for security issues at this level. Starting again and building a new Internet from the ground up, one with better attention paid to security management, is quite arguably a non-starter. From the basic details discussed earlier, this might not seem to be the case. But, again, perhaps the best way to think about this is in terms of the levels of analysis. With the history and brief detail of the development of the Internet offered earlier, we have shown what amounts to a topographical view of the digital world from on high. If one were asked to describe the international system, one might start by talking about the existing borders of the world and how people – and money and militaries and information – move across them. That is essentially what we've done here. We have described the borders and the transmission functions of the digital world. But we have not yet fundamentally described where many enduring vulnerabilities come from, particularly not in a way that allows us to move on to topics more specifically associated with the study of cyber conflict – the shape of cyber attacks, the actions of defenders and the prospects for conflict mitigation. To do that, it is necessary to think of yet more fundamental issues bound up in the design and development of information technologies.

The security of information and information systems

If the core function of new information technologies is to enable more efficient communication, then the first challenge in designing such systems is in how to customize information transmission mechanisms to ensure security. In other words, designing information technologies requires building ways for people to send information to whomever they want without publicly broadcasting that information and while preventing malicious third parties from intercepting what is being communicated. This is an age-old problem and isn't unique to the digital age. How does one make sure that only those you intend to read your messages can actually read them? How do we prevent public transmission of information that is intended to be private?[17]

This is an issue of information security. Functionally, this is a question of **cryptography**, **code making** and **code breaking**. Broadly, cybersecurity (in the innumerable ways that it manifests) is problematic and worrisome on two fronts. First and foremost, information technologies undergird an increasingly complex global system of interactions within and across traditional national borders. ICTs are the wiring of critical societal functions, and attacks on that wiring can naturally lead to deleterious outcomes, from the trivial (i.e. stealing small amounts of money) to the extreme (i.e. subverting an element of a country's command-control infrastructure for nuclear weapons contingencies). Second, it is never going to be perfectly possible to design ICTs wherein the core function of systems can be assured. That function has to do with access to information. When different parts of the global information infrastructure talk to one another, they have to make sure that only those with authorization to see different information or use different features are able to do so. All security issues within the networked world, thus, can be boiled down to the issue dual of *authentication* and *authorization*. In essence, is someone who they say they are, and do they have the right to access what they're trying to access? These issues manifest at two levels – the mathematical and the functional (i.e. how we implement information security when designing technology).

This subsection deals with the former; the next deals with the latter, computer and network security. Chapter 4 then further discusses the historical context of information security challenges for states in greater depth and provides a framework via which it is possible to trace the development of computerized and networked information systems for national security purposes over the course of the past century.

Traditionally, the general shape of the authentication problem with the digital world has been illustrated in game theories and mathematical proofs. One common game, for instance, involves a scenario wherein several generals and their armies have laid siege to a city.[18] They cannot take the city except with the aid of the other two generals. In order to coordinate their attack effectively, they must send couriers to each other to relay relevant orders. However, they do not know if one or more counterparts are actually traitors. If one is, then the communication might be misleading. How can they ensure that their peers are telling the truth? How, in essence, can the generals ensure that their system is relatively **fault tolerant** – i.e. able to withstand the possible betrayal of incoming information and be best positioned to determine false intention? Here, there are two basic concerns. How can the privacy of information be ensured, and how can the receivers of information ensure that said information has not been altered?

Though such concerns are not unique to the digital age, they apply in a fundamental sense. In designing information systems, we must attempt to ensure a degree of fault tolerance. Mostly, we want to authenticate usage by authorized users in as secure a manner as possible. But how is this done? The basic idea is that we figure out how to scramble the contents of our information transmissions in such a way that only our affiliates and our intended recipients know how to unscramble them. If Rupert wants to talk to Rose without others knowing the content of their messages, they need only design a secret code that transforms the original message into gibberish and then – through application of the same code known to the recipient – translates that gibberish back into meaningful text. In this scenario, Jimmy and Betsy use an algorithm they have previously agreed upon to translate messages – called the **plaintext** – into gibberish – called the **ciphertext** – that only they can resolve.

The problem with this is twofold. First of all, the algorithm used to translate plaintext into ciphertext and back again must be robust. Human beings have a great many reasons to want to break the encryption of others' communications and are, eventually, quite clever. Even the most robust algorithms for encrypting information are usually mathematically imperfect. The reality is that a sufficiently powerful computer or the efforts of a human analyst in dissecting a given encryption approach might spell doom for a particular given method of secure communications. Second, symmetrical encryption is problematic because it does not provide for a way to eject untrusted individuals from the pre-existing secure communications agreement. If Rupert and Rose decide they don't like Willie, then there is no realistic way of ejecting him from the system without entirely changing the system. Therefore, given the usual challenges of human relationships, symmetrical encryption approaches to securing information will enduringly fall apart as time goes by.

So, how might we design fault tolerant systems that ensure the transmission of information securely over time? The answer to this question can be found in answering the question outlined earlier of the generals trying to coordinate an attack. How might we answer the question: how can you be sure that a courier has not been intercepted or that a counterpart general is not a traitor? The answer to doing so does not rest with information protocols. Even if generals that receive messages from their peers request further instructions or commit to action based on a contingent event, the level of trust cannot exist such that a successful attack would be mounted. But what if the courier involved was a lifelong friend or family member? What if the courier spoke a secret codeword known only to a select few? Or what if the courier spoke at the same time as a corroborating message – perhaps a specially colored smoke – was sent from their general's camp as corroborating evidence? In short, the answer to a range of information security issues lies in the appropriate use of *both* public and private knowledge when designing effective information systems.

The simplest form of this solution to the basic issue of common knowledge of algorithms is called **symmetric encryption**. With symmetric encryption, all users of a system hold a **key** that allows them unique access to information sent from a specified other party. Much like how lots of people have the same design of lock on their front doors, or many banks utilize the same vault lock systems, algorithms are designed well and used for everyone (using a given service or system). However, a personalizable key is necessary to open messages that are sent to you from another person. In essence, you and another person have agreed upon a key that both can use to solve the algorithm for only your communications. The key will be different for your communication with others. This is one of the most common ways of encrypting information found in the world today. However, again, there are distinct problems. For one, the number of keys a single individual must hold – particularly if that individual is running a large company, for instance – can be immense. And though symmetric encryption allows one to exclude individuals from communications that have left an organization or are thought to be traitors (or what have you), their keys need to be destroyed at that point. If keys are willfully discarded and discovered by others, the whole encryption scheme falls apart.

But there must be further steps that, mathematically speaking, we might take to ensure information security, right? Of course! In general, the rule of thumb on information security is that defenders and security designers have a massive advantage over potential malicious attackers in that they can mathematically stack the deck – from a probabilistic point of view – against someone being able to fool defenders or crack encryption in a reasonable amount of time. This is done by yet further employing private knowledge to act as a force multiplier to ensure that authentication is almost always possible and that privacy of information is ensured. Cryptographic solutions to information security challenges constitute a broad set of approaches to sharing information that involve different authentication mechanisms and one-way mathematical functions – calculations that are easy to make one way, such as the multiplication of prime numbers, but not the other, such as the division of the result to equal the initial set – wherein detailed knowledge of the sender of information is required in order to authenticate incoming information.

These solutions take several different forms that can generally be thought of, again, as focused on authenticating access to information systems by authorized personnel and ensuring the actual privacy of information. In other words, some solutions just assure readers that the message they're reading is the exact same one sent to them from the known sender (i.e. not intercepted and altered). Others ensure the privacy of the content of data packets themselves.

One simple method of authenticating the contents of incoming information packets is for recipients to study a **message authentication code**. The notion behind these is simple. A mathematical function is used with the shared private key by individuals in the know, which produces a distinct code. When the message is received, the recipient simply uses his or her own version of the same key to compute the code and compare it with the incoming one to see if they match. If they do, the message has not been tampered with. This is a simple method of authenticating information, as intercepting parties should be unable to alter *both* a message and its code accurately without access to the secret key. **Hash functions** do a similar thing. A one-way mathematical formula is employed that produces a short, unique code in place of a much longer one. The formula cannot be reversed (i.e. the hash value of a couple dozen digits cannot be worked backward into the original information) and is complex enough that the chances of two large pieces of data producing the same hash are zero, probabilistically speaking. Hash functions are massively useful, as they provide simple ways for users to translate information into verifiable (and short!) code pieces that would require the initial data to replicate. Interception is functionally improbable in the most extreme sense.

The general notion behind hash functions – i.e. that we can use mathematical functions that are easy to compute one way but really (really, really, really!) difficult to reverse – also serves to solve the problems with symmetric encryption outlined earlier. Using such functions, we have developed **asymmetric encryption** procedures. In essence, users still use common algorithms to protect their data but split the job of the key into two keys. Using these one-way mathematical functions, they create one key for encrypting messages and another for decrypting them. They publish the encryption message for the entire world to see, thus ensuring that anybody can send them encrypted information (a useful thing in a world beyond the imprecise scenario wherein only intra-organization encryption is desired). But only the recipient can unlock messages and, of course, there is no dissemination of the private key (i.e. no potential breaches from unscrupulous partners). This kind of encryption is also often called **public key encryption** and is remarkably similar in format to **digital signature encryption**. Whereas asymmetric functions used to ensure data privacy focus on exclusive access to private keys that can decrypt data, digital signature schemes only seek to ensure that authentication is possible. They are not about ensuring the actual privacy of information sent between parties. Thus, these schemes switch the role of the public and private keys. Information senders can encrypt messages using their private keys and provide a public key to everybody that can decrypt them. Information decryption is easy, but only the sender could possibly have encrypted the message to begin with (as only they have the private key and, thus, the ability to "sign" the message).

Finally – most simply, but perhaps most notably – one solution to information security issues is just to use incredibly long digit requirements when requiring different encryption standards. If passwords are extremely long, they are exponentially harder to crack with **brute force attacks** (i.e. computer-based attacks where an attacker simply tries every possible combination of letters and numbers until they get it right) than are short ones. Mathematically, this actually breaks down to an advantage for defenders based on extremely simple actions. Just one additional letter in a passcode (a key) multiplies the difficulties involved for attackers trying to guess the right one and force entry (aka the breaking of encryption). And this advantage in incremental steps with cryptography for defenders is actually suggestive of a broader dynamic reality in that the various techniques and approaches outlined in this section are rarely employed in a vacuum. Rather, designers of systems most often employ different protocols wherein users must employ different techniques in steps in order to authenticate, prove their authorization and interact with information. This multi-technique protocol approach to information security stacks the deck against would-be attackers in powerful ways.

But if the deck can be stacked against attackers, why is information security such a pressing concern? Why do attackers still attack? A big part of the answer is the same as it is at other levels of analysis – that the users and designers of information systems are human. In just 70 years, humans have become very capable of designing advanced computer systems (or computers that can help design newer computers). Moore's Law states that processing power potential will enduringly double roughly every 18 months.[19] At that rate, computers will exist in some years' time to whom today's basic key length protections will present a limited challenge. Whereas today it might take a computer thousands of years to brute force attack a robust 1,024-bit key, in 20 years it may only take a few days or hours. And particularly with quantum computing power in sight as a real possibility, prior encryption procedures may largely go by the wayside and have to be rethought given unprecedented access to processing power. In short, cryptography is about making information attacks probabilistically unfeasible. Out into the future, this means constant vigilance and new development as computers will inevitably manage to catch up to today's standards and make attacks far more feasible.

Beyond processing power considerations, information security is so pressing a concern because of the element of human usage of information systems. This manifests in two ways. First, in building protocols, we often resort to using less complex authentication and authorization schemes than we should in order to provide for a degree of practicality. Some of what has been described here constitutes processes that take some time – real-time minutes or hours. In applying these practices to the design of systems used by banks or militaries, designers must consider operational imperatives. Real-time delays might blunt the function of organizations or put what is considered to be an unreasonable demand on operators. Thus, a compromise will be instituted that makes things slightly less secure (but still pretty darn secure!) in exchange for enhanced time management abilities.

Second, it should be reasonably obvious to anybody reading this chapter that the topic of keys and algorithms can proxy for (and, indeed, is primarily intended to proxy for) the more specific topic of computer passwords and security systems. We discuss the latter – computer and network security systems – further later in this chapter. In thinking about keys as passwords, however, it has likely occurred to you (the reader) that key security is a major challenge. The most common passwords in the world today remain "p@ssword," "12345," "guest" and so on. Even with genuinely robust passwords, we as users are encouraged to use information in the construction of passcodes that will help us remember our credentials. Among even the best thought out passwords then, there is a pattern of human input that can be taken advantage of by attackers. Some kinds of brute force attacks, for instance, employ a hierarchical list of likely words based on a study of how people tend to construct passcodes. Others employ web resources to try to learn about potential victims and then use personalized information to break encryption more quickly. And personal information in the content of messages can allow attackers to circumvent encryption procedures as well. The famous case of the codebreakers at Bletchley Park deployed to crack Germany's Enigma encryption scheme during World War II is a case in point on how a combination of minor design flaws and deft intelligence work can lead to defunct information security practice.[20]

Implementing it all (or "the security of networked computers")

Though information security challenges are partially about prospective future computational capabilities, the primary problems we face stem from issues with implementation. Designing information systems of various kinds that are truly secure is a serious challenge. Partially, this is because of the ways in which humans use information systems. But there are other problems that should be addressed too.

This section addresses these "computer" security issues. The word computer is entered in quotation marks because the security of information systems from an implementation and design perspective does not fully pivot on the study of individual computers. Rather, the "computer" challenges that we care about exist and must be discussed in a layered fashion. There is the security of individual computers used by multiple individuals with different degrees of authorization. There is the security of computers vis-à-vis authenticating any of those users' access credentials. There is the security of computers connected to network connections. And there is the security of networks themselves. There are distinct design issues that computer engineers and scientists have to grapple with at each layer. Here, we face similar issues as we do with cryptography insofar as the human element is often to blame for the general vulnerability of information systems. But with computer security, the technological solutions to information security issues themselves present as less probabilistically powerful than do the mathematical solutions to basic cryptographic problems. Computer security is fundamentally a function of man-made systems attempting to solve organizational problems. Thus, there are almost infinitely more avenues for malicious actors to forge paths.

But what is actually involved in computer security? In essence, computer security is about the implementation of information technology in such a way as to ensure three things – the **integrity** of information, the **confidentiality** of information and the **availability** of information. Collectively, these three requirements are known as the **CIA Triad** and are the fundamental elements of **information assurance**, the practice of assessing and managing risks related to the threat of information attack.[21] In reality, there are two other requirements that are less generally considered to be main pillars of information security in this vein – the **non-repudiation** and the **authenticity** of information.

Maintaining the integrity of information is a reasonably simple notion – it means that actions must be taken to ensure that unauthorized users do not alter information stored in computer systems or transmitted between systems. Hand-in-hand with this, the information assurance requirement of confidentiality simply means that steps must be taken to ensure information privacy. That is, those without authorization must not be allowed access to information, period. This is actually simpler than integrity, as the point here is simply about access and not focused on alteration of information. Maintaining availability of information systems, the third leg of the information assurance triad, is where computer security turns to more purely functional considerations. Maintaining availability simply means that information systems need to continue to function as intended by designers and operators, regardless of malicious efforts to compromise information security. Finally, the two lesser-known requirements of information assurance are related to the ability of users to know that information is accurate *when it is being sent from one party to another.* Authenticity differs very subtly from integrity insofar as information should be able to be verified as sent by the specified recipient. Integrity, by contrast, is simply about keeping information – which is often stored and used but not sent – unaltered (at least by those without authorization). Non-repudiation is more broadly about standards for ensuring that authenticity can be universally acknowledged. In short, non-repudiation is the ability to validate the authenticity of communications between two users (or systems) in a public setting.

Box 2.3: Information assurance as a metaphor

In international and national security studies, thinking about systems – from political systems to conflict environments – as kinds of information systems isn't such a bad idea. The functionality of much of what is involved in the national security enterprise can be best understood by looking at the mechanisms of information transmission, information privacy and the mitigation of uncertainty related to information inherent in a given system.

A good example is that of political systems, particularly democratic systems of governance and public participation. As we discuss in later chapters, the digital age has augured in new means and modes by which states are attempting to interfere with foreign political systems for strategic gain. Between at least 2013 and 2018, the Russian Federation was widely accused of using cyber means to enable broad-scoped subversive espionage campaigns against Western political systems in Europe and North America. The most prominent of these, interference in the U.S. presidential election campaign season in 2015–2016, included a broad range of cyber attacks alongside manipulation of new digital content systems (i.e. social media platforms and other online content distribution systems). Putting aside the objectives and effects of such a campaign (which are discussed in later sections), it is easy to see the intended effects of Russian efforts if we use information assurance as a metaphor for how the system – in this case, democratic systems of discourse and politics – works.

Simply put, we might analogize democracies as functioning through the existence of different mechanisms that modulate how information is treated when it is in the public domain (i.e. when facts are being reported, debated and discussed among the public). Experts, legislators, free news media organizations and more all engage with information and present it to civil society in such a way that relatively prudent public policy decisions might emerge in voting and in representative governance. *Modern cyber-enabled interference campaigns aim to degrade the ability of those mechanisms to work properly so that democratic processes become less effective.*

What's the value of the information assurance metaphor here? Easy! By manipulating the content of the information environment by producing fake news and by maligning entrenched political interests via questionable sources, foreign influence campaigns aim to degrade faith in the **integrity** of available information by violating the **confidentiality** of private information and making **non-repudiation** extremely difficult. The goal, in short, is the manifestation of what is called **Byzantine faults** in the system. This term emerges from the problem of the generals (typically portrayed as the **Byzantine Generals' Problem**) trying to communicate with one another that appears throughout this chapter. Given that the generals don't know who might be a traitor, their system isn't just flawed – there is clear potential for faults to exist without their knowledge. With influence operations against democratic systems, this is the primary goal.

These aspects of security manifest across design and implementation considerations over every element of the apparatus of computer security, from the internal programming design of personal and specialized computers through the functionality of networks. In reality, each element of the information assurance triad (and the two additional elements) can be boiled down to access control or "how our technology can be set up to allow for different customizable types of access regulation." At the level of the personal computer, the primary concern is in regulating access to information and the ability to alter information across different people using a computer. This was a major problem until the early 2000s when personal computers became so ubiquitous that expectations regarding usership of computers changed (i.e. it became far less common for multiple users to share a computer). Before that time, security models – of which the **Bell LaPadula model** is perhaps the most broadly known – were developed wherein integrity and confidentiality were maintained largely via a process of status-to-classification comparison. In other words, the level of permission that a user had (e.g. "Confidential" or "Top Secret" clearance) was compared to the classification of a document or process to determine whether or not that person could undertake a particular action *as different from another user.*

That said, these concerns have re-manifested in the decentralization of computer functions across network systems such that multiple users now commonly access parts of the same computer(s). The main challenge with internal computer security from a software design perspective is in keeping the core security programming of a computer simple. With operating systems, the **kernel** is the main section of hardware, firmware and software that is both trusted and considered to be vital to system function. The challenge for computer engineers is in preventing the kernel from becoming too expansive, as only by keeping the kernel tight can the possibility for manipulation of user access privileges be realized.

Then, there is the problem of authenticating user efforts to access individual computers (or distributed computers). This is different from essential computer security issues because there doesn't exist an assumption that at least some degree of authentic access is allowed to users. Rather, the concern is that malicious actors might try to access a computer without the proper authorization to do so. Naturally, the main question here is: how might we ensure that access allowed is access authorized ahead of time? The answer lies with information security and the key that authorized users are given. In other words, it lies with passwords or equivalent entry credentials. As described earlier, the challenge with designing systems to be secure at this level is simply one of determining how to prompt the construction of better keys. Most systems require users to adhere to certain rules when doing so, such as having a certain number of characters or a certain diversity of characters (i.e. at least two numbers and a non-alphanumeric symbol). Many more additionally take security steps to prevent unauthorized access, such as locking down after a certain number of unsuccessful attempts. And yet other systems employ passcodes that are of a different nature than simple alphanumeric sequences. Biometric systems – wherein a person's fingerprint or retinal information, for instance, are the passwords – are increasingly common in mobile computing product development and in private industry. The problem with all of these, of course, is that they rely on human input and are relatively simple to fool for the dedicated attacker.

Moving beyond the internal function and design of computer systems, we quickly run into issues of computer security in the digital age. Issues of access control on specific computers have been grappled with for more than half a century. Beginning in the 1960s and 1970s, however, designers and engineers increasingly had to adapt security procedures to deal with the threat of network-based attacks. Specifically, they had to adapt their work to deal with the threat of malicious software, or **malware**, which was designed and prepackaged to disrupt the normal function of computer systems.

Understanding malware is relatively simple insofar as we might think about malware as several categories of techniques developed to subvert access controls and violate the requirements of information assurance in several ways. Broadly, malware includes **viruses**, **worms**, **Trojan horses** and a host of other related code applications for achieving malicious effects that are less easy to categorize. Of these, worms are perhaps the simplest to understand. Though they can be designed to employ sophisticated break-in techniques, worms are essentially pieces of code that first break into a computer and then use any available connections to try to spread to other computers. If successful, the worm clones itself and is replicated across a network of computers, disrupting functioning. By contrast with worms, which do not necessarily hide their actions, viruses are pieces of malicious code that co-opt whatever system they enter and attempt to spread their influence to achieve a pre-set outcome. Viruses can work in a number of different ways but typically either target-specific programs or subvert systems via direct interaction with a computer's underlying programming language. Trojan horses are pieces of software that are either designed to be or introduced as a malicious element of otherwise seemingly benign computer programming. As was true with their namesake, Trojans are highly seditious and virulent tools for stealthily entering a system and subverting access controls to enable malicious actions.

Box 2.4: Viruses, worms, Trojan horses . . . what do they really do?

Though the technical design of given malware instances can be highly complex, it is relatively easy to understand the function and general workings of malicious programs designed to achieve harmful effects in computer systems. Functionally, we might think of malware as being employable for two purposes. Some malware is designed with delivery in mind, and some is focused on achieving a malicious effect. In any malware attack, some malicious code is wrapped up in a delivery mechanism of some kind. The "wrapper" is designed to elude security measures in networked computers so that the **payload** can be delivered.

It is reasonably commonplace to label all malware as computer viruses. In reality, viruses are malware which, much like their biological equivalents, spread from computer to computer in reference to some condition. Where a biological virus might spread between people off the back of poor hygiene practices or through accidental exposure to the bodily fluids of an infected person, computer viruses spread off the back of users' actions (such as the sending of an email). What a virus does is not determined by its ability to infect different computers, but rather by what the payload it is carrying is designed to do. Indeed, viruses are such a ubiquitous threat to computer users and network operators because they are merely the courier for a potential universe of malicious code.

Worms and Trojan horses are, similarly, mechanisms for delivery malware with a more specific purpose. Worms are a subset of viruses that are often held apart because they do not require human action to spread. Worms are more virulent and purposeful (but also often more detectable) than viruses because they are capable of self-propagation. Trojan horses, by contrast, are simply pieces of delivery software aimed at entering computers not through subterfuge, but by misdirection. Much in the way the original Trojan horse enabled Greek soldiers to enter Troy via trickery, Trojan horse programs are designed so as to convince victims that they are legitimate. Then, after a user downloads the seemingly legitimate program, the payload is released.

Generally, computer viruses employed by a dedicated attacker are difficult to combat. Viruses are designed to mimic the biological process of adaptation and mutation over time. Specifically, computer viruses are designed to **self-modify** in order to fool antivirus software, which most often works by simply scanning files for evidence of known malware signatures. In the simplest sense, viruses are able to rewrite themselves by simply jumbling and reordering the subroutines that constitute the code determining their movement. More pressingly, more complicated instances are able to employ cryptographic techniques in order to evade detection. Encrypting the **payload** malware hidden within the virus delivery mechanism is a common counter to antivirus defenses that involves the program transforming the payload code into ciphertext to be decrypted just prior to delivery. The trick here is that the *decryption module* must inevitably remain outside the encrypted section of the software. Antivirus programs look for such tools, though sophisticated viruses contain **polymorphic** capabilities that allow them to rewrite the code of the decryption module for each new computer. In rarer instance (particularly rare because this swells the size of the file in question), viruses are sometimes given **metamorphic** capabilities that allow the program to entirely rewrite itself with each new computer. The result is a program that achieves the same effect over time but never contains the exact same code sequences.

And finally, information assurance issues exist in questions of network security. Network security is different from computer security insofar as this is the first place where our discussion of digital security intersects with the earlier discussion of the shape of the Internet. Here, security concerns are largely about the security of packets of data that enter a given network. Ensuring the integrity of those packets is critical for the task of ensuring that no malicious behavior takes place within a network. And the threats of packets are significant. Common attacker stratagems include the use of **packet sniffers** to read the content of packets being sent to a given network (so as to extract password information or modify contents to redirect traffic to a malicious web address, an action called **IP spoofing**).

At the same time, network security involves maintaining the functionality of the network itself, often in the face of persistent opposition. **Denial of service** attacks are commonly employed to disrupt the function of networks and, thus, the operations of computers within those networks. Denial of service attacks take a range of formats but, in essence, the point is to take up so much bandwidth that a network server is unable to continue legitimate operations. One common type of such attacks (called a **SYN flood**) takes advantage of the handshake protocol described earlier in this chapter to do this. A handshake is extended and the network server puts aside a small amount of runtime (memory) in order to handle the new incoming request. But the victim's acknowledgment is never acknowledged. After enough of these requests, the victim runs low on memory and has to stop legitimate operations to handle what has become a massive and seemingly unending volume of incoming traffic.

Distributed denial of service attacks do pretty much the same thing, though the method of disrupting legitimate network operations is incoming traffic from a large host of malicious computers. These computers are often themselves compromised by malicious actors (often without the knowledge of the user) and directed against a specific network target. Those compromised computers are called **zombie computers**.

Network security is a tricky beast. Defensive technological principles emphasize two sets of activities: manning the perimeter and monitoring traffic within networks to try and spot odd activities. The next few chapters deal with computer network exploitation/attack (CNE/CNA) and defense (CND) in more detail. But it's important to note here that network security is as inherently challenging as are other elements of computer security and for similar reasons. The networked world is complex, and there exists a plethora of connections in computer-based human activities that constantly offer avenues for the violation of information assurance principles – for information theft, modification and more. In truth, computer defense and network defense are not doomed tasks, of course; the defender has a broad range of tools available to them that makes the job of potential hackers extremely difficult. Again, these are discussed over the next few chapters. But where the mathematical realities of information security favor the defender in a probabilistic sense, the realities of computer security (broadly writ) are a more balanced affair. Technology design is simply not refined to such a degree that malicious action is either unthinkable or unfeasible.

What's vulnerable in the information age?

So what is actually vulnerable in the information age? The previous sections of this chapter have highlighted a number of different issues with the construction and function of the digital world. Most of these can be boiled down to the human factor. More specifically, most of these can be boiled down to two human-derived facts. First, humans create and implement information technologies. As such, there are inherent insecurities in the imperfect designs we promote. Refinement occurs but is often outpaced by innovative advancement in the construction of better technology. Thus, at the same time, humans are providing both the flaws in information systems and, despite positive developments in achieved in parallel, the mechanisms for taking advantage of them. Second, humans have to use information technologies. As creatures of pattern and habit, we favor interactions with technology that balance convenience with security. Thus, we inevitably use bad passwords and systematically endorse the use of methods that fall short of ideal security efficiency.

This said, it is not accurate enough to blame all that is vulnerable about the digital world on humans. Rather, as a number of scholars have in their treatments of cybersecurity issues, it would be more accurate to split the digital world up into layers.[22] Specifically, discussion of cybersecurity as it intersects with traditional topics of international security – from interstate conflict and the function of military-intelligence apparatuses to the use of ICT by non-state actors – is arguably best served by generalizing on the content outlined so far in this chapter and identifying the broad contours of the cyber threat ecosystem.

Scholars and practitioners operating at the intersection of cybersecurity issues and national security processes tend to break up cyberspace into four layers. They do so with an eye to describing the distinct threat modalities faced by actors from the full range of foreign threats. The first of these is the **physical layer** of the networked world. The physical layer is constituted of the real-kinetic components of the Internet and other network infrastructure. More specifically, this layer is constituted of fiber optic cables, microwave receivers, physical computers (i.e. personal computers, servers) and other kinds of connecting wiring. Of particular significance in this layer are the submarine cables that crisscross the world's oceans and carry data packets from one side of the globe to the other with almost no delay. These cables are particularly important considerations for national security planners because destruction of a relatively small number could significantly hamper Internet functionality (and, thus, the functionality of economic sectors). Submarine cables could be mined or attacked by naval vessels in the event of war. Perhaps more worryingly, these cables are vulnerable at their "landing zones" where they come on shore. Many such points of transition between sea and land are insecure and, though meaningful disruption would require large-scale attacks on numerous landing points, would make easy targets for a dedicated non-state attacker.

The second layer of cyberspace is the **logical layer**. The logical layer is constituted of those systems and procedures that dictate how packets are sent from one part of the networked world to others. As outlined earlier, this principally includes the Domain Name System and various protocols that are critical to expected functionality of the web for everyday citizens of planet Earth. However, the logical layer also includes tools used – both legitimately and illicitly – by denizens of the digital world to access networks in a specialized fashion. Browsers like The Onion Router (TOR) that are designed to allow access to secure sites and to anonymous web traffic to a degree are used by people the world over to securely (and, admittedly, often criminally) interact with others. TOR is actually the result of DoD funding,[23] but other similar tools are the work of non-state actors and private companies around the world. Dynaweb is an example of one such non-government-developed product that constitutes an important part of the logical layer for dissident non-state actors around the world.[24] Developed by exiled members of Falun Gong, an outlawed spiritual exercise group in China, Dynaweb, allows individuals in repressive digital environments to circumvent state controls and access the global Internet more freely than would otherwise be the case. Both TOR and Dynaweb are tools that enable certain kinds of specialized sociopolitical activities and, as such, stand as strong examples of vulnerable elements of the logical layer of cyberspace.

The **information layer** of cyberspace constitutes the actual content of the digital world, from the languages underwriting systems' designs to the text, imagery, and multimedia content of different databases. We might break this layer down further into two sub-layers of the networked world – the *syntactic layer* and the *semantic layer*. The syntactic layer is constituted of the informational design components of information systems and, as such, might also be said to contain elements of the logical layer. The semantic layer, in contrast, is entirely constituted of the above said content that computers are built to accommodate and make secure. This description of the information layer is brief, largely because of the attention paid to security implementation earlier and because of the attention given to the topic of computer network attack and defense in chapters to come.

The final layer of cyberspace is the **user layer**. Naturally, this layer is not technologically constituted. But if cybersecurity is the security of socio-technical systems emerging from or affected by the information revolution, then any map of cyberspace must necessarily include the primary category of sociopolitical inputs to the other layers. That category is defined by human approaches to using computer systems. Much of what security researchers and practitioners are concerned about at this layer falls under the moniker of **cyber hygiene**, or those actions taken by individual users to maintain a healthy position with regard to computer usage (i.e. to maintain effective security habits). From the perspective of the attacker, the most viable approach to intrusion is most often through the user layer. **Social engineering** occurs when attackers attempt to trick users into surrendering credentials (i.e. passwords and usernames) or valuable information that might make the job of forced entry into digital systems easier (for instance, information commonly incorporated into passwords such as birthday information, the name of a pet, family names or a home address). Social engineering can occur in a range of formats, but by far the most common techniques include **phishing**, **spearphishing** and **waterholing**. Phishing involves the large-scale distribution of false messages in an effort to have victims follow a link to a malicious website (or to download a piece of malware). After a link is clicked, users invariably are asked to offer personal information or are unwittingly made to authorize access privileges for some downloaded malware. Spearphishing differs from phishing only insofar as messages are more intensively customized and sent to a few targets (as opposed to a massive number). Waterholing, by contrast, flips the phishing scenario by placing malware or tricks for naive users on commonly frequented websites. In doing so, attackers are "staking out the watering hole" in the same way that predators might do in the wild.

Box 2.5: What is cyberspace?

Given what we know about the design and function of the networked world, what exactly is cyberspace? To some degree, this is an unfair question. There is no consensus position on what cyberspace actually is, nor any agreed-upon standard of judgment for settling any debate on the subject. Some regard cyberspace as a technical system of interconnected computers; others think about the thing metaphorically as a medium in which human interactions occur in non-physical ways.

The first question on the nature of cyberspace that many ask is quite simple: "Is cyberspace the same thing as the Internet?" Here, there *is* some consensus that the Internet *cannot* be synonymous with cyberspace for a couple of reasons.

The Internet is constituted of the physical and logical layers of cyberspace. In essence, the Internet is a transit system for digital communications that has a clear underlying infrastructure and, in the logical elements that enable information transmission, a well-defined functional design. But does the Internet include the information and user layers of cyberspace? Though it is harder to tease apart from the vision of the Internet that most people hold in their minds, the information layer is not specifically linked to the function of the networked world. Prior to the ARPANET, information was stored in computers around the world in much the same way that it is today – it just wasn't accessible for anyone without physical access to a given computer. Indeed, it was the digitization of information and of control systems in computers that prompted many to support the development of network features within the DoD community. So while the information layer is intrinsically linked to the Internet, it preceded and clearly exists separately from it. The same might be said of the user layer. Again, it is hard to tease apart the people that use the Internet, given that the Internet is designed to accommodate broad-scoped human interactions. But users are not, in their utilization of networked computer systems, a critical element of how the Internet functions. Thus, the user layer of cyberspace is not intrinsically a part of the Internet.

So what is cyberspace? It is worth thinking about this in terms of how national defense establishments have conceptualized the thing. In the United States, specifically, the DoD has regularly described cyberspace as the **fifth domain of warfare**. This means that, alongside land, sea, air, and space, the U.S. service branches consider cyberspace a unique domain in which security operations can occur.

From this point of view, cyberspace clearly includes more than just what functionally constitutes the Internet. After all, the target of offensive cyberattacks is often information stored on servers, internal computer processes that control an industrial system or the ability of malicious hackers to act online. While cyberattacks do principally involve manipulation of the logical design of information and network systems, the scope of security operations inevitably includes things not defined by how networks work (i.e. computer-stored information and the ability of users to engage others online). Thus, while it is certainly possible to debate where cyberspace ends and other domains start, it is obviously the case that cyberspace is bounded most specifically by the way in which networked ICT cedes humans the ability to engage one another in a manner categorically divorced from how humans interact in other domains. It is defined by human usage of networked computer systems and so includes the infrastructure of the Internet, the underlying design characteristics of information systems, information contained within those systems and human agency reflected in those systems.

Of interest, the DoD, which envisions cyberspace as being constituted of layers in this fashion, does not specify the user layer in its definition of the domain. Nevertheless, any survey of the basic principles of information and computer security – like the brief sections presented in this chapter – demonstrates clearly that the human factor is the most significant determinant of variable risks and of threat profiles in the cyber world. Humans not only have to design and implement information security requirements, but we are also the primary users and manipulators of the information systems constructed in the world to date.

Beyond design: network externalities as an underlying source of insecurity

This chapter has placed significant emphasis on the role that humans play in being responsible for enduring insecurities in the design of computer systems and the function of the networked world. An important theme in chapters to come, however, is that human agency in the design and use of information technologies extends beyond static instances. That is to say that humans do not simply design individual systems and use them in a vacuum. In reality, it is the context of human development and deployment of new information technologies that arguably has the most to say about the nature of enduring vulnerabilities in digital affairs.

Much of what we are saying in previewing this theme has to do with network externalities. An externality is simply a consequence or a side effect of the activities of human inventions, social and cultural practices, and institutions. With cyberspace, the context of the Internet's development and subsequent global adoption matters a great deal. For reasons discussed in Chapter 4, much of the early development of computers and the subsequent employment of cyber instruments for national security purposes was concentrated within the U.S. and U.K. intelligence communities. As Chapter 3 discusses in part, the rise of industries based on the Internet in the e-commerce/dotcom boom era prevented the widespread adoption of robust security management practices and, to this day, incentivize mediocrity in solutions to broad-scoped cybersecurity problems.

From bytes to fights

The following chapters in this book deal with topics in computer network attack and defense in greater detail, before moving on to deal with substantive issues of theory, policy and the empirics of cyber topics in international affairs. In particular, chapters to come pay particular attention to the notion of the **Advanced Persistent Threat** (APT). APTs are often construed as particularly virulent pieces of malware or specific country threat groups (i.e. foreign intelligence agencies or military units), but the label really just signifies sophistication in design of malicious actions taken in cyberspace. Specifically, an APT is an attempt by a dedicated attacker to broach information systems beyond what we might think of as normal hacking activities. APTs are worthy of note as a final point here because it should be noted that many threats manifest across every layer of cyberspace. Indeed, the layers of cyberspace outlined earlier are cognitive aids and are not truly separate. Insofar as the construction of computer systems automatically involves the simultaneous development of informational, logical and user-focused platforms, truly sophisticated cyberattacks will threaten each layer in service of a particular objective. Thus, the next chapters, in focusing on specific offensive and defensive topics in cybersecurity, look at the flip side of the information security coin and assess the socio-technical shape of security threats rather than (as has been the case here) the technological basis of insecurity. From there, this volume aims to introduce a broad range of topics related to cyber conflict between countries, the use of ICT by non-state actors, the history of national cybersecurity experiences around the world and the shape of governance of the digital domain at the highest level.

Notes

1 For perhaps the best use of this metaphor to describe critical infrastructure, see Hunker, Jeffrey. "Policy Challenges in Building Dependability in Global Infrastructures." *Computers & Security*, Vol. 21, No. 8 (2002), 705–711.

2 See Leiner, Barry M., Vinton G. Cerf, David D. Clark, Robert E. Kahn, Leonard Kleinrock, Daniel C. Lynch, Jon Postel, Larry G. Roberts, and Stephen Wolff. "A Brief History of the Internet." *ACM SIGCOMM Computer Communication Review*, Vol. 39, No. 5 (2009), 22–31; and Leiner, Barry M., Vinton G. Cerf, David D. Clark, Robert E. Kahn, Leonard Kleinrock, Daniel C. Lynch, Jon Postel, Lawrence G. Roberts, and Stephen S. Wolff. "The Past and Future History of the Internet." *Communications of the ACM*, Vol. 40, No. 2 (1997), 102–108.

3 Salus, Peter H., and G. Vinton. *Casting the Net: From ARPANET to Internet and Beyond . . .* Reading, MA: Addison-Wesley Longman Publishing Co., Inc., 1995.

4 Though it is generally held to be apocryphal that military concerns about computers as targets that were uniquely vulnerable within the nuclear deterrence complex led to the funding of ARPANET, it is certainly the case that a number of military reports through the 1950s and 1960s consistently cautioned against overreliance on static, centralized computer control systems to regulate key military functions.

5 Perhaps described best, initially, in Cerf, Vinton, and Robert Kahn. "A Protocol for Packet Network Intercommunication." *IEEE Transactions on Communications*, Vol. 22, No. 5 (1974), 637–648.

6 Also see early descriptions in Cerf, Vinton, and Bernard Aboba. "How the Internet Came to Be." *The On-line User's Encyclopedia: Bulletin Boards and Beyond*. Reading, MA: Addison-Wesley, 1993; and Cerf, Vinton G., and Edward Cain. "The DoD Internet Architecture Model." *Computer Networks (1976)*, Vol. 7, No. 5 (1983), 307–318.

7 Cerf and Kahn. "A Protocol for Packet Network Intercommunication." 1974.

8 Braden, Robert, Ted Faber, and Mark Handley. "From Protocol Stack to Protocol Heap: Role-Based Architecture." *ACM SIGCOMM Computer Communication Review*, Vol. 33, No. 1 (2003), 17–22.

9 Berners-Lee, Tim, Roy Fielding, and Henrik Frystyk. *Hypertext Transfer Protocol – HTTP/1.0*. No. RFC 1945. 1996.

10 See Mockapetris, Paul, and Kevin J. Dunlap. "Development of the Domain Name System." *ACM*, Vol. 18, No. 4. (1988).

11 For one of the best narrative descriptions of the "terrain" of the Internet, see Blum, Andrew. *Tubes: A Journey to the Center of the Internet*. New York: Ecco, 2012. Additionally, for narrative description of the relationship between diverse commercial infrastructures and services in the context of cyber conflict, see Greenberg, Andy. *Sandworm: A New Era of Cyberwar and the Hunt for the Kremlin's Most Dangerous Hackers*. Anchor, 2019.

12 For perhaps the best historical narrative treatment of the development of the Internet and its lack of security features, see Kaplan, Fred. *Dark Territory: The Secret History of Cyber War*. New York: Simon & Schuster, 2016. Also see Carlin, John P., and Garrett M. Graff. *Dawn of the Code War: America's Battle Against Russia, China, and the Rising Global Cyber Threat*. Hachette UK, 2018; and Ware, Willis H. "Security and Privacy in Computer Systems." In *Proceedings of the April 18–20, 1967, Spring Joint Computer Conference, ACM*, 1967, pp. 279–282.

13 See Mockapetris and Dunlap. "Development of the Domain Name System." 1988.

14 See Hill, Jonah Force. "A Balkanized Internet? The Uncertain Future of Global Internet Standards." *Georgetown Journal of International Affairs* (2012), 49–58; Hoffmann, Stacie, Dominique Lazanski, and Emily Taylor. "Standardising the Splinternet: How China's Technical Standards Could Fragment the Internet." *Journal of Cyber Policy*, Vol. 5, No. 2 (2020), 239–264; and Healey, Jason. "The Five Futures of Cyber Conflict and Cooperation." *Georgetown Journal of International Affairs* (2011), 110–117.

15 For more on cloud computing, see Armbrust, Michael, Armando Fox, Rean Griffith et al. "A View of Cloud Computing." *Communications of the ACM*, Vol. 53, No. 4 (2010), 50–58; Mell, Peter, and Tim Grance. "The NIST Definition of Cloud Computing." (2011); and Armbrust, Michael, Armando Fox, Rean Griffith et al. "Above the Clouds: A Berkeley View of Cloud Computing." Vol. 4. Technical Report UCB/EECS-2009–28, EECS Department, University of California, Berkeley, 2009.

16 Though we discuss the subject in some detail in later chapters, for more on Internet governance, see Mueller, Milton L. *Networks and States: The Global Politics of Internet Governance*. Cambridge, MA: MIT Press, 2010; DeNardis, Laura. *The Global War for Internet Governance*. New Haven, CT: Yale University Press, 2014; DeNardis, Laura, Derrick Cogburn, Nanette S. Levinson, and Francesca Musiani, eds. *Researching Internet Governance: Methods, Frameworks, Futures*. MIT Press, 2020; and Kahler, Miles, ed. *Networked Politics: Agency, Power, and Governance*. Ithaca, NY: Cornell University Press, 2015.

17 For perhaps the best book on this entire topic, see Schneier, Bruce. *Secrets and Lies: Digital Security in a Networked World*. New York: John Wiley & Sons, 2011.
18 Lamport, Leslie, Robert Shostak, and Marshall Pease. "The Byzantine Generals Problem." *ACM Transactions on Programming Languages and Systems (TOPLAS)*, Vol. 4, No. 3 (1982), 382–401.
19 Schaller, Robert R. "Moore's Law: Past, Present and Future." *IEEE Spectrum*, Vol. 34, No. 6 (1997), 52–59.
20 This is a primary point of discussion in Chapter 4.
21 See Perrin, Chad. "The CIA Triad." www.techrepublic.com/blog/it-security/the-cia-triad/488 (2008).
22 A tact taken variously in, *inter alia*, Choucri, Nazli, and David D. Clark. "Integrating Cyberspace and International Relations: The Co-Evolution Dilemma." (2012); Clark, David. "Characterizing Cyberspace: Past, Present and Future." *MIT CSAIL, Version* 1 (2010): 2016–2028; and Russell, Alison Lawlor. "Strategic Anti-Access/Area Denial in Cyberspace." In *Cyber Conflict: Architectures in Cyberspace (CyCon), 2015 7th International Conference*. IEEE, 2015, 153–168.
23 Dingledine, Roger, Nick Mathewson, and Paul Syverson. *TOR: The Second-Generation Onion Router*. Washington, DC: Naval Research Lab, 2004.
24 Global Internet Freedom Consortium. "FreeGate." (2012). www.internetfreedom.org/FreeGate.html.

Further readings

Blum, Andrew. *Tubes: A Journey to the Center of the Internet*. New York: Ecco, 2012.

Carlin, John P., and Garrett M. Graff. *Dawn of the Code War: America's Battle Against Russia, China, and the Rising Global Cyber Threat*. Hachette UK, 2018.

Ferguson, Niels, Bruce Schneier, and Tadayoshi Kohno. *Cryptography Engineering: Design Principles and Practical Applications*. New York: John Wiley & Sons, 2011.

Jasper, Scott. *Strategic Cyber Deterrence: The Active Cyber Defense Option*. Rowman & Littlefield, 2017.

Leiner, Barry M., Vinton G. Cerf, David D. Clark, Robert E. Kahn, Leonard Kleinrock, Daniel C. Lynch, Jon Postel, Larry G. Roberts, and Stephen Wolff. "A Brief History of the Internet." *ACM SIGCOMM Computer Communication Review*, Vol. 39, No. 5 (2009), 22–31.

Mitnick, Kevin D., and William L. Simon. *The Art of Deception: Controlling the Human Element of Security*. New York: John Wiley & Sons, 2011.

Perlroth, Nicole. *This Is How They Tell Me the World Ends: The Cyber Weapons Arms Race*. Bloomsbury Publishing USA, 2021.

Salus, Peter H., and G. Vinton. *Casting the Net: From ARPANET to Internet and Beyond*. Reading, MA: Addison-Wesley Longman Publishing Co., Inc., 1995.

Schneier, Bruce. *Secrets and Lies: Digital Security in a Networked World*. New York: John Wiley & Sons, 2011.

3 Cyberspace and international relations

Cyberspace is a fact of daily life in the 21st century. It is the unusual political episode or diplomatic incident that occurs without feeling the modifying influence of the digital world. Cyberspace affects the real-kinetic dynamics of everyday global society at almost every level – virtual networks aid in the spread of information about a particular issue, while different computer systems serve as a direct path between the functions of different national or subnational institutions – and, in doing so, has a very real impact on policy, politics and security.

It seems reasonable to say that cyberspace is the most significant emergent issue for the **International Relations** (IR) field in the new century. This is largely because new information and communications technologies (ICTs) broadly and *systematically* affect the contours of human interactions and institutions. If cybersecurity is the security of socio-technical systems pertaining to the information technology revolution and ICTs have had an impact on almost every function of human societies, then cyber issues are roughly synonymous with the full gamut of topics in IR.

More specifically put, the digitization of global infrastructure has changed the ways in which information is accessed, controlled and transmitted across every type of interaction – economic, social and security. Traditional political phenomena are increasingly influenced or changed entirely by the nature of cyberspace's "special" characteristics. Interstate security calculations, for instance, must cope with distinct asymmetries of capability and knowledge in a world where networked computer systems underwrite most societal functions and where attribution is difficult. Moreover, the rise of digital technologies has laterally produced changes in the content of international society. Large populations of global constituents systematically access and utilize new means of communicating with each other. And information, due to the rise of social media and related platforms, is presented across borders in meaningful (and potentially vulnerable!) new ways. These dynamics, and the institutions that are emerging from different national and international efforts to adapt, constitute new challenges of empirical understanding for scholars and students of both IR and public policy.

In this chapter, we consider the importance of cyberspace and cybersecurity for the IR field. In particular, we attempt to place the rise of cyberspace in an appropriate context as a systematic development that continues to impact upon the dynamics of world politics in a myriad of ways. First, we describe the historical circumstances surrounding the rise of cyberspace and the current state of cyber affairs. Then, we ask why cyberspace matters for international security and IR theory. What are current debates? And what challenges – both empirically and epistemologically – do scholars face with future efforts to describe the cyber-infused world? We then conclude by describing IR's main paradigms and discussing their applicability to today's digitally augmented world.

DOI: 10.4324/9781003246398-3

The rise of a global cyber ecosystem

Advances in ICTs have defined major shifts in sociopolitical and economic affairs in the modern international system for centuries. From the telegraph to space technologies, new ICTs have consistently been the cause of changes in the setup of economic and political systems that ultimately impact upon the power of states, the shape of industries and more. Digital technologies are no different. Indeed, the rise of cyberspace arguably constitutes the most significant set of changes to the contours of IR and the core function of global processes ever. The digitization of infrastructure has, since the late 1980s, involved complete transformations in the way individuals interact with society writ large, in the way that society itself cultivates and treats major issues, and in the way that institutions and authorities comport themselves on issues of diplomacy, business and security.

This section describes the history and trajectories of advances in information technologies and the rise of cyberspace. It then describes the shape of the system – an international system with cyber characteristics – that we see today, including the institutional and technical architecture of computer networks and the political capital behind it.

The rise of cyberspace in world politics: history and circumstances

How have information technologies come to impact the processes of world politics in fundamental ways? What has been the trajectory of cyber developments and how have political entities – from governments to social institutions and international organizations – considered them over time? The phrasing of these questions is somewhat misleading. Pundits, analysts and scholars alike tend to think of the "cyber phenomenon" as a cohesive set of developments that have gradually become a major set of issues for global society, when in reality there is significant disconnect between IR's focus on "cyber" and the broad-scoped impacts of information technologies on the contours of world politics.

The Internet – and, as a result, cyberspace – has been around since the 1960s.[1] As the last chapter describes, the first networked computer systems emerged in the 1960s and 1970s from the research and innovation of a series of U.S. government programs, notably the predecessor of the modern Internet – called **ARPANET** – designed by the U.S. **Advanced Research Projects Agency (ARPA)** in the late 1960s. The development of the ARPANET has its roots in several spheres of national security, notably the postwar expansion of the U.K. and U.S. intelligence communities and the subsequent computerization of key defense establishment functions through the 1950s and 1960s. These are described in greater detail in Chapter 4.

Despite the fact that the Internet and its direct predecessors have been around since the 1960s, it would be entirely unreasonable to assume that cyberspace has had a concentrated and regular impact on all aspects of world politics over that period of time. In reality, it would be far more accurate to say that different theaters of international affairs have experienced distinct and relatively unique periods in which information technologies have both changed the fundamental functionality of a given set of processes and prompted a range of geopolitical responses. Three sets of such developments – and the sociopolitical events that accompanied them – bear particular mention.

In the realm of national and international security, the story of information technology adoption and integration has, broadly speaking, been one of punctuated political equilibrium. For intelligence agencies, militaries, and other cogs in the traditional apparatuses of state security, ICT adoption through the 1970s, 1980s and 1990s generally took place in much the same fashion as it might have for private industry. New technology was adopted as it became cost efficient or particularly effective. Except for where institutional tribalism stymied progress, better security mechanisms were constructed and implemented as threats – technological and sociopolitical – became apparent.

Despite this general condition, however, the modern apparatuses of state cybersecurity functions have tended to emerge from the unique ways in which different countries have been forced to deal with incipient cyber challenges. In summarizing the history of such institution-building in the United States, Healey (2013) points out that the national experience with cyberspace as a national security concern has largely been shaped by "realization episodes," in which government at the highest level has been forced to address cybersecurity as a result of a specific threat. This kind of reactionary pivot to cyber policy inevitably disrupts the organic adaptation of elements of national security apparatuses to new challenges and opportunities, and forces the formation of unique policy regimes and perspectives. Both Chapters 4 and 8 expand upon this in their description of different national experiences with computers and new network technologies.

By contrast with the tumultuous history of orientation toward cyberspace in the national security setting, the story of information technology adoption in global economic terms has been relatively smooth. Quite apart from the reactionary fits and starts that have characterized the emergence of focus on cyber in other areas, the integration of ICT to economic ventures of all stripes has occurred regularly and in a more deterministic fashion than has been the case with national security institutions since the 1980s. Internet-based services and technologies have not only revolutionized technical aspects of global society – such as the tools of information access and transfer available to consumers the world across – but also made for remarkable economic transformations at the national and regional levels.

Network technologies have enabled countries like India, China, Brazil and many others to expand massive industrial sectors in support of information externalities – i.e. to e-commercialize broad sectors of national industry in order to meet rising consumer demand for Internet-based products and services. Likewise, network-based information technologies have lowered the barriers to development and welfare for countries and individuals around the world, as education, medical services and more have increasingly cultivated viable online characteristics. Today, policy that addresses cyberspace in terms of the massive economic impact that information technologies have had on global society includes focus on necessary adaptations to international law and interstate treaties, the distribution of economic assets in line with technological advances, and the evolution of issues related to ICT's impact on the world economy, including immigration.

And beyond the realms of national security and global economics, we might broadly consider the shape of information technology adoption in terms of global society itself. Global economics and international security dynamics are highly relevant elements of any conversation about the shape of global society, of course. But there *have* been unique events and distinct effects on the social and cultural dynamics of world politics beyond those that emerge from economics and information security.

The development of the Internet has essentially transformed global social processes at two distinct points. First, the emergence of a public facing set of information access and dissemination tools in the 1990s revolutionized not only industrial or government functions but also the dynamics of distinct social phenomena. As one prominent literature in political science points out, Internet technologies in the 1990s and 2000s became "liberation" tools for social movements of all stripes around the world. Protest, revolution and citizen advocacy have taken on new forms in the digital age by enhancing the ability of civil society to coordinate and visibly apply political pressure. Information technologies have also affected the way that information is presented and social processes – including political participation – take place in response. The Internet age has seen the intensification of media effects on populations, such that the specific contours of media information reporting and "spin" have a direct impact, in various ways, on the preferences and behaviors of citizenry.

On top of this, second, the recent emergence and proliferation of social media platforms around the world has continued to transform social processes in meaningful ways. Social and political communities are increasingly taking on virtual characteristics. This is significant for two reasons. First, virtual society is dramatically different from pre-digital society in a number of respects, including that there is greater mobility of social interests, fewer barriers to information sensationalism and a more diverse range of sources for sociopolitical learning. Second, not all social development in the Internet age has happened uniformly. The experience of virtualization of society differs greatly across countries, with unique local conditions – including government regulation, prevailing social conditions and enduring political dynamics – setting the scene for social adaptation to life in the digital era.

"Cyber" means different things across vastly different aspects of world politics. Information technologies have had unique effects on different political, economic, social and security processes in the global system such that understanding policy and issues related to "cyber" anything means comprehension of a tangled web of disparate but interlocking topics. This reality is no less the case for security researchers just because security at some level is the focus of our interests. As parts of this chapter and much of the content of later chapters show, understanding the contours of strategic calculations in digital affairs or the dynamics of international cooperation on cybersecurity involves knowledge of how information technologies tie together issues related to – among other things – international law, human and civil liberties, military effectiveness and parochial politics.

The political architecture of cyberspace

Though discussed elsewhere in this book (partially in Chapter 2 and more fully in later chapters), the question of who runs the digital world is particularly relevant to any consideration of cybersecurity as world politics. In particular, understanding who runs the digital world is important for trying to understand what rules of the road – informal or formal – exist with regard to the regulation and governance of cyberspace, as well as what rules might exist in the future. Where does the political capital lie across cyber issues, and how does this setup relate to the organizational and strategic dynamics of world politics more broadly?

In truth, the management of cyberspace relies on a great number of institutions. Some of these institutions have been established specifically to govern aspects of the digital world, while others are either old or new parts of the regulatory apparatuses of different authorities in world politics – state and local governments, international organizations and nongovernmental organizations.[2] Though the latter category of institutions tend to have purview beyond just matters of cyberspace governance – i.e. they are concerned with the digital aspects of issues that go beyond cyberspace, such as human rights, civil liberties, human security, social activism and so on – all of these weigh in on some subset of the same group of cyber management issues, including the development and management of infrastructure, the function of computer systems and the services linked to cyberspace.

As suggested in the last chapter, the governance of the Internet largely has to do with the regulation and coordination of the logical elements of the innumerable network components that make up the global Internet apparatus. The key difference between network ICT and previous telecommunications technologies (like the telephone or telegraph) lies in how information is transmitted between end nodes. With older technologies, information was sent through a set circuit (i.e. one device location was physically connected to another for the purpose of exchanging information). With digital technologies, information is sent in packets across a network of nodes. Data packets contain pieces of information and address headers that contain information about the intended recipient of communications. The path of packets across the Internet is determined sequentially by different computers that receive data packets, read the address headers and then forward information down what is understood to be the most efficient route. As such, determination of responsibilities for registration of different network addresses and standards for best practices (for development of new technologies, etc.) is the main function of most Internet governance entities.

At present, the governance of cyberspace and the functions related to it in the international system take the form of what is called a **multi-stakeholder model**. This model is described further later, but essentially describes a reality in which various actors, *not just states*, have a say or role in how the domain is governed. This setup, to which there is some opposition – and which, again, will be described further later – has emerged from the circumstances of cyberspace's early development in the United States. In essence, the U.S. government's early institution of a private nongovernmental entity – called the **Internet Corporation for Assigned Names and Numbers** (ICANN) – to regulate the Internet's Domain Name System (DNS) set the stage for massive private actor authority over the core functions of public-facing computer networks. Alongside a variety of affiliated organizations, the ICANN – which was "internationalized" in 2016 to allow for more representation of interests by foreign states and interested organizations – remains today as the main functional agent for network maintenance and regulation in world politics.

More broadly, a range of different entities and interested parties are driving the global agenda on the governance of cyberspace. Foremost among these are the actions of non-profit organizations that adjudicate and advise on standards for network technology implementation and adaptation. The **Internet Engineering Task Force** (IETF) has self-awarded purview (by consensus among the world's engineers and scientists) over the function of protocols for network technologies. The IETF, in essence, takes upon itself to request feedback from relevant parties around the world and then works to adapt existing standards to keep up with advancing technical realities. By contrast with the IETF's focus on developing awareness of different issues through consensus and community operations, the **Internet Society** (ISOC) actually oversees technical standards in a more direct way. In particular, the **Internet Engineering Steering Group** (IESG) offers direct oversight capacity for developers globally, while the **Internet Architecture Board** (IAB) oversees the operations of the IESG itself. These sub-units of ISOC together constitute the primary means of coordinating standards and practice on non-crisis development issues for the international community.

Second among governance efforts is that of the United Nations. Though the global agenda on cyberspace has been taken up by a diverse host of interested parties since the late 1990s, the UN has, in recent years, encouraged a series of initiatives under the auspices of its different agencies aimed at streamlining coordination on and progress regarding rules of the road for cyberspace regulation and policy. The World Trade Organization has, in particular, been at the forefront of advocacy for better adoption of favorable region-specific practices under international law. The UN is also behind the **World Summit on the Information Society** (WSIS), which is constituted of a series of meetings and documents on the subject of global development and information technologies. The main idea behind WSIS is that ICTs are responsible for increasing information content accessibility around the world and that sustainable economic development across a number of sectors benefits from stability and favorable regulation of cyberspace. WSIS is notable in that it has called for and received non-binding agreement from many countries for agreement on unfettered global access for all on projects related to education, e-governance, social services and research. Thus, though responsibility for cyber governance and for coordination on regulatory issues in all veins continues to rest on the diffuse motivations of a host of actors in world politics, the UN has emerged as perhaps the most important cog in international efforts to meaningfully move toward a progressive regime for information technology management.

By comparison with more general efforts to achieve meaningful progress on streamlining the processes of global governance of cyberspace, attempts to obtain international coordination on security issues reflect an interesting dichotomy of interests in which there are at least somewhat effective mechanisms in place to monitor non-state cyber instabilities and very few formal pillars on coordination on the subject of interstate cybersecurity. With regard to the former, it is clear that there is significant international agreement on the responsibilities of governments to protect their citizenry from the negative externalities associated with increased societal exposure to network technologies, namely crime, political extremism and terrorism online. To be fair, the landscape of mechanisms put in place across countries in this regard is a mélange of uncoordinated institutions that function much as police organizations in different countries might – by coordinating inefficiently but as needed about different incipient threats. Nevertheless, there is some degree of central planning in international security efforts to mitigate the effects of cybercrime and non-state militancy. A number of countries maintain a **Computer Emergency Response Team** (CERT) (there are more than 200), which takes on the threefold task of coordinating responses to cyber emergencies, promoting better general cybersecurity and building for more robust service options within a given jurisdiction. CERTs are additionally able to coordinate internationally through the CERT Coordination Center and, though national experiences differ in the approach adopted, tend toward the mission of building better relationships with the private sector for the purposes of information sharing. In this way, CERTs have emerged as a standard component part of procedures across countries for coping with particularly egregious situations of criminal or extremist disruption and intrusion. Moreover, CERTs are increasingly proving their value as a relatively impartial interlocutor between private industry and governments.

By contrast with efforts to coordinate on issues related to crime and low-level disruption from non-state sources, there are thus far only limited formal features of the coordinative landscape for interstate security cooperation. Though cooperation and recognition of norms of behavior through codification of appropriate rules of the road for cybersecurity are desirable in the future, the reality is that major powers in world politics have barely moved beyond formal assurances regarding interference online. China and Russia have signed a pact stating that they will not interfere with each other in cyberspace, and the United States has obtained a similar assurance from Beijing that national processes will not be targeted online. However, there are few formal aspects to such agreements at present, with most bilateral interactions between states on cyber issues taking the form of memoranda of understanding. Though we cover international norms and law on cyberspace in a later chapter, it is clear that there still exists uncertainty for states in linking emerging norms and viable formal practices for interaction on a number of fronts. Specifically, there are a range of issues bound up in constructing acceptable treaty structures for future cyber international relations, and the domain's special characteristics – particularly the attribution problem – only exacerbate state concerns about the ability to validate, verify and enforce binding cyber agreements.

In the next section, we describe how scholars and analysts have been theoretically and descriptively studying cyberspace across a range of issues. In particular, we discuss the current state of IR scholarship on the cyber phenomena and engage with the problem of streamlining conceptual approaches for producing better knowledge on the contours of world politics in the digital age.

Why cyberspace matters for IR theory

IR theory is the examination of world politics and society wherein researchers apply different frameworks to try to explain patterns of human behavior across individuals, organizations, countries and supranational entities (like the UN or the European Union). The IR field can be roughly divided according to a number of theoretical debates about how the world works. Some scholars, for instance, hold that the world can best be explained via an understanding of how humans assign meaning to different objects and arrive at unique identities. Others, by contrast, broadly argue that relationships in international affairs entirely emerge from and are shaped by specific disparities in power between human institutions. Yet other debates pivot on disagreement over the role of economics, types of government, features of different national societies and the constitution of norms. Regardless of the flavor, however, the bottom line is that IR theories allow us to describe different logics of outcome for world affairs depending on what forces we find most relevant and impactful.

Major advances in information technologies and the rise of cyberspace are of special interest for scholars in the IR field largely because of how systematically their effects can be felt in world politics. Almost all societal functions are connected digitally. Information is available for access or transmission instantaneously and in a far less secure fashion than might have been the case in eras past. Moreover, changes to the function of international society and what might be called the mechanical substrate of the global system are reflected in actor behavior and organization. We exist at a time, thanks to the rise of digital technologies, in which myriad different actors possess preference sets molded by the dynamics of the new interconnected world and in which political organization has been forced to take new forms to cope with the challenges of that world. These developments have significant implications for scholars of world politics and, as is the primary focus of this book, analysts of national and international security.

By far the most cited and studied set of developments in the IR field emerges from what might be called the digitization of global infrastructure or the process by which most societal functions – from various aspects of the global economy to military systems and tools of local governance – have, since the 1980s, gone digital. This process is the reason most commonly listed by scholars of cyberpolitics and security as to why IR should pay attention to cyber developments.[3] The digitization of infrastructure has exposed global society to new opportunities and vulnerabilities, and has significant implications for the calculations of political actors of every kind.

For scholars of international security, in particular, the digitization of international society portends new threats and vulnerabilities on a number of levels. Information and the systems that run our society (particularly the more "important" parts, such as military or government systems) are at risk of infiltration, exfiltration and broad-scoped disruption in an unprecedented fashion. Sensitive information is at risk of digital theft not only for private citizens and industry but also for the intelligence community and the developers of military hardware.[4] And specific military and vital national systems, such as air defense platforms or nuclear reactor computer networks, might be vulnerable to disruption and exploitation by any number of aggressors. In addition, the costs of entry for actors seeking to threaten state security, whether one is talking about other states or non-state actors like terrorists or "patriot hackers," are exceedingly low.[5] The know-how and hardware needed to execute reasonably sophisticated cyberattacks aimed at intrusion and disruption are, in most instances, well within the means of a dedicated individual or small group. Thus, with cyber, the balance of capabilities is – perhaps for the first time in history – uniquely and systematically stacked against the state.[6]

Of course, cyberspace matters for security and other scholars in IR not just because of the systematic way in which it links different societal functions and produces new types of interactions. Rather, it also matters because of the great number of special characteristics that must be considered when controlling for the impact of information technologies. Not only are the costs of entry low for actors of all stripes, but also the actions undertaken in cyberspace benefit from a diminished threat profile. The "attribution problem," which describes the difficulty faced by the victims of cybercrime and aggression in identifying the real-world source of cyber intrusions, forces decision-making and strategic planning to take place in a range of new ways.[7] Deception has become a major guiding component of both defensive and offensive strategies, at least for state actors, and authorities face new challenges stemming from attribution difficulties. Moreover, the nature of cyber "weapons" as benefiting from secrecy in development and execution introduces a range of new strategic calculations to be considered by policymakers and officials. Unless defenders use deception to effectively derail attacks, cyberspace presents as an offense dominant domain where the incentive for competitors is to hit first and commit early to parallel actions.[8] This dynamic is further compounded by the relative difficulties involved in determining the effects of cyber actions against foreign computer systems.[9]

Cyberspace also matters for IR scholars and for IR theory more broadly because of the secondary effects of the digitization of societal functions.[10] In particular, the massive global adoption and integration of information technologies has had major impacts on both the behavior of different actors – from individuals to states and international organizations – and the way they self-organize. Information technologies, in introducing new modes of information access and control, have significantly affected the preference sets of actors concerned with a diverse range of economic, social, and political issues. As one set of literature points out, problem solving and decision-making in the digital age often occur in a "networkized" fashion,[11] with individuals and the institutions they build reacting to challenges by offering solutions with distinct network characteristics. Social mobilization, for instance, is approached in a much broader and more complex fashion than it might have been in eras past, and security challenges are problematized as crosscutting issues affecting the entire portfolio of actor interests.

Box 3.1: Skeptics and revolutionaries: debating the impact of cyberspace on international security

The existing literature on cyberspace and international security principally focuses on a debate about the importance of cyber developments for stakeholders in international affairs. Specifically, the debate revolves around the significance of information technology development for state efforts to govern and secure national security imperatives. Broadly speaking, there is a disagreement between scholars on the extent to which cyberspace portends complex and challenging obstacles to security and stability for actors of all stripes in the international system. One side of the debate holds that information technologies are the next revolution in military affairs (i.e. they are forcing military planners and practitioners to fundamentally rethink their approach to security issues) and that the massive transformation of societal functions at nearly every level to a digital format augurs widespread vulnerabilities for actors of all stripes.[12] These vulnerabilities are particularly diverse and numerous for advanced states, where ICT integration occurs and guides industrial function on a much larger scale than it might elsewhere in the world.[13] The suggested global dynamic for such scholars is one of mismatched capacity – smaller states and non-state actors hold disproportionate powers of intrusion and disruption against a diminished great power's ability to regulate effectively.[14]

The other side of the debate on the significance of cyberspace for international security, however, eschews this reading of emerging global dynamics and argues that much of the analysis of threats from online oversimplifies matters.[15] In particular, aggression online never takes a violent form, and there are a great number of considerations to take into account when analyzing state-level vulnerabilities.[16] Incentives for would-be attackers and exploiters of all forms are limited by the realities of strategic circumstances, regardless of the bounds of technical possibility, thus producing a global landscape possessed of new technological dynamics but fundamentally unchanged basic dynamics.[17]

This stark division among scholars of cyberspace and IR appears in the range of debates that have, in recent years, attempted to explain the realities of politics and state-level security in the digital age. In particular, advocates of both perspectives dominate discussion over (1) how important cybersecurity should be in national security policymaking and rhetoric, and (2) what effect cyberspace has on patterns of conflict and interaction in the international system. In the former discussion, a range of prominent officials and scholars have consistently argued that cybersecurity has quickly emerged as the most serious challenge for modern society. From Leon Panetta's claim of a possible "cyber Pearl Harbor" to the arguments made by Tom Donilon and others that the core functions of society – particularly critical infrastructure and intellectual property – are prone to disruption at any time, the public sphere in the West is awash with the belief that cyberspace is a revolutionary development that has altered the conduct of national security in its entirety. By contrast, a small but growing number of scholars argue that the nature of cyberspace itself is used to overblow assessments of threats to national security. If cyberspace is a crosscutting domain that augments actors' abilities and introduces new vulnerabilities to society at every level, then it is analytically disingenuous to lump all security issues under a cyber moniker.

Rather, research and policymaking must attempt to parse out the implications of information technology adoption over existing, traditional issue areas. In so doing, it is argued, we might see that cyberspace's destabilizing effects are highly contextual and that, in many instances, new inter-connectedness just means new dimensions to old problems, such as terrorism, interstate competition and global economics.

With regard to the effect of cyberspace on patterns of conflict and interaction in the international system, it is important to understand how scholars have differently used phrases like "cyber warfare," "cyberwar," and "cyber conflict."[18] Scholars have used *cyber warfare* consistently as a catchall term to describe the use of network technologies during conflict. In referencing cyber warfare, one might as easily be talking about cyber aspects of traditional conflict scenarios – such as war between great powers or a military intervention – or intentional low-level disruption caused by either state or non-state belligerents outside an official state of war. **Cyberwar**, by contrast, has been used not only by scholars[19] but also by policymakers, pundits, and private sector spokesmen to describe a condition of interstate conflict that might take place *only* (or at least primarily) online, while **cyber conflict** has most often been used to describe any conflictual interactions via cyberspace that *don't* meet the threshold of interstate warfare (formal or otherwise).

The question of the likelihood of cyberwar itself has largely been the content of the debate between the cyber skeptic and cyber revolutionary camps. Rid's seminal article – and the book that followed[20] – points out that nothing about conflictual cyber interactions is violent and that the vast majority of incidents involve espionage, sabotage, vandalism or other defacements. And while sabotage might be used on a massive scale – many of the more alarming lines of rhetoric in punditry and analysis consistently bring up the question of critical infrastructure security, for instance – any advantages brought to a foreign state aggressor would be temporary. Thus, cyberwar is a fiction outside the scope of traditional conflict scenarios that include kinetic campaign aspects.

However, in introducing this point about the myth of cyberwar, the works of Rid, Gartzke and others in this vein have consistently demonstrated that the same cannot be said at all of cyber conflict. Indeed, it is the crux of the counter-cyberwar argument that cyber conflict happens consistently in the international system in *low-intensity* terms – meaning espionage, vandalism by non-state proxies and more. Given this, what might be said of the effect of cyber conflict on state behavior in international relations? Do low-level provocations by one state tend to produce aggressive foreign policy responses by others? And how do the special characteristics of cyberspace – including attribution problems and the difficulty in determining what the "use of force" might look like online – play into this?

These questions lie at the heart of ongoing study into conflict in the fifth domain and form in part the basis of later chapters' discussion of cyber conflict across numerous settings in world affairs.

Information technologies have also spurred new institutional development in world politics. Two dimensions of these developments highlight the need for more study of advances in ICT development by IR scholars. First, global ICT adoption has and is continuing to produce institutions concerned with the governance, regulation and contestation of cyberspace. These include international organizations concerned with legal and normative aspects of cyber regulation, as well as national and local institutions concerned with the reconciliation of meaningful cybersecurity practices and the welfare of different communities.[21] Second, information technologies, in changing the dynamics of interaction and information usage in the world writ large, have introduced the need for existing institutions at every level of society to adapt rules and operating procedures to the contours of the digital age. Doing so might entail the introduction of new processes, the reinterpretation of existing doctrines or the layering of reimagined approaches on top of existing procedures. These developments, occurring as they are systematically within global society, portend major changes in the directionality and the outcomes of political behavior in international affairs and present compelling explanatory challenges to the IR field.

Realism, liberalism, constructivism: IR's paradigms and cyber conflict

Just how useful is IR theory when it comes to explaining patterns of and potential for cyber conflict? Off the bat, it seems reasonable to think that IR theories are going to be of great help in unpacking the impacts of the information revolution on world politics. Since the 1970s, IR theory – whether paradigmatic theory that attempts to explain entire modes of human interaction, or what some have called "mid-level" (aimed more specifically at explaining unique political phenomena)[22] – has developed conceptual approaches and borrowed from other social science fields such that students of international affairs can draw on a range of perspectives to explain the world around them. It seems intuitive that IR theories will be adapted, in time, to provide explanations of the contours of the digital world.[23]

At present, however, it would be disingenuous to suggest that political scientists are possessed of macro theories that link new cyber developments to the fundamental behaviors of states, institutions or individuals. *Broadly, we argue that IR theory, in the broadest sense, is ill-suited to describing digital politics but that core concepts across the IR field are of immense utility in problematizing new dynamics of human interaction in the information age.*

In reality, existing IR theories and approaches to different conceptual problematics suffer with cyber from a problem of imprecision on two fronts. First, recent theoretical and analytical work on cybersecurity and digital politics has suffered from the imprecise and sometimes clumsy application of concepts and principles previously used to describe things such as strategies for nuclear conflict. Terms like "offense dominance," "deterrence" and even "power" have been used to describe the manner in which cyberspace's unique characteristics constitute a set of strategic dynamics for both state and non-state actors.

Offense dominance, for instance (and which we return to in later chapters), describes a situation in which strategic logic favors the first use of a particular weapon type due to losses in effectiveness at a later stage of conflict.[24] The basic logic of offense dominance in cyberspace, an assumption found in numerous scholarly and pseudo-academic texts stretching back to the 1990s, references the logic of zero-day exploits in explaining why the incentive will always be to strike first.[25] If a vulnerability is patchable and a capability able to be mitigated with even a small amount of advanced information, then full effectiveness of a cyber arsenal can only be realized in conflict under conditions in which an actor is the first mover. The problem with this use of an offense- versus defense-dominance framework for understanding strategic calculations online, of course, is that the unique characteristics of cyberspace also influence actors' behavior in other ways. Cyber "weapons" are often ineffective, a fact that is difficult to determine for both attacker and defender. Cyber tools often also accomplish things other than their intended purpose, making it difficult for a defender to interpret situations. And deception also awards a number of abilities to defenders to reroute or otherwise mitigate incoming threats, a fact, again, which might be difficult to fully comprehend for would-be attackers. Knowing these things, the strategic calculation for attackers becomes significantly more complex than the simplistic and dichotomous offense/defense dominance label might suggest.[26] Overreliance by scholars and strategists on such imprecise terminology thus carries with it significant risks for misinterpretation and premature prescriptions, and the field is only recently beginning to address the need for greater nuance in conceptual treatments of cyberspace.

Second, the IR field has so far struggled with describing both the broad-scoped impact of the information revolution on international affairs and the effect thereof on the various aspects of state behavior in theoretical terms. One need go no farther than the field's main paradigmatic perspectives on world politics to see how mainstream IR theories both provide useful lenses for viewing cyber developments and yet fall short when it comes to applying their broad assumptions to the description of human interactions online. The many forms of realism, liberalism and constructivism – IR's three main theories – not only hold insights useful for scholars interested in cyberspace, as do other theoretical perspectives, but also largely fail to function in their entirety when faced with issues of cyber interaction in international affairs. The task presently before political scientists thus lies in parsing relevant conceptual foundations apart from scholarly bluster such that our theories accurately describe developments in world politics in the digital age.

Realism

Realism is fundamentally about power. In reality, there are several realist theories that each rely on different justifications and formalizations of a relative cohesive set of assumptions about how the world works. Realism has its roots in the *realpolitik* thinking that dominated the class of statesmen, military thinkers and scholars focused on imperialist politics through the 19th century and into the twentieth. Realism was formalized in the writings of men like E. H. Carr and Hans Morganthau, as well as in the geopolitically focused work of George Kennan, Herman Kahn and others. **Classical realism**, as the first set of articulations of the theory is referred to, essentially holds that IR is an inherently tense business because humans are naturally flawed creatures. Specifically, humans always tend toward the search for greater power, the result of which is that peace only ever results from balanced power relationships that can effectively deter the conflictual excesses of any one actor.

By contrast, **neorealism**, which attempts to justify the assumptions of realist thinkers via better grounding in logic and empirical observation of the nature of relationships between countries, essentially holds that the state is a unitary actor that exists in an anarchic global system.[27] Since anarchy is the underlying condition of IR, state behavior inevitably rests on the assumption that strategy must be self-help – i.e. since there is no enforcing authority above states, state actions must necessary be focused on securing the conditions for survival and welfare.[28] This condition is made worse by the fact of the problem of other minds, essentially the surety in life and in world affairs that it is impossible to truly know the intent of peer competitors.

Box 3.2: How relevant is realism in a world of powerful non-state actors?

A particular challenge for anyone interested in adapting the tenets and assumptions of realist theories to explaining conflict dynamics in the information age is in explaining the role and significance of non-state actors. As Chapter 11 describes in greater detail, the cyber conflict ecosystem is in many ways better defined by those non-state actors that operate therein than by states that employ cyber tools for conflictual purposes. This is largely because, in addition to the immense crowd of terrorist organizations, criminal syndicates, social movements and hacker irritators that have moved their operations to the web, most state-related acts of cyber contention are executed under the cover of deniability. In other words, when states hack, the direct culprit is often non-state hackers under the hidden employ of the state or state agencies that actively hide their identity. In cyber conflict, because full attribution of digital actions to those politically responsible is inherently difficult, it is both common and relatively simple to bypass the direct consequences of aggression found in other domains.

For realism, this is particularly problematic not because such a dynamic excludes the utility of core concepts linked with the school of thought but because it throws a wrench in the works of the predictive elements of the thing. Later chapters break down different parts of realist dogma as they relate to conflict in the fifth domain, including *anarchy*, the **security dilemma**, the **offense-defense balance**, **coercion** and the nature of *power*. These concepts bear significant deconstruction and debate for various reasons. Neorealism is defined as a paradigm, however, not just by those conceptual elements but also by the implications thereof for state behavior. The two main variants of neorealism both hold that anarchy and the problem of other minds drive states toward self-help behaviors. While this is relatively uncontroversial, they also note that states respond to perceived threats via balancing actions. While a neorealist might argue that cyberattacks regularly fail to elicit balancing actions in interstate relations because they don't often hit the threshold of threat credibility, there are no provisions in the theory to explain broad-scoped commitment on the part of many states to tactics that emphasize reliance on non-state proxies and don't directly affect the military power of competitors. Realism, in other words, does not present as a compelling framework by which we might explain the choices of states and non-state actors to engage in cyber conflict.

When it comes to cyberspace, realism's main problem lies in the common critique that its structural (and most popular) variants are overly parsimonious. With cyber, as with all things, this means that states are likely to adopt policies and take actions to enhance capacity and strategically mitigate threats.[29] The problem of being overly simplistic is perhaps more obvious with cyberspace than it is with any other empirical phenomenon. Certainly, realism holds important insights for understanding how states might think about the digital component of national security, but a focus on unitary state processes and systemic mechanisms as primarily responsible for compelling state action simply doesn't gel with the reality of cyber interactions. Given that cyberwar is unlikely[30] and low-intensity cyber conflict is the norm of digital interactions in the contemporary international system[31] – points we return to in later chapters – the relationship between realism's primary explanatory assumptions and the incentive to engage in cyber conflict is not clear. Moreover, as later chapters discuss, it seems to be the case that particular cyber aggressions emerge from the unique interests of institutions within state security communities and that cyber conflict is often linked to the existence of a range of parochial, political circumstances. In short, with neorealism, the focus on unitary state agency seems to suggest few real methods for accounting for variation in behavioral outcomes with regard to cyber, even if (as we discuss them in depth in later chapters) many core realist concepts have utility for our study of cyber conflict.

Liberalism

Of the three main schools of thought in IR, **liberalism** in its various modern formats might be of most use to the research program on cyberspace and international security. Two iterations bear particular mention. First, the variant of liberalism presented by Andrew Moravscik and extended by others fits most closely with the evidence regarding patterns of interaction and effect in cyber international affairs.[32] Liberals explicitly reject the idea that everything pivots on power politics and promote the notion that international cooperation can emerge as a self-interested output of institutional politics, often aided by international institutions and the development of positive norms of behavior. Though theoretically undeveloped as yet, the initial evidence we have that states are relatively restrained when it comes to responding to major cyber speaks to the tenets of modern liberalism in that state behavior might be expected to emerge from configurations of political capacity and interests at both the domestic and interstate levels. The abilities and inclinations of private actors matter a great deal when it comes to determining propensity for conflictual interstate relations,[33] when it comes to norm development in international relationships, and when it comes to predicting deviations from established patterns of interaction.

Seeing international affairs in a cyber-augmented world as being constituted of private actors – states and otherwise – trying to maximize bounded self-interest makes a great deal of sense, as one of the main theses of the literature on cybersecurity in world politics holds that information technologies have fundamentally altered the abilities of actors at every level of global society to achieve informational outcomes disproportionate to their kinetic abilities.

Box 3.3: Why no international cooperation? attribution as the ultimate spoiler

Proponents of liberalism within the IR field advocate that stakeholders in world affairs tend toward cooperation over time. There are a broad range of reasons offered for this, from the formation of shared cultural ties that often come with (additionally restraining) trade relationships to the logical need to mitigate the otherwise harmful excesses of life in an anarchical world. Given that, at this juncture, there are remarkably few constraining arrangements in existence that serve to restrain the conflictual tendencies of cyber actors, we might reasonably ask liberals to explain themselves.

A number of concepts and arguments strongly linked with liberalism are relevant here. Liberals regularly argue that international organizations serve a series of purposes in helping to develop global compacts on trade, finance, social issues and the prevention of war. International organizations help build common cultures of understanding about what is and is not permissible in IR. Moreover, they incentivize cooperation among states that otherwise would have to compete on a playing field that is often uneven. They also offer enforcement mechanisms for those actors that break the rules and, as Chapter 12 discusses, help sustain positive norms of behavior.

When it comes to international conflict, the most relevant element of international law is quite arguably **the law of armed conflict** (**LOAC**, also often called **the law of war**). Simply, the LOAC is constituted of all parts of international public law that deal with conflict. Major elements include the provisions of the Geneva and Hague Conventions and those articulations of just right to wage war included in the United Nations charter (which dictates when member states can engage in conflict and when the international community should intervene to prevent uninhibited warfare). In most cases, these formalizations of mutual understanding about how war should be conducted link with various **just war theory** traditions.

The direct application of the LOAC to the fifth domain is not a simple proposition. There is significant debate, for instance, about how to protect targets traditionally held as off-limits during wartime, such as hospitals or schools, in a cyber conflict. Given that the Internet enables a broad range of medical services beyond those simply contained in hospital facilities, simply labeling certain IP addresses with a special "do not attack" label will not protect all national or organizational medical functions. Likewise, it is unclear as to the degree to which the LOAC should protect hackers that are caught by states and held as responsible for major disruptive attacks. Some national actors use proxy actors extensively in cyber campaigns to ensure deniability. These non-state agents are not terrorists that have disavowed a recognizable political entity, but they are likely to be specifically disavowed by a state sponsor interested in maintaining its own neutrality. So, are they combatants to be protected by international law (as a soldier hacking as part of a national cyber military service would be) or not?

Another useful concept for thinking about cooperation and the Internet from the liberal perspective is that of the **tragedy of the commons**. The tragedy of the commons emerges in situations where there is some resource that is available to everybody but controlled by nobody, such as public lakes where you might want to go fishing or public land where you might graze cattle. The tragedy itself is that individuals are incentivized to take all of the resource they can because it is free to them (i.e. to let their cattle graze as much as possible on public land), despite the fact that this kind of self-interested action will reduce the usability of the resource for everybody over time (i.e. there will be less grass for others' cows to graze on).

Is there a tragedy of the Internet commons? After all, the Internet is a resource that is not **rivalrous** (meaning that my consumption of Internet bandwidth doesn't prevent others from getting online) and is only minimally **excludible** (meaning that it's possible to prevent someone from using it). In terms of international cooperation, it simply isn't possible – short of a major disruptive attack, as will be discussed in later chapters – to prevent other countries from using cyberspace. From the perspective of the individual, it is possible that a gatekeeper – an ISP or an authoritarian government, for instance – could prevent you from getting online. But in general, Internet usage is so common now in world affairs that the thing is only minimally excludible.

That the Internet is a **common pool resource** in this way is not inherently problematic for attempts to have countries cooperate on cyber conflict issues. After all, numerous treaties have existed over the centuries governing what countries can and cannot do militarily in space, on the high seas, etc. All states have interests in mitigating threats of misuse of public goods that could hamper trade or encourage criminal enterprise. Moreover, it is often quite possible to coerce or shame states into good behavior. On the one hand, it is possible to call out a country for abuse of the commons. On the other hand, abuse of the commons is most often met by a balancing action, such as a Russian militarization of spacecraft in response to U.S. efforts to station ICBMs in orbit.

Herein lies the difference between the Internet and other public goods that have been successfully regulated by international agreement. Much as is the problem with applying international law to cyberspace, challenges of attribution prevent the development of effective verification capabilities. Simply put, it is not always – or even often – possible to satisfactorily identify when a state is using the Internet inappropriately. Moreover, the Internet seems more truly non-rivalrous than other domains of warfighting like space of the oceans in that one country's ability to act offensively online does not naturally limit or balance the ability of others to do so. The result is a digital age tragedy of the commons wherein international cooperation is difficult because there are strong incentives to abuse the resources juxtaposed against no incentives (i.e. no social or political costs) to stop.

Second, the neo-institutional or neoliberal perspective on world politics bears mention as a potentially useful framework for understanding how international cooperation might emerge on cybersecurity and other digital issues.[34] Though grounded in the behavioralist language of structural realism, neoliberalism historically has no quarrel with the idea that states are not the only important actors in international relations. Moreover, the neoliberal approach emphasizes the manner in which iterative relationships can compel complex learning and the formation of unique configurations of either cooperative or conflictual interactions. Taken with the more traditional take on liberalism updated by Moravscik and others, neoliberal approaches appear to gel well with research to date on the empirics of the cyber phenomenon in world politics. Institutional learning and strategies matter, as do parochial political dynamics. If anything, the challenge for constructing liberal theories of cyberpolitics moving forward will be in the details. In particular, liberal theories often make claims about the directionality of political relationships (the hierarchy of local, national and state interests, for instance) that present as clumsy in the digital age. A substate actor's political capacity in using cyber tools need not be employed via the prism of state configurations in the digital age, a fact that opens up new possibilities for cause and effect beyond the pale of interstate politicking.

Constructivism

Finally, as the research program on cyberspace and international security continues to develop, it seems likely that researchers will increasingly turn to **constructivism** to flesh out and explain patterns of interaction and political behavior in world politics in the digital age. Constructivism, broadly writ, holds that the environment in which political action takes place is social and that the social setting of interactions can essentially *provide* states and other actors with their core preferences.[35] This differs dramatically from the perspectives of realism and structural liberalism, where core conceptual assumptions about the sources of state power are based on the condition of anarchy in the international system. For constructivists, "anarchy is what states make of it," meaning that social conditions in world politics compel actors to construct different perspectives on the anarchic condition of the system.[36]

Box 3.4: Ideas, the Internet and insecurity

The global proliferation of Internet technologies has naturally aided a worldwide transformation of the processes of information diffusion and identity creation. Specifically, the information revolution has led to the development of unique new tools for information dissemination and to the construction of novel information environments wherein humans now access, consume and have framed content in ways not seen before.

Constructivist modes of thinking about world affairs are quite arguably the most relevant for anybody interested in understanding the *new* sources of contention and conflict in the digital age. This is quite simply because so much new contestation is bound up in these broad-scoped changes to the global information environment. From the Arab Spring to the Umbrella Movement protests in Hong Kong in 2014, information technologies have been important enablers of social conflict and have defined the course and perception of activist campaigns around the world. Likewise, the rise of the Internet has prompted significant international focus on issues of infrastructural policy in ways that would have seemed ridiculous in eras past, simply because of the way that ICTs have rewired most core functions of global society. Net neutrality, for instance, is a hotly debated issue primarily because making sure network operators aren't allowed to treat different kinds of data based on their economic self-interest is an obvious guarantor of ideational freedom in democratic societies. And much of what is increasingly being thought of as part of the toolkit of cyber contention in the form of information warfare conducted by subversive state agents is, as later chapters discuss, only logically significant when one considers the security and political stability of IR from an ideational perspective.

There's a problem of imprecision with constructivism that is, in many ways, no different from the broader critique of the school of thought. Simply put, constructivism – the idea that "ideas matter" – fails to provide agent-centric mechanisms for explaining and predicting patterns in international affairs. Nevertheless, constructivism provides some interesting insights and has likely applications for theoretical treatments of cyberspace in world politics in the future. The school lends itself to understanding how institutions and states learn from the conditions of their interactions. In particular, a constructivist perspective – which emphasizes the role that constraining norms play in changing state behaviors over time – lends itself to future studies of international cooperation on cyber issues.

Box 3.5: Future shock (or "from speed comes vulnerability")

As Chapter 4 and onward repeatedly note, any discussion of cyber conflict and the information age is invariably an old discussion. Just because the most recent information revolution has arguably been the most dynamic and unprecedented in human history, does not mean that there have not been other information revolutions or that other new information technologies have not fundamentally changed the way the world works. Famously referenced with some regularity by political scientists, the Gutenberg printing press is an example of a revolutionary ICT that, among other things, helped expand literacy in Europe, standardized conceptions of time and language and generally fed the rise of the nationalistic state in Europe. The development of papyrus paper and simple written shorthand in Egypt spurred the rise of the first major bureaucratic empires. And so on.

Humankind, in short, has regularly experienced the tumult and catalyzing force of information revolutions. Inasmuch as we often associate these revolutions with great progress on a civilizational scale though, it is worth noting that truly revolutionary ICTs have often been key drivers of great catastrophe and conflict in human history. As a series of thinkers, notably the scholar Paul Virilio, point out, new information technologies dramatically speed up the functions and interactions inherent in human society. And yet, we humans don't really speed up to match, at least in terms of how we fundamentally operate. The result is quite often a mismatch between techno-political societal progress and the development of the human condition. With great speed comes significant disruption to traditional identities and relationships, as well as a diminished understanding about the way the world around us works. The shadow of the future, as some scholars might put it, grows shorter because technological progress reflected in societal advancements injects uncertainty into the human experience. At these times, constructivists – or post-structuralists, post-materialists, etc. – suggest that we are significantly more prone to "future shock" where an inability to reconcile prevailing conditions with uncertainty about the future produces vulnerability to conflict.

In addition, constructivist theories are perhaps the only set of perspectives that might accommodate psychological evidence put forward on the impact of cyberspace on the preferences of individuals and institutions in the digital age (i.e. on how people think and behave differently as a result of changes in the global information environment). Much recent work has pointed to the way in which the information revolution has caused a sea change in the way that global constituents – from individual citizens to state organizations – undertake problem solving and strategic planning. Today, it is increasingly the case that these tasks take place in a "networkized" fashion, with actors formulating responses to sociopolitical events in unprecedented ways. Constructivism, alone among the three main schools of thought in IR, provides interesting insight as to how such developments might affect world politics.

Cyberspace and international relations: islands of study

To this point, this chapter has considered international relations in broad terms. Specifically, we know that cyberspace is meaningful for world affairs and for the IR field. The question has been: to what degree is the IR field and its theory relevant to the study of cyberspace-related phenomena? To some degree, of course, the answer is highly relevant. There is a clear need for researchers to look past limiting assumptions about human interaction and the function of international politics and security. Clearly, the paradigmatic views on international affairs espoused by the great debates of the 1980s and 1990s impact upon today's questions about world affairs in the digital age in an incomplete fashion. Many core assumptions and concepts are useful for guiding scholars as they attempt to produce new knowledge. Others, particularly as they pertain to modes of conflict and international organization in eras past, need to be updated or entirely jettisoned in favor of concepts that better reflect the new normal of the Internet age.

Today, the state of affairs for the IR subfield that has oriented on issues of digital security – which we broadly refer to as the cyber conflict studies field in this book – has constituted a large amount of what's often referred to as "middle range theory," which is essentially an approach to research that assesses empirical data and then infers from there to make generalizable statements about the state of the world (or, more accurately, one particular dimension thereof). At this time, this subfield within IR might be broken up into four constituent areas of substantive focus. These are (1) cyber conflict, (2) espionage and counterintelligence, (3) Internet controls and cyber repression and (4) cyberspace governance. We discuss each here in brief. There are also lesser areas of research that don't fit neatly into these arenas of research. These include work on the epistemology of digital developments, narrow work on terrorism and crime that abuts more common research on cyber conflict writ large and research that grapples with questions of communication, propaganda and disinformation in the Internet age. All of these areas, of course, are inherently interdisciplinary. We describe them as a subfield of IR in this chapter and there is a large amount of organizational and topical truth to that statement. But research in each of these veins clearly links to work undertaking by researchers in related areas of the social sciences. Indeed, some of these areas are more heavily located under the banner of other disciplines like sociology or communications studies. The immense subfield focused on disinformation and misinformation that clearly links to issues of digital security of concern for IR scholars, is one such example.

Cyber conflict

We use the term "cyber conflict studies" in this book to describe the broad subfield within IR that has formed to study issues of international security, politics and information technologies. However, there are distinct questions and areas of focus that more reflect the emphasis made by many researchers on *conflict* in the age of the Internet than others. Cyber conflict research has its roots in several pieces of scholarship that emerged to draw broad lines around the emerging phenomenon of the Internet as a national security imperative in the late 1990s and early 2000s.[37] The first groundswell of serious IR research, however, occurred after 2010, with several authors orienting on the core question of what engagement in cyberspace actually looked like[38] and what it constituted in terms of a sociopolitical threat to states and their societies. It is from this early work that the evolutionary versus revolutionary camps described earlier in this chapter were initially defined,[39] as some voices argued we simply saw a continuation of technological advances linked to warfighting and others saw a groundbreaking shift toward new paradigms of human interaction for purposes as conflict. Issues discussed by such voices included the nature and likelihood of so-called "cyberwar" and the condition of cyber operations as a tool for warfighting versus espionage, sabotage and subversion. We take up each of these questions later in this book.

Today, the cyber conflict studies component part of the broader subfield functions as somewhat of a core of assumptions, definitions and terminology. Some of this is for the better, with initial work on cybersecurity in world politics making clearer the parameters and types of effects being studied by student of international affairs. As we discuss in later chapters, however, some of these assumptions and definitions are ill-fitting – with old having been grafted onto the new phenomenon without necessary updating, in some cases – and still being debated.[40] Today, research in this subfield has seemingly landed somewhere between the evolutionary and revolutionary camps, seeing cyber conflict as a major issue but not the same thing as traditional conflict. Cyber conflict has not, according to many, revolutionized military affairs but it may have brought about other revolutionary effects.[41] As such, ongoing work centers on the relationship between cyberspace and national power,[42] military capabilities and the various functional vulnerabilities of national societies as they pertain to strategic competition.

Espionage and counterintelligence

Given that so many IR scholars have accepted that cyber conflict is not warfighting in the traditional sense, it should be unsurprising that many research programs have shifted their context and objectives to align more closely with the intelligence studies field.[43] Several empirical studies, which we broach in later chapters, have shown that most cyber activity is undertaken – at least by state actors – for purposes of either short-term or long-term espionage. Espionage is obviously an age-old human activity, but cyberspace and evolving web technologies have made information retrieval possible in a truly revolutionary fashion. Intelligence gathering can be undertaken remotely and can occur rapidly with few operational resource commitments. There is no small amount of plausible deniability associated with cyber espionage and there are immense logistical advantages to be accrued by actors interested in the immense amounts of new data being produced by non-governmental actors, particularly private industry and domestic political actors.

At present, most espionage research within the growing cyber conflict studies subfield centers on questions of preferences. Specifically, when and why do states choose to conduct different kinds of cyber espionage? When are counterintelligence operations warranted? How do state actors guide and control proxy actors that hack on their behalf? To what degree are the outputs of cyber espionage actually useful to belligerents? On this last front, there is no doubt that espionage leads to at least targeted gains for those willing to employ cyber operations along such lines. However, there is great pushback against the idea that immense volumes of data can be handled and processed in a fashion conducive to, for instance, spurring domestic innovation. Chinese theft of intellectual property has clearly led to narrow advances in military technology and industrial techniques, for instance, but there is less evidence to suggest that the large troves of information stolen over the years from Western competitors has produced economic activity that wouldn't otherwise have occurred.[44] So is cyber espionage truly a substantial evolution of traditional espionage activities or is it just more of the same?

Internet controls, authoritarianism and cyber repression

The third area of focus within the cyber conflict subfield focuses on issues of Internet controls and cyber repression. This is the area of the subfield that most heavily focuses on the domestic politics of nations, with strong ties to research programs within IR focused on social movements,[45] protest activities,[46] rebellion and insurrection[47] and media control.[48]

As with the other areas of the field, the central question for many is the degree to which the Internet has revolutionized versus just evolved pre-digital dynamics of the relationship between the state and society. There is a broad literature that argues that the web is a boon to activists and democratists across global society. Authors like Jared Diamond have famously labeled the Internet a "liberation technology"[49] and more recent research has upheld the reality that the web helps dissidents and protesters overcome conventional logistical challenges to organization, build collective agreement on needed social change without formal mobilization and generally hold authorities accountable. Cyber attack is a focus of some research in this area, with a few studies describing certain techniques – such as distributed denial of service (DDoS) and virtual private networks (VPNs) – as useful for either drawing attention to a cause or permitting dissenters a means of getting around government surveillance or information lockouts.[50]

On that point, of course, the field has inevitably oriented on the opposing argument, namely that the Internet has been as useful for upgrading state powers of social control as it has been for providing democratizing capacity to social movements. Gohdes makes the argument that some governments specifically design their web architecture – both technically and organizationally – with surveillance in mind.[51] This applies particularly to authoritarian regimes, of course, but is relevant in some democratic or semi-democratic states. Control of web access allows countries to leverage information outages to political gain, with some governments like Syria even using controlled Internet outages to hide state-backed violence against dissidents, minorities and other troublesome parts of the domestic population. And, more than just a simple control method, this authoritarian upgrading via the co-option of the web has taken on a sophisticated form in countries like China and those attempting to copy Beijing's approach. Here, web technologies are weaponized against the population, with an immense apparatus of censors guiding civil discourse and preventing the organization of dissent behind the scenes, all the while more targeted cyber tools – spyware, for instance – is employed to spy on or entrap known troublemakers for public shaming, imprisonment or worse.[52]

The governance of cyberspace

A final major category of research within the cyber conflict studies field is that focusing on cyberspace governance. Research in this vein focuses on the complex system of actors and architectures – again, both technical and organizational – that help manage the global functionality of the Internet. Politically, the focus of much work centers on the provision of web services as a public good versus a more restricted commodity. This, of course, ties directly into the questions of multilateral versus multistakeholder perspectives previously outlined, with research often orienting on the processes that lead to competing models for solving a host of regulatory, technological and geopolitical challenges.[53]

Subsets of this research arena focus on the function of international organizations in governing the Internet, as well as the international law surrounding the regulation of different Internet "territories" and the actual objectives of the global Internet governance regime.[54] On this last point, researchers and analysts debate what the goals of cyberspace governance activities should be. Should the provision of an open Internet across the entirety of civil society be the stated goal of such efforts?[55] Should the self-determination of peoples or nations to select what their evolving web experience should be override such an objective?[56] And how does all of this link to the aforementioned areas of cyber conflict research, particularly the actions of multiform actors that use web technologies to attack one another, to spy on one another, to affect political interference and to spread influence? These are obviously questions of law, as well as of the development of appropriate norms of web use, development and misuse.[57]

Where to next?

Cyberspace and information technologies affect global society at virtually every level. From the infrastructure of security, economic and sociopolitical functions to the ways in which global constituents consume information and approach problems, the maturation and continued evolution of the cyber phenomena have significant implications for the dynamics of world politics.

There are a great number of important issues to be considered in the nascent cyberpolitics field. What might future international cooperation on conflict avoidance online look like? How will the international community deal with cyber extremism? How do human and civil rights tie into processes of human interaction and information availability in the digital age? At stake in studying these issues and in answering critical questions about what might determine variation in outcomes along different lines, is a better understanding of the phenomenon itself. Better knowledge produces better real-world outcomes in policy and in norms of coordination.

At present, the state of scholarship on cyberspace and IR is relatively immature but is growing fast. This is not to say, of course, that there is not important and significantly insightful work already published on the cyber phenomenon and its impacts on international affairs. But scholars have only in recent years moved to an analytic footing that emphasizes the interdependent complexity of issues that link across traditional categories in an unprecedented fashion. And overreliance on imprecise or underdeveloped theoretical constructs continues to impede progress on research related to cyberspace.

Moving forward, scholars and analysts will need to continue to recognize that a great number of sociopolitical factors impact upon the already unique situations that different actors – from states to international organizations and domestic political groups – find themselves in in the digital age. In order to apply IR theories successfully as claims about the ontology of IR in a cyber-infused world, there is a great deal of work needed on fleshing out those non-traditional areas of focus in the field that are most impacted by digital developments. The rest of this book is designed so as to give students of world politics interested in cyber conflict a foundation from which such work might emerge.

Notes

1 For perhaps the best description and analysis of the history of cybersecurity in international politics, see Healey, Jason, *A Fierce Domain: Conflict in Cyberspace 1986 to 2012*. Vienna, VA: Cyber Conflict Studies Association, June 1, 2013.

2 For additional overviews of the various aspects of the landscape of cyber governance in world politics, see Ferwerda, Jeremy, Nazli Choucri, and Stuart Madnick, "Institutional foundations for cyber-security: Current responses and new challenges." Working Paper CISL 2009–03. Massachusetts Institute of Technology, Composite Information Systems Laboratory, September 2010; and Choucri, Nazli, *Cyberpolitics in International Relations*, Cambridge, MA: MIT Press, 2012.

3 For perhaps the earliest examples of this, see Deibert, Ronald J., "Black code: Censorship, surveillance, and militarization of cyberspace." *Millennium Journal of International Studies* 32, no. 2 (2003): 501–530; Goldman, Emily O., "Introduction: Information resources and military performance." *Journal of Strategic Studies* 27, no. 2 (2004): 195–219; and Eriksson, Johan and Giampiero Giacomello, "The Information Revolution, Security, and International Relations: The (IR)relevant Theory?" *International Political Science Review* 27, no. 3 (July 2006): 221–244.

4 For a review and description of arguments in this vein, see Whyte, Christopher, "Power and predation in cyberspace." *Strategic Studies Quarterly* 9, no. 1 (Spring 2015): 100–118.

5 See Adams, James, "Virtual defense." *Foreign Affairs* 80, no. 3 (2001): 98–112; and Nye, Joseph, *Cyber Power*. Cambridge, MA: Belfer Center for Science and International Affairs, Harvard Kennedy School, May 2010.

6 Nye, *Cyber Power*, 2010.

7 For perhaps the best descriptions of the attribution problem, see, among others, Hurwitz, Roger, "Depleted trust in the cyber commons." *Strategic Studies Quarterly* 6, no. 3 (2012): 20–45; and Lindsay, Jon R., and Erik Gartzke, "Weaving tangled webs: Offense, defense and deception in cyberspace." *Security Studies* 24, no. 2 (2015): 316–348.

8 Lindsay and Gartzke, "Weaving Tangled Webs." (2015).

9 Gombert, David C., and Martin Libicki, "Cyber Warfare and Sino-American Crisis Instability." *Survival* 56, no. 4 (2014): 7–22.

10 Perhaps the best work to date that interfaces with the question of social processes and national security processes is Manjikian, Mary M., "From Global Village to Virtual Battlespace: The Colonizing of the Internet and the Extension of Realpolitik." *International Studies Quarterly* 54, no. 2 (June 2010): 381–401. Beyond national security, a diverse range of works interface with questions related to the contours of world politics and the societal impacts of the rise of cyberspace, including, among others, Whyte, Christopher, "Dissecting the Digital World: Old Questions, New Answers." *International Studies Review*, Forthcoming; Cotnoir, Amy, Angelos Stefanidis, Arie Croitoru, Andrew Crooks, M. Rice, and Jacek Radzikowski, "Demarcating New Boundaries: Mapping Virtual Polycentric Communities through Social Media Content." *Cartography and Geographic Information Science* 40, no. 2 (2013): 116–129; Steven Shaviro, *Connected, Or What It Means to Live in the Network Society*. Minneapolis, MN: University of Minnesota Press, 2003; Galloway, Alexander R. and Eugene Thacker, *The Exploit: A Theory of Networks*. Minneapolis, MN: University of Minnesota Press, 2007; Dreyfus, Hubert, *On the Internet: Thinking in Action*. London: Routledge, 2001.

11 Galloway and Thacker, *The Exploit: A Theory* (2007).

12 See, for example, Gregory J. Rattray, *Strategic Warfare in Cyberspace*. Cambridge, MA: MIT Press, 2001; Borg, Scott, "Economically Complex Cyberattacks." *IEEE Security and Privacy Magazine* 3, no. 6 (November/December 2005): 64–67; McConnell, Mike, "Cyberwar is the New Atomic Age." *New Perspectives Quarterly* 26, no. 3 (Summer 2009): 72–77; Clarke, Richard A., and Robert K. Knake, *Cyber War: The Next Threat to National Security and What to Do about It*. New York: Ecco, 2010; Junio, Timothy J., "How Probable is Cyber War? Bringing IR Theory Back into the Cyber Conoict Debate." *Journal of Strategic Studies* 36, no. 1 (2013): 25–133; Peterson, Dale, "Offensive Cyber Weapons: Construction, Development, and Employment." *Journal of Strategic Studies*, Vol. 36, No. 1 (2013), 120–124; and Lucas Kello, "The Meaning of the Cyber Revolution: Perils to Theory and Statecraft." *International Security* 38, no. 2 (2013), 7–40.

13 Brenner, Joel, *America the Vulnerable: Inside the New Threat Matrix of Digital Espionage, Crime, and Warfare*. New York: Penguin, 2011; Nye, Joseph S. Jr., "Nuclear Lessons for Cyber-security?" *Strategic Studies Quarterly* 5, no. 4 (2011): 18–36.

14 Adams, "Virtual Defense," p. 98; Nye, *Cyber Power*, pp. 1–4; and Nye, Joseph S., *The Future of Power*. New York: Public Affairs, 2011, Chapter 5.

15 See, broadly, Thomas Rid, "Cyber War Will Not Take Place." *Journal of Strategic Studies* 35, no. 5 (2012): 5–32; Betz, David. "Cyberpower in strategic affairs: neither unthinkable nor blessed." *Journal of Strategic Studies* 35, no. 5 (2012): 689–711; Liff, Adam P., "Cyberwar: A New 'Absolute Weapon?' The proliferation of cyberwarfare capabilities and interstate war." *Journal of Strategic Studies* 35, no. 3 (2012): 401–428; Libicki, Martin C., *Conquest in Cyber-space: National Security and Information Warfare*. New York: Cambridge University Press, 2007; Lindsay, Jon R., "Stuxnet and the limits of cyber warfare." *Security Studies* 22, no. 3 (2013): 365–404; Gartzke, Erik, "The myth of cyberwar: Bringing war in cyberspace back down to earth." *International Security* 38, no. 2 (2013): 41–73; and Lindsay, Jon R., "The impact of China on cyber-security: Fiction and friction." *International Security* 39, no. 3 (2014–15): 7–47.

16 Rid, Cyber War Will Not Take Place. pp. 7–16; and Rid, *Cyber War Will Not Take Place*, Chapters 1, 2.

17 Much scholarship in this vein argues that cyber revolution rhetoric and analysis is exaggeratory in nature and emerges from a range of alternative motivations on the parts of governments, officials, military institutions, industry actors, and more. Among others, see Ohm, Paul, "The Myth of the Superuser: Fear, Risk, and Harm Online." *University of California Davis Law Review* 41, no. 4 (2008): 1327–1402; Dunn Cavelty, Myriam, "Cyber-terror: Looming threat or phantom menace? the framing of the US cyber-threat debate." *Journal of Information Technology & Politics* 4, no. 1 (2008): 19–36; Brito, Jerry and Tate Watkins, *Loving the Cyber Bomb? The Dangers of Threat Inflation in Cyber-security Policy*. Arlington, VA: Mercatus Center, George Mason University, 2011; Lawson, Sean, "Beyond Cyber-Doom: Assessing the Limits of Hypothetical Scenarios in the Framing of Cyber-Threats." *Journal of Information Technology & Politics* 10, no. 1 (2013): 86–103; Schneier, Bruce, *Liars and Outliers: Enabling the Trust That Society Needs to Thrive*. Indianapolis, IN: Wiley, 2012; and Evgeny Morozov, "Cyber-Scare: The Exaggerated Fears Over Digital Warfare." *Boston Review* 34, no. 4 (2009). http://bostonreview.net/archives/BR34.4/morozov.php.

18 Other descriptions of these terms appear in Choucri, Nazli, *Cyberpolitics in International Relations*. Cambridge, MA: MIT Press, 2012; Singer, P.W., and Allan Friedman, *Cyber-security and Cyberwar: What Everyone Needs to Know*. New York: Oxford University Press, 2014; Valeriano, Brandon, and Ryan Maness, *Cyber War versus Cyber Realities*. New York: Oxford University Press, 2015; Libicki, Martin C., *Conquest in Cyberspace*, 2007; and Reveron, Derek S., ed., *Cyberspace and National Security: Threats, Opportunities, and Power in a Virtual World*. Washington, DC: Georgetown University Press, 2012.

19 For the history of the use and conceptualization of this term, see Gartzke, Erik. "The myth of cyberwar: Bringing war in cyberspace back down to earth." *International Security* 38, no. 2 (2013): 41–73; Rid, Thomas, "Cyber war will not take place." *Journal of Strategic Studies* 35, no. 1 (2012): 5–32; Rid, Thomas, "Think again: cyberwar." *Foreign Policy* 192 (2012): 80–84; and Mahnken, Thomas G., "Cyber War and Cyber Warfare." In Kristin M. Lord, and Travis Sharp, eds., *America's Cyber Future: Security and Prosperity in the Information Age*. Washington, DC: Center for a New American Security, 2011.

20 Rid, Thomas, *Cyber War Will Not Take Place*. Oxford, UK: Oxford University Press, 2013.

21 For perhaps the best descriptions of such institutions, formal or otherwise, in this vein, see Choucri, Nazli, *Cyberpolitics in International*, 2012.

22 See, for instance, Jackson, Patrick, and Daniel Nexon, "Paradigmatic faults in international relations theory." *International Studies Quarterly* 53, no. 4 (2009): 907–930.

23 To date, few works have undertaken this task. See Junio, Timothy, "How probable is cyber war? bringing IR theory back into the cyber conflict debate." *Journal of Strategic Studies* 36, no. 1 (2013): 125–133; and Russell, Alison, *Cyber Blockades*. Washington, DC: Georgetown University Press, 2014. Other examples might be found in Reardon, Robert, and Nazli Choucri, "The role of cyberspace in international relations: a view of the literature." *Paper Prepared for the 2012 ISA Annual Convention* (April 1, 2012).

24 See Lynn-Jones, Sean M., "Offense-defense theory and its critics." *Security Studies* 4, no. 4 (1995): 665.

25 For a good description of this logic, see Adam P. Liff, "Cyberwar: A New 'Absolute weapon?' the proliferation of cyberwarfare capabilities and interstate war." *Journal of Strategic Studies* 35, no. 3 (2012): 401.

26 For a full explication of these dynamics, see Gombert and Libicki, "Cyber Warfare and . . ." (2014). For a discussion of the role of such uncertainty in driving decision making and strategy on cybersecurity, see David D. Clark and Susan Landau, "Untangling Attribution." In *Proceedings of a Workshop*

on Deterring Cyberattacks: Informing Strategies and Developing Options for U.S. Policy. Washington, DC: National Academies Press, 2010, 25–40; and Alexander Klimburg and Heli Tirmaa-Klaar, *Cyber-security and Cyberpower: Concepts, Conditions, and Capabilities for Cooperation for Action within the EU*. Brussels: European Parliament Directorate General for External Policies of the Union, Policy Department, April 2011.

27 For seminal overviews of the neorealist perspective and arguments, see Waltz, Kenneth, *Theory of International Politics*. Reading, MA: Addison-Wesley Pub. Co., 1979; Keohane, Robert O. *Neorealism and Its Critics*. New York: Columbia University Press, 1986; and Mearsheimer, John J. *The Tragedy of Great Power Politics*. New York: W.W. Norton & Company, 2001.

28 An overview of particular focus on the nature of power and power strategies might be found in Mearsheimer, *The Tragedy* (2001).

29 Ibid., pp. 57–60.

30 Rid, *Cyber War Will . . .* (2012).

31 See, among others, Valeriano and Maness (2015).

32 For an overview of liberal theories on the nature of foreign policymaking, see Fareed Zakaria, "Realism and Domestic Politics: A Review Essay," Myths of Empire: Domestic Politics and International Ambition by Jack Snyder, *International Security* 17, no. 1 (1992): 177–198; Moravcsik, Andrew, "Taking preferences seriously: A liberal theory of international politics." *International Organization* 51 (1997); Levy, Jack S., "Domestic Politics and War." In Robert I. Rotberg and Theodore K. Rabb, eds., *The Origin and Prevention of Major Wars*. New York: Cambridge University Press, 1988; Rosecrance, Richard, and Arthur A. Stein, eds., *The Domestic Bases of Grand Strategy*. Ithaca, NY: Cornell University Press, 1993; and Hagan, Joe, "Domestic Political Explanations in the Analysis of Foreign Policy." In Laura Neack et al., eds., *Foreign Policy Analysis: Continuity and Change in Its Second Generation*. Englewood Cliffs, NJ: Prentice Hall, 1995.

33 Putnam, Robert, "Diplomacy and domestic politics: the logic of two-level games." *International Organization* 42 (1988): 427–460.

34 For seminal works on the neoliberal perspective, see Keohane, Robert and Joseph Nye, *Power and Interdependence: World Politics in Transition*. Boston, MA: Little, Brown and Company, 1989; Lebow, Richard Ned, "The long peace, the end of the cold war, and the failure of realism." *International Organization* 48, no. 2 (1994): 251–252; Keohane, Robert O., and Lisa L. Martin, "The Promise of institutionalist theory." *International Security* 20, no. 1 (1995): 39–51; and Keohane, Robert O., *After Hegemony: Cooperation and Discord in the World Political Economy*. Princeton, NJ: Princeton University Press, 1984.

35 For an overview of the constructivist perspective, see Checkel, Jeffrey T., "The constructivist turn in international relations theory." *World Politics* 50, no. 2 (1998): 324–348.

36 Wendt, Alexander, "Anarchy is what states make of it: the social construction of power politics." *International Organization* 46, no. 2 (1992): 391–425.

37 Notably Arquilla, John. *The Advent of Netwar*. Rand Corporation, 1996; and Arquilla, John, and David Ronfeldt. *Networks and Netwars: The Future of Terror, Crime, and Militancy*. Rand Corporation, 2001.

38 Rid, Thomas. "Cyber war will not take place." *Journal of Strategic Studies* 35, no. 1 (2012): 5–32; Lindsay, Jon R. "Stuxnet and the limits of cyber warfare." *Security Studies* 22, no. 3 (2013): 365–404; Gartzke, Erik. "The myth of cyberwar: Bringing war in cyberspace back down to earth." *International Security* 38, no. 2 (2013): 41–73; Barzashka, Ivanka. "Are cyber-weapons effective? Assessing Stuxnet's impact on the Iranian enrichment programme." *The RUSI Journal* 158, no. 2 (2013): 48–56; and Valeriano, Brandon, and Ryan C. Maness. "The dynamics of cyber conflict between rival antagonists, 2001–11." *Journal of Peace Research* 51, no. 3 (2014): 347–360.

39 Langø, Hans-Inge. "Competing academic approaches to cyber security." In *Conflict in Cyber Space*, pp. 23–42. Routledge, 2016.

40 The authors of this book have discussed this dynamic elsewhere, particularly in Whyte, Christopher. "Dissecting the digital world: A review of the construction and constitution of cyber conflict research." *International Studies Review* 20, no. 3 (2018): 520–532.

41 Van Puyvelde, Damien, and Aaron F. Brantly. *Cybersecurity: politics, governance and conflict in cyberspace*. John Wiley & Sons, 2019.

42 For instance, Inkster, Nigel. *China's cyber power*. Routledge, 2018.

43 For instance, Lindsay, Jon R. "Cyber conflict vs. Cyber Command: hidden dangers in the American military solution to a large-scale intelligence problem." *Intelligence and National security* 36, no. 2

(2021): 260–278; Akoto, William. "Accountability and cyber conflict: examining institutional constraints on the use of cyber proxies." *Conflict Management and Peace Science* 39, no. 3 (2022): 311–332; and Canfil, Justin Key. "The illogic of plausible deniability: Why proxy conflict in cyberspace may no longer pay." *Journal of Cybersecurity* (2020).

44 Gilli, Andrea, and Mauro Gilli. "Why China has not caught up yet: Military-technological superiority and the limits of imitation, reverse engineering, and cyber espionage." *International Security* 43, no. 3 (2018): 141–189.

45 Pierskalla, Jan H., and Florian M. Hollenbach. "Technology and collective action: The effect of cell phone coverage on political violence in Africa." *American Political Science Review* 107, no. 2 (2013): 207–224.

46 Shandler, Ryan, Michael L. Gross, and Daphna Canetti. "Can you engage in political activity without Internet access? The social effects of Internet deprivation." *Political Studies Review* 18, no. 4 (2020): 620–629.

47 Van Laer, Jeroen, and Peter Van Aelst. "Cyber-protest and civil society: The Internet and action repertoires in social movements." In *Handbook of Internet crime*, pp. 248–272. Willan, 2013.

48 Dragu, Tiberiu, and Yonatan Lupu. "Digital authoritarianism and the future of human rights." *International Organization* 75, no. 4 (2021): 991–1017.

49 Diamond, Larry, and Marc F. Plattner, eds. *Liberation technology: Social media and the struggle for democracy*. JHU Press, 2012.

50 Asal, Victor, Jacob Mauslein, Amanda Murdie, Joseph Young, Ken Cousins, and Chris Bronk. "Repression, education, and politically motivated cyberattacks." *Journal of Global Security Studies* 1, no. 3 (2016): 235–247.

51 Gohdes, Anita R. "Repression technology: Internet accessibility and state violence." *American Journal of Political Science* 64, no. 3 (2020): 488–503.

52 King, Gary, Jennifer Pan, and Margaret E. Roberts. "How censorship in China allows government criticism but silences collective expression." *American Political Science Review* 107, no. 2 (2013): 326–343.

53 DeNardis, Laura, and Mark Raymond. "Thinking clearly about multistakeholder internet governance." In *GigaNet: Global Internet Governance Academic Network, Annual Symposium*. 2013.

54 Nye, Joseph S. *The Regime Complex for Managing Global Cyber Activities*. Vol. 1. USA: Belfer Center for Science and International Affairs, John F. *Kennedy School of Government*. Harvard University, 2014.

55 Holman, JoAnne, and Michael A. McGregor. "The Internet as commons: The issue of access." *Communication Law and Policy* 10, no. 3 (2005): 267–289.

56 Davidow, William H. *Overconnected: The Promise and Threat of the Internet*. Open Road Media, 2011.

57 Schmitt, Michael N., ed. *Tallinn manual 2.0 on the International Law Applicable to Cyber Operations*. Cambridge University Press, 2017.

Further readings

Betz, David J. *Cyberspace and the State: Towards a Strategy for Cyber-power*. London and New York: Routledge, 2017.

Choucri, Nazli. *Cyberpolitics in International Relations*. Cambridge, MA: MIT Press, 2012.

Goel, Sanjay. "National cyber security strategy and the emergence of strong digital borders." *Connections* 19, no. 1 (2020): 73–86.

Healey, Jason, ed. *A Fierce Domain: Conflict in Cyberspace, 1986 to 2012*. Washington, DC: Cyber Conflict Studies Association, 2013.

Horowitz, Michael C. "Do emerging military technologies matter for international politics?." *Annual Review of Political Science* 23 (2020): 385–400.

Kello, Lucas. "The meaning of the cyber revolution: Perils to theory and statecraft." *International Security* 38, no. 2 (2013): 7–40.

Manjikian, Mary McEvoy. "From global village to virtual battlespace: The colonizing of the internet and the extension of realpolitik." *International Studies Quarterly* 54, no. 2 (2010): 381–401.

Stevens, Tim. *Cyber Security and the Politics of Time*. Cambridge, UK: Cambridge University Press, 2015.

Valeriano, Brandon, and Ryan C. Maness. *Cyber War Versus Cyber Realities: Cyber Conflict in the International System*. New York: Oxford University Press, 2015.
Valeriano, Brandon, Benjamin M. Jensen, and Ryan C. Maness. *Cyber Strategy: The Evolving Character of Power and Coercion*. Oxford University Press, 2018.
Van Puyvelde, Damien, and Aaron F. Brantly. *Cybersecurity: Politics, Governance and Conflict in Cyberspace*. John Wiley & Sons, 2019.
Whyte, Christopher, A. Trevor Thrall, and Brian M. Mazanec, eds. *Information Warfare in the Age of Cyber Conflict*. Routledge, 2021.

4 Exploit

From signals intelligence to cyber warriors

In the chapters to come, we unpack a host of topics related to modern-day manifestations of cyber conflict. However, it would be inexcusable to start our foray into warfare in the fifth domain without first contextualizing the rise of cyberspace since the 1960s. To do this, we need to go back almost a full century to describe the manner in which intelligence-focused developments fueled – at first gradually and then rapidly – the construction of new technologies that are today at the heart of our topic. Today, conversation about cyber conflict often pivots on issues of military strategy. While this is a worthwhile perspective on what the phenomenon has become, the reality is that computers and network systems have been built and used at the cutting edge primarily by intelligence communities. First, these technologies were the tools for spying, instruments for breaking codes and analyzing vast amounts of information. Then, these technologies became the targets and victims of espionage, as intelligence organizations realized that secrets stored in machines could be stolen as easily (if via a somewhat different method than was traditional) as if they were stored on paper in a filing cabinet. And finally, use of these technologies has become the means of spying itself, a set of exploitative methods where machine interactions are the entire basis of stealing information, deceiving adversaries and understanding enemies. In short, though the focus of our book must inevitably turn to questions of computer network attack (CNA) and disruption beyond intelligence matters, the roots of such a focus are unquestionably the substance of espionage – of computer network exploitation (CNE). This chapter briefly offers a history of these issues and developments.

The unbreakable marriage of computers and espionage

The starting point of any attempt to understand the origins of cyber conflict lies in understanding computers. This might seem an obvious thing to say, but here we would note that use of the term "computer" to refer to a mechanical, electronic device of some kind is something that has only come about since the 1960s. Traditionally, computers were simply people that computed. Anybody whose job involved the analysis and computation of patterns in information might have, at one point in time, been generically referred to as a "computer." Traditionally, computational vocations came in two main flavors. First, "computers" might have included anybody who used mathematics and other relevant acquired knowledge to deconstruct and assess patterns in coded information, such as the scrambled communications of state enemies. Second, we might have labeled someone a "computer" if they were involved in the complex analysis of information via some dumb process. To understand this statement, it is necessary to know something of the

DOI: 10.4324/9781003246398-4

conceptual design of modern computers. In turn, we first require some historical context reaching back to the years following World War I.

As any student of international relations will tell you, the story of great power politics from Westphalia (and backward) through the present day is the story of strategic gambit and maneuver involving nation states, their vassals and agents, wherein actions are the end result of both geostrategic circumstances and parochial national interests. Intelligence organizations have classically and enduringly been a critical element of state security apparatuses whose job is to inform national leaders on the condition and intentions of foreign foes (and those of allied actors too!). Intelligence, essentially information collected, processed and analyzed so that some kind of inference about world affairs is possible, comes in a number of flavors. **Human intelligence** (HUMINT) is information gathered directly from human sources, often via the actions of special operatives and informant networks. **Imagery intelligence** (IMINT) is derived from representations of objects reproduced electronically or via other optical means, such as visual photography, radar or infrared sensors, lasers or electro-optics. **Measurement and signature intelligence** (MASINT) is derived from scientific and technical information that is used to characterize specific targets of national interest. MASINT encompasses a broad set of disciplines, including those that fall under the nuclear, seismic, chemical and materials sciences. **Open source intelligence** (OSINT) emerges from publicly available information, such as that found in newspaper reporting or social media postings. And **geospatial intelligence** (GEOINT) involves the collection and study of imagery and mapping data produced through an integration of imagery, imagery intelligence and geospatial information (i.e. data that comes from satellites, reconnaissance aircraft).

Our story begins with the systematization – meaning the development of institutions given the mission of collecting all relevant signals information for national security purposes – of the last category of intelligence (**signals intelligence** (SIGINT)) in the years during and following World War I.[1] The Great War, as it was referred to following the cessation of hostilities, occurred several decades following what we might think of as the last great information revolution. Across the world, Europe's colonial empires and independent countries alike had come to rely on the telegraph and thereafter the telephone for the transmission of information critical to international commerce, government, war and social processes. At the outset of the war, the British Empire in particular held sway over much of the physical infrastructure of global telegraphy in the form of close patronage relationships with the British firms that directly managed the thing.[2] This made signals intelligence, which is simply the interceptions of signals passing between people (via machines, inscription, etc.) for the purposes of extracting strategically useful information, a more promising prospect than it had been at any point in history up until then. Shortly after the war began, Britain instituted a censorship regime wherein individuals placed at key telegraph exchange points around the world (given the title of "censor") would review massive volumes of information and weed out relevant data that could be processed further by British intelligence.[3] Britain would even quickly move to try and force the Central Powers to communicate entirely via British-overseen infrastructure. They did so at the beginning of the war by cutting German telegraph cables laid under the English Channel. The British effort was only somewhat successful on this front, largely due to the fact that radio technology rapidly came of age during the war, but the seeds of what would become signals intelligence – and, much later, cyber exploitation – were laid in such broad-scoped efforts to mine the world's communications systems for useful data.

Britain's efforts during the Great War were not unique in their shape, though they did perhaps stand apart in scope. During the war years, both Allies and Central Powers spent large sums of money developing systems that would effectively secure communications for military and high-level political purposes. Each major power involved in the war leveraged access to intellectuals and defense establishments that had, in the prewar years, made rapid advances in the mathematics field specifically related to cryptography. The result, quite simply, was that many of the Great Powers' ciphers were broken by the midpoint of the war.[4]

Following the war, as Europe rapidly rebuilt and the United States receded from the world, there was a veritable renaissance of intellectual and industrial enterprise on a number of fronts. That this happened is not necessarily a great surprise, given the incredible number of military and intelligence personnel that re-entered civilian life at universities and in business having been exposed to both other great minds and practical challenges that required solving. Part of this renaissance of sorts was, particularly in Western and Central Europe, a surging interest in cryptography – in the theory and practice of sending (and intercepting) secret messages. The result of this interest was a series of rapid advances in the development of more secure ciphers, many of which relied on physical devices to work. One such advance, for instance, was the development of the **one-time pad**, a programmable disc of sorts that could be used to encrypt information to be sent to a recipient elsewhere in the world.[5] The device, which shared its encryption cipher with a partner pad, was unique and immensely useful for two related reasons. First, the jumbling of letters that occurred each time a message was sent would be truly random. Pads were built two at a time and physically distributed to those who wanted to communicate in secret. Each pair of pads would have a different, randomly decided key. Second, no code would ever be reused. The pads were truly "one-time use," designed to be discarded after use and never reused. There was no mathematical chance that a key would ever be reproduced at a later stage because of the immense number of possible key combinations that the manufacturers had to choose from. Thus, there would be no pattern or logical system that could be targeted by foreign spy organizations to "break the code."

Enter Enigma

A problem with some advances, like the one-time pad, was that they were innovative but fundamentally inefficient in their design. With the pads, as is generally the challenge with most private key encryption schemes, the problem was that the paired pad needed to be physically distributed to those in need of the ability to communicate. Key distribution on a scale required for securing military communications during wartime, quite simply, was out of the question. After all, the distribution of thousands of keys is certainly possible. The billions of paired pads needed to support tens of thousands of daily communications from hundreds of military units deployed over expansive theaters of war did not seem so.

For Germany, whose military leaders were shocked to learn in the 1920s and early 1930s that their ciphers had been so thoroughly deconstructed by the Allies during the World War I, a solution to the pressing challenge of communications' security came in the form of a now-famous series of machines called **Enigma**.[6] Developed by a German engineer called Arthur Scherbius in the late 1910s and updated considerably over the next two decades, Enigma was an electro-mechanical device that promised unheard of encryption power. Interestingly, the German military was reluctant to purchase the device for a long time, as it was incredibly expensive to produce when first introduced. The inflection point was the discovery of the extent to which German communications had been previously tapped. One source of this information, ironically, was a book published by future Prime Minister Winston Churchill in the 1930s.[7] Another source, one which would worry many other countries besides Germany (notably Japan) in the years before World War II, was a book called *The American Black Chamber*.[8] The book was published by a disgruntled former government employee called Herbert Yardley and detailed much about the inner workings of U.S. efforts to systematically dismantle foreign encryption schemes.

Enigma – as well as the efforts of Polish,[9] French and British analysts[10] through the 1940s to break it – stands (alongside a device called the Tunny) as one of the most significant drivers of processes that would lead to the computer revolution at the mid-point of the 20th century. Enigma was a remarkable device. Prior to the German military's purchase and deployment of its own Enigma machine variants in the 1930s and 1940s, the device was largely targeted for sale to banks and other financial institutions that were uniquely interested in making sure that their transmissions were secure from prying eyes. The upside of this commercial genesis of the Enigma in the interwar European experience was that the head of British intelligence's codebreaking section was able to simply walk into a company office in Germany just prior to the start of World War II and purchase one of the Enigma devices.[11] Unfortunately, unlike the one-time pads where the device itself constituted the key needed to decrypt enemy signals, possession of the machine relayed no more advantage to the Allies than perfect knowledge of how seemingly improbably the task before them – of breaking the Enigma code – was.

The Enigma device works on a simple mechanical concept, namely that someone can type in a message and immediately have it scrambled. The internal wiring of the device is triggered when somebody types a letter into a keyboard sitting atop the device casing. From any distance, Enigma looks remarkably like a typewriter given its standard keyboard and boxy design. However, the device does not print text on a page, but rather scrambles the words input into the thing as they are entered. When a key is typed, it creates an electrical circuit wherein the input signal – a letter – is switched over and over again before outputting a final alternate letter. For example, if "B" is typed and "F" were to be the output, the user would type in "B" and "F" would appear in the form of a bulb that would light up atop the machine, then to be recorded for message transmission via radio, telegraph, etc.

The reason that possession of the Enigma machine would not itself give someone the ability to read all Enigma-encoded messages lies in the fact that the inside of the device is configurable. Specifically, the input letter would be switched in reference to a series of moving internal rotors. These circular parts would each switch the input letter for a new one as part of the electrical circuit being produced; indeed, they would do so twice, as the circuit looped back through the rotors after an initial pass because of the use of a reflector that bounced the current back toward the light panel (i.e. the bulbs that would light up to show the user what letter their input letter had been switched for). For each letter typed into the machine, the rotors would move forward one position. Thus, no letter would be switched for the same alternative letter twice in a row unless the result came about through pure luck.

This was the immense power of Enigma. The fact that the internal rotors required configuration meant that the device itself merely functioned as a sort of padlock for scrambling messages. In order to unlock the message (i.e. the decrypt the ciphertext),[12] you didn't just need the lock. You needed the "key," which in this case was knowledge of the starting positions of the rotors inside the machine itself. Without that information, messages typed into a recipient's own Enigma machine would remain scrambled. The

German military added a plugboard to its own initial versions of the machine, which allowed for the additional switching of letters inside the device based on the toggling of a pair of switches. Later versions added additional rotors and more complicated plugboards. The idea with each addition, of course, was that each new element multiplied the challenge of breaking the code exponentially. And Enigma was logistically secured from compromise via robust operational planning on the part of the German military, with instructions on rotor settings distributed on dissolvable paper and changed daily so that even capture of a "key" would not compromise operational security for more than a few hours. Some German operators even began their radio transmissions by sending seemingly random numbers in the clear to their counterparts. The idea was that the counterpart operators would turn their rotors that random number of times before commencing Enigma transmission. This meant not only that the onus was on the enemy (i.e. the Allies) to have guessed the daily set of starting positions but also that they would necessarily have to intercept all radio transmissions over and above.

Enigma was a seemingly insurmountable problem for Allied intelligence in the late 1930s and into the early 1940s. The standard configuration of the device used by the German navy – the "Naval Enigma" – was set up in such a way that any brute force attack on the code (i.e. an effort to break the cipher by simply guessing all possible combinations of machine settings) had a chance of success of about 1 in 159 million million million. Some variations on the device worsened those odds to 1 in billions of billions.[13] Quite simply, it would take thousands of "computers" thousands of years to break even one day's worth of coded signals unless they were improbably lucky. Obviously, some answer other than a large-scale human effort to guess the right "key" each day was required.

We go through the details of Enigma's inner workings to reinforce the significance of the relationship between information security and what would ultimately become the technological foundations of modern network-based computer systems. At the same time, it is necessary to know some detail of the challenge facing the British intelligence community in 1940, prior to the United States' entrance to World War II and when the threat of invasion seemed imminent on an almost daily basis. It was in that environment that the now-famous members of the British intelligence establishment, notably men like Alan Turing, Bill Tutte and Tommy Flowers, who were gathered to Bletchley Park in England in the late 1930s to serve Her Majesty's government in codebreaking endeavors, devised techniques for cracking mechanically enabled cryptographic innovations like the Enigma machine.[14]

The simple answer to Enigma was that "computing" – as an analytic activity – needed to evolve beyond human capabilities in much the same way that cryptography had been enhanced via technological augmentation. Specifically, a machine had to be constructed to allow for greater ease of brute force attacking the Enigma encryption. That machine was called a *bomba* and, of particular interest, it was not actually developed initially either during World War II or at Bletchley Park. Rather, the first bomb (or "bomba") was developed by Polish codebreakers in 1937 working to break an initial employment of Enigma during the Spanish Civil War.[15] They shared their accomplishments with their French and British counterparts shortly thereafter, and their work became the basis of more sophisticated bombs (so called probably because of the constant ticking noise they made as they worked) at both Bletchley Park and Arlington Hall (the U.S. cryptanalytic center set up in the 1940s). The bomb's design was quite simple. It was a recreation of the inner workings of the Enigma machine designed to crank through possible combinations at incredible pace, only stopping its incessant clicking when it landed on the right "key."

But even the development of bombs by British intelligence was not in itself enough to break Enigma. The volume of possible keys was still too great to work through in a single day before the German code changed. Those at Bletchley Park needed some way of limiting the number of possible combinations in a given day. The answer to the whole thing came down, as it often does, to the human factor. Researchers realized that two features of human efforts to create secure communications with Enigma actually allowed for the elimination of millions of possible settings combinations from the get-go. One of these features was design-based; the other was in the Germans' actual usage of the machine.

The first flaw that gave researchers an advantage stemmed from the specific nature of the marriage of mathematics and technology inherent in the machine's construction. The inner workings of Enigma did what all good encryption algorithms do as they are set to work to jumble information in a fashion both reversible and difficult to reverse – it simulated complexity so as to drastically lower the odds of brute force discovery while systematically allowing for key-based decryption. However, the specific design of Enigma meant that any letter entered could never be transformed into itself, an outcome that should obviously be mathematically possible given the process of randomly swapping one letter of the alphabet for another.

The second flaw, one familiar to most computer users today, stemmed from the fact that the use of Enigma by German operators invariably took on some amount of the personality of the operator themselves. In truth, the fact of Enigma was that even with knowledge about the limitations of the machine factored into the design of the attempt to brute force break the Enigma's code, one still needed to know something of what was already inside a message in order to decode the entire thing. Today, an attempt to guess somebody's password is dramatically more likely to succeed if some personal information is known about the target – their birthday, pet's name, mother's maiden name and so on. If some detail of how a user persistently interacts with an information system (e.g. sets passwords) can be uncovered, then that leads to a systematic vulnerability across everything that user touches. If that detail manifests at the level of an organization – say, because a company issues default usernames like "admin" or "firstname_lastname@company.com" to its employees, some of whom will be too lazy to change their settings – then that organization is systematically at risk. In the case of Enigma, this last scenario manifested in the regular usage of certain terms in German communiques, from regular weather reports where Bletchley staff could confidently assert that formulaic words (i.e. "Begin weather report: . . .") would appear to the standard "Heil Hitler" end-of-message sign-off.

Box 4.1: Systematic vulnerability and backdoors from Enigma to Huawei

Later in the book, we evoke in greater detail a story about how the first offensive action undertaken by the British Empire in World War I was actually about information and communications security. Very shortly after the official declaration of war, a British ship dredged up German telegraph cables off the northern coast of the European continent and cut them. The effect was a massive advantage for Great Britain in terms of infrastructural control of global communications.

British operators were able to more effectively plan the intercept of German and Austro-Hungarian communications. This allowed British forces to more effectively counter Central Power moves, though largely outside of Europe proper, and to thwart secret plans to import resources in circumvention of the Allied embargo.

The case of Enigma is similar to the case of British attacks on German telegraph systems in that the whole thing revolves around exploiting a systematic vulnerability. Because of specific flaws in the way the device was designed and used, the Allies gained a systematic ability to compromise a foreign military system. And indeed, following the war, Allied governments made sure that Enigma machines found their way into the hands of dictators around the world, particularly in Africa. Given that Ultra and the breaking of Enigma were kept secret for more than 30 years, this spread of the machines essentially constituted a backdoor into foreign communications that the U.S. and British governments exploited to great gain.

Today, we have enduring concerns about the design and origins of technology that ultimately underwrites global communications and the various societal functions based on the function of global information infrastructure. Perhaps the best example of national security concerns about backdoors manifesting in policy is the enduring resistance in the United States to allowing the import of technology produced by Huawei, the largest Chinese manufacturer of network infrastructure. In short, the enduring concern is that something in Huawei's design might give the Chinese government a functional weapon against global Internet infrastructure in the event of future conflict. Great Britain, though it has allowed Huawei imports in recent years and has come to an arrangement with Huawei based entirely on this concern. In order to allay fears that there is something in the source code of Huawei devices, which now make up about a third of the world's physical routing infrastructure, the company has set up a secure Ministry of Defence room where British technicians can observe and dissect the source code. Nevertheless, concerns remain that the rise of non-Western technology behemoths – as well as the submission of Western firms to foreign oversight practices that might compromise design knowledge in exchange for market access – pose an immense security threat.

Taken together, British cryptanalysts were regularly able to eliminate enormous quantities of possible rotor settings for the machine. Then, bombs were able to calculate the proper key in a matter of hours each day by cranking through combinations faster than even an army of human "computers" ever could. Enigma's breaking, which by some estimates led to the early cessation of hostilities by as much as two years, thus involved the ushering in of a new kind of intelligence where the tools of the game were machine-based, not human. It was the start of a new era of signals intelligence. And yet, the bombs and the work on Enigma would not be the direct progenitor of the digital machines that came after the war. That honor fell to a device developed later in the war to address an even more daunting encryption challenge than Enigma.

The Tunny and the Colossus

Before World War II, the renaissance of mathematics and mechanical advancement referenced earlier not only saw the production of cryptanalytic innovations that would be applied in aid of the allied war effort but were also the first real steps toward the digital age. Though it should be noted that a number of scientists in the West, in Central Europe and in Russia notably worked on similar projects in the interwar years, credit is often and rightfully given to Alan Turing as the father of modern computers (not to mention artificial intelligence). This is largely because of his being the first to conceptualize the modern computer's design.

Turing published a now-famous paper in 1936 called "On Computable Numbers."[16] It is quite dense. Even given this, however, it is surprisingly readable for non-specialists in those sections where the author deigns to momentarily put aside mathematical formula to describe his names. The paper is about human intelligence and the notion that machines might be made to simulate such intelligence. Up until this point in history, humanity had made a vast number of machines built to accomplish narrow goals with some efficiency. Circuit switches for telephones, for instance, efficiently and dumbly enabled the rapid connecting of phone calls from their first widespread usage in the early 20th century. The problem with such narrowly focused machines, of course, was that they could not be redeployed to be useful for other tasks. A human "computer" was enduringly more useful than a machine simply because the human brain is intelligent and intelligence is defined, to some greater or lesser degree, by an ability to apply skills to any problem.

Turing's paper asks whether or not it was possible to design a machine that mimicked this flexibility in humans, an adaptability that he saw as roughly synonymous with intelligent behavior.[17] He designed a machine in his paper. This machine would work on a fairly simple set of principles. The main principle that is critical to understanding how it worked is the deconstructability of information to dichotomous decisions. That is, every decision, no matter how complex or nuanced, can be reduced to a series of yes-or-no calculations. Turing drew, quite accurately it has turned out, his notion of human intelligence emerging from complex combinations of simple calculations from his studies of microbiology. In principle, his notion here reflects the function of neural networks (i.e. the "thinking" substrate of the human brain) in that they are basically complex sets of nodes that receive information, interpret and then send on a new piece of information based on some set of rules. The cumulative result is human consciousness. Turing was not entirely novel in suggesting this as the basis for simulating intelligence beyond the human brain; after all, "computers" of the repetitive action kind (e.g. the armies of censors that would scan letters during World War I who decided to pass or flag one by responding to simple rules) had been doing this for years. But Turing was the first to acknowledge how, with existing technologies, a machine might be made that entirely automated the process.[18]

Turing's machine, which existed for him only in his mind and on paper, was set up to receive information via two sets of paper tapes. Tapes with a single piece of data would be fed into the machine slide-by-slide for analysis. These tapes contained the information that was the focus of the machine's work. The other set of paper tapes would include the instructions from which the machine would take a simple action. The whole thing worked by a simple "if this, then that" logic wherein outputs from the machine depended entirely on the thing's "state of mind" (meaning the disposition of the machine, determined by the instructions, to make dichotomous decisions in a single way). Turing had described, in other words, not just the first programmable computer system but also the first *reprogrammable* computer system with general application. The task before the machine did not matter so much as the possible programming permutations that the thing's design allowed for. The implications were immense, not least for signals' intelligence.

Toward the mid-point of the war, following the breaking of the Enigma code by cryptanalysts at Bletchley Park, new and puzzling communications were intercepted by British intelligence. They were not encrypted by Enigma and proved an immediate, unbreakable challenge for Allied intelligence. The messages were encrypted by something called a Tunny device, the use of which was reserved for the communications of German high command.[19] The Tunny device was significantly more advanced than the Enigma and far more ergonomic. Typing in a message did not result in light bulbs blinking with letter outputs. Rather, the messages typed in would be encrypted and sent directly from inside the machine along wires to the destination. The Tunny worked by translating input letters into bits of information (i.e. into binary representations) that would then be scrambled via the use of 12 rotating wheels that added additional letters to the message. All the recipient would have to do would be to add the same coded letters to the received ciphertext to reveal the plaintext within. The encryption value of the Tunny was, if it is possible to believe, astronomically higher than the ubiquitous Enigma, with estimates ranging from the device being between one million and ten million times harder to break.[20]

The key to breaking Tunny was not some prediction of regular usage patterns or a design flaw of the machine as it was with Enigma. For one thing, the usage patterns of the German high command were much more careful and the relatively limited usage of the devices prevented usage predictions based on regimented daily provision of keys in the same fashion. For another, the Allies had no Tunny machines. Rather, the answer to Tunny came from a series of Bletchley researchers – Tiltman, Tutte and Newman, initially – who realized that there might be mathematical ways to reveal the correct key that did not require possession of one of the machines. Tutte, in particular, realized that incorrect keys would mark an incoherent distribution of outputs when run against encrypted messages but that correct keys would produce clustered distributions, what was referred to as a statistical bulge. Mathematically, if it could be possible to set up a process in which hundreds of millions of calculations could be performed very rapidly, it would be possible to find indicators of the right keys.

A machine like the one Turing described in 1936, which had never before been built or seriously designed with real-world technological limitations in mind, was the obvious answer, and a man called Tommy Flowers stepped up to the plate with a design in mind. He and a team of 50 others used telephone circuits to construct a fast-"thinking" device that could be programmed via use of a light plate to make simple decisions about whether or not input data conformed to a specific condition. If the statistical bulge existed, the machine's work was done. If not, it kept going.

The machine was called Colossus and was the first legitimately digital "computer" ever built. Colossus went to work barely months after Flowers and his team began construction of the device to great success. Moreover, Colossus was adapted a number of times to tackle similar kinds of challenges. It was the generally focused, programmable machine that Turing had envisioned in 1936 and allowed the Allies even greater advantages in the closing days of the war. While breaking the Enigma helped the Allies save countless lives and maneuver with greater strategic effect in the latter years of the war, the breaking of the Tunny machines allowed for information deceptions of the most sophisticated kind. The example most often cited relates to the D-Day landings in 1944, when Tunny intercepts were used to verify that information transmitted to German high command by double agents about fake Allied landing targets had been believed. It was largely confirmation of this fact via Tunny communiqués that persuaded General Eisenhower to launch the invasion when he did.[21] As significant as such wartime advantages were, however, it was Colossus's demonstration of general-purpose computing at work that did most to shape the coming computer revolution.

Cold War computing and the move from narrow to general

The cooperation of the intelligence communities of the United Kingdom and the United States both during and immediately following World War II drove and shaped the computer revolution of the 1950s and 1960s, a period when major commercial manufacturers made incredible advances in electronic computer technologies.[22] Even before the war began, a tenuous trust was established that blossomed thereafter into an unprecedented relationship between two foreign powers that had, 100 years earlier, still regularly risked at least low-level warfare. The initial deal between the intelligence branches of both countries was fairly simple. The United States would take the lead against Japanese cryptographic and other espionage efforts with Britain in support; the flip was true for Britain against Germany. An early trade of information – details on Japan's Purple cipher against data on how researchers at Bletchley Park had cracked the Naval Enigma – cemented the relationship.

Following the war, both countries began to wind down their intelligence networks while at the same time grabbing all of the German equipment and personnel they could. And yet, it would only be a few short weeks and months before the governments of both countries decided that the future would be a broad-scoped expansion of signals intelligence work, rather than a recession of wartime efforts. The General Communications Headquarters (GCHQ) was set up jointly as a successor to Bletchley Park. The new focus was the Soviet Union, a foe that the British were much more used to facing off against than were their American cousins. Thus, in the early few years of what would become the Cold War, the United Kingdom roared ahead and took some amount of precedence in the relationship.[23]

That position of pre-eminence would not last long. Britain's historical experience with combating Russia aside, U.S. security services excelled in playing catchup. More importantly, U.S. technological efforts would soon surpass those of Great Britain, largely as a result of industrial strength. Companies in the United Kingdom would quickly move to help GCHQ develop newer and better versions of the Colossus-style general-purpose electronic computer. By the time the last Colossus machines were decommissioned in 1959, several dozen more powerful successors were in operation.[24] In part, these computers were built to combat new and novel Soviet cryptographic techniques. One major focus of their work was Project Venona, an effort to exploit a flaw in the way the Soviets had produced a series of one-time pads – in essence, they had accidentally created about 35,000 duplicate pads – that left them vulnerable to decryption.[25] The work was slow and did not provide real-time intelligence, but with the help of new general-purpose computers, GCHQ and the new **National Security Agency** (NSA) were able to uncover significant information on traitors within Western security services.

Where the United Kingdom lost out was in the industrial might of U.S. companies, several of which were given exclusive contracts to aid the NSA in the construction of new high-tech computer systems. Whereas the United Kingdom required strict adherence to secrecy laws from a limited set of contracting companies that begin to build machines under government supervision, the United States surreptitiously shopped around with the result that innovation in computer systems was nested in private enterprise. This meant, on top of the greater moneys thrown their way by U.S. intelligence organizations, more rapid updates and user-friendly designs than U.K. companies were incentivized to build. The result was that, several years into the Cold War, the computer centers at both the NSA and GCHQ were filled almost entirely with IBM machines. The U.S. primacy in the intelligence relationship was thus established not so much by ingenuity or adaptation to the substantive realities of Cold War espionage, but rather by the greater resources and dynamism of U.S. industry.

Enter the Internet

The upshot of such a dynamic in the early years of the Cold War, as Chapter 2 discusses briefly, was a proliferation of computer systems across *both* private industry in the West and most elements of state national security apparatuses. Entire commercial sectors grew up around the use of increasingly sophisticated computers, and the non-intelligence sections of both the United States and allied defense establishments adopted computers in their Cold War efforts to deter Soviet aggression.

In some ways, the story of how the Internet came into being is almost a side story to the broader narrative about how the intelligence world shaped the conditions that would lead to the appearance of what we are now calling cyber conflict in the international arena. What's almost as important as the technical details of what projects were funded and new methods of information transmission developed is the shift around the midpoint of the 1960s in the role of computer systems in national security. Whereas before computers were used as tools of spycraft, now they were both used as tools of statecraft (more broadly defined than spycraft) and had become targets of the same.

Chapter 2 describes the nature of the technological development that manifested as the ARPANET in the late 1960s.[26] The ARPANET was the product of U.S. research funding of ideas that had been percolating in scientific communiqués in the transatlantic scientific community since at least 1962, when a scholar at MIT called J. Licklider hypothesized a "Galactic Network" concept.[27] In the simplest sense, the main motivation behind the development of packet-switching technologies and the design of an open architecture networking environment was the desire to more effectively share resources. On such a network, communications could be made simpler and more accessible to the average user. The result of the ARPANET's development – and its various expansion stages that ultimately opened the nascent network to more and more institutions in academia and industry – was, eventually, the modern Internet.

Important in all of this, the example of the ARPANET as a novel and innovative method for improving efficiency in the use of computers by large organizations – or, in ARPANET's case, connected universities and laboratories spread across the United States – was not lost on the U.S. military and intelligence communities. Defense interest in the greater efficiencies involved in networked computer operations led to great private sector investment in the relevant technologies and helped fuel the expansion of Internet infrastructure through the 1970s and 1980s. Interestingly, however, interest in the Internet would be more pronounced outside of the NSA. As a result, the trajectory of Western national security establishments toward the practice of cyber warfare would not be particularly smooth, but rather defined by the pathologies of the intelligence community.

Stunted development, crypto protection and delayed realization

The lack of interest in the Internet among those at the NSA and at GCHQ is relatively easy to understand. In the world of spycraft, computers were used to store and analyze incoming information. Communication of that information was itself a closely guarded enterprise, and the Internet was nowhere near developed enough that Western intelligence forces would be incentivized to seriously adapt their cryptographic skills to web-based forms of signaling.[28] The upshot was that, while militaries and research departments were investing immense resources in efficiency-improving projects based on this new network platform, the spooks at Fort Meade were generally ignoring the whole thing.

Box 4.2: Crypto as weapons of war: should security software be exported?

From the 1970s through at least the late 1980s, the United States and Western European defense communities were engaged in what many have called the "crypto wars." Far from being a conflict – cold or otherwise – against foreign actors, this period was actually characterized by growing hostility between the scientific and intelligence communities in the West around the subject of secret communications. As we describe later, the simple shape of the conflict was disagreement over the right of the government to have a monopoly on encryption and a rapidly developing set of academic approaches to securing information transmission in increasingly sophisticated ways.

Two academic researchers in particular took up the mantle of questioning what they saw as the "Big Brother" practices of the NSA in monopolizing information security. They would, though they eventually came to an understanding with the government in the interests of a shared feeling of patriotic obligations, be instrumental in the development of public key (asymmetric) encryption techniques. This has, in the intervening years, led to a revolution in the way that the public is now able to enjoy the benefits of cryptography in everyday applications.

For the government, the development of public key encryption techniques (described in Chapter 2) is reflective of a broader set of enduring concerns about the nature of information technologies (particularly cryptographic ones). Given that asymmetric encryption empowers individuals and makes the task of breaking encryption far more difficult (when implemented effectively, of course), the task of law enforcement and intelligence agencies has been made harder. And yet, law on the books in most Western countries permits the government certain abilities to ensure that interception of communications is possible for national security and law enforcement purposes. Should this not extend to asymmetric encryption systems? In other words, should governments be able to mandate backdoors be built into applications that employ a public key security system? Or does the public have the right to security for its information beyond government authority?

Beyond the traditional tension between national security oversight authority and public concerns over information privacy, governments must also consider the value of security software and techniques to foreign competitors' military and intelligence services. In the mid-20th century, cryptographic techniques were regularly considered to be instruments of national security and so the export of new knowledge in the space was illegal. From the 1970s onward, this position waned with the explosion of academic and commercial interest in cryptography beyond the intelligence sphere. And yet, governments are still regularly concerned about the proliferation of information security know-how around the world and the security implications of such spread. The NSA maintains informal advisory processes for coordinating the publication (or non-publication) of new techniques with academic sources where national security implications are pronounced.

More significantly, there is a series of import-export control agreements that consider certain information security applications and techniques to be instruments of conflict. In particular, the **Wassenaar Arrangement on Export Controls for Conventional Arms and Dual-Use Goods and Technologies** is a multilateral arms control treaty between 42 countries that has increasingly seen the addition of software to the lists of prohibited trade items. Originally focused on traditional weapons or war and unique products like chemicals, the Arrangement has been updated in recent years to include software that encompasses intrusion command-and-control features and zero-day exploits. Though the logic behind such a move is simple insofar as the signatories seek to limit the spread of potent techniques for malicious behavior, the wording of the ban on these items is broad enough that it seems to ban products that legitimately incorporate such features. These include penetration testing products made by cybersecurity companies and intrusion software that doesn't provide execution features.

In short, both the domestic and international cases of government oversight of information security technologies serve to illustrate an enduring tension between national security imperatives and the desires – economic, social and political – of non-government sectors of global society. Should governments treat these technologies as weapons or instruments of war (or crime) and act to restrain them appropriately? We'll assume the answer here is yes. But to what degree should governments be able to affect oversight?

Of more concern to spies at the NSA and at GCHQ in the 1970s was the gradual loss of control being experienced over the tools of signals intelligence.[29] In addition to computers proliferating throughout U.S. and U.K. industries, the mathematical art of cryptanalysis was gradually becoming a topic of great focus among non-government researchers. Specifically, academics in California and in the Northeast were beginning to publish new and innovative work on how to secure information. Whereas in the past such researchers would often be visited by the NSA with an offer of consultancy and a non-disclosure agreement to sign, many younger mathematicians believed that the public had the right to greater information security in an age of rapid computerization. Moreover, many researchers objected to what they saw as the NSA's attempt to securitize a public good at the expense of civil liberties.[30] A series of "wars" thus broke out between the signals intelligence wing of the U.S. Department of Defense and a broadening community of researchers focused on "democratizing crypto" through the 1970s and into the 1980s.[31] Though a compromise system of sorts – wherein academics could enter into dialogue with the NSA to ensure patriotic adherence to common sense national security arguments with actual censorship – was eventually put in place, these spats largely came to define the NSA's approach to the technologies of the information age until something changed in the mid-1980s.

Maturation and exploitation in the age of cyber conflict

In the 1980s, the United States faced its first true crises of a cyber nature. These are described in much greater detail in the later chapter cataloguing notable episodes in the history of cyber conflict. Briefly, however, a series of events convinced the NSA and its counterparts in allied countries to finally give the Internet its due as a revolutionary development in the ongoing information revolution.

Among those episodes, like the Cuckoo's Egg and the Morris Worm, a common feature was the use of the Internet to launch intrusive computer network attacks for the purpose of stealing information. With the end of the Cold War, this method of approach – which had been on the minds of SIGINT operators for two decades already – was regularly and officially recognized by several directors of the NSA as a further evolution of the computerization of the intelligence business that was started at Bletchley Park some 50 years prior. Computers had become tools and then targets of espionage. Now, computers were themselves the means of spying on one's enemies, and the Internet was the medium in which spy operators would lurk. As a result, signals intelligence work that centered on networked computer systems – which would, of course, be an increasingly large section of the SIGINT portfolio of both the NSA and GCHQ – became less about the use of computers to allow for exploitation (based on information that one decoded) of enemies in the real world. Rather, use of computers *was* the exploitation that was the ultimate end goal of signals intelligence. The NSA thus evolved rapidly following the end of the Cold War from an organization largely focused on informing others within the U.S. defense community to a frontline operator directly engaging with the United States' enemies.

Structurally, this meant the rapid retooling of the NSA's operational divisions. "Group A" – i.e. those parts of the NSA focused on the Soviet Union – diminished in stature and resources while other divisions, notably the Information Assurance Directorate, rose to prominence given their traditional role in developing technological and mathematical solutions to SIGINT challenges.[32] There was, of course, entrenched opposition from those who objected to a reimagining of the NSA toward digital age operations. But such opposition gave way to the experiences of the NSA in supporting national security missions on a number of fronts. Significantly, NSA operators aided U.S. military efforts during the First Gulf War by conducting what was then often referred to as counter-command/control operations to disrupt Iraqi communications. There, the NSA was authorized to engage in operations designed to compromise computer systems and networks to aid national security objectives in line with the SIGINT mission. They were authorized, in other words, to engage in computer network exploitation.

Box 4.3: Going for the "whole haystack"

Following the events of September 11, 2001, in the United States, the U.S. government underwent a major retooling of elements of the homeland and national security establishment. One major development was the installation of a Director of National Intelligence (DNI) between the President and the heads of individual intelligence agencies. The idea was to coordinate the transmission of intelligence from the community to the rest of government more effectively than had previously been the case and to allow for a voice that could effectively lobby the executive branch.

The second DNI, Mike McConnell, is well known for (among other things) supporting the upscaling of the NSA's mission to capture and interpret signals for intelligence purposes. Early in his tenure, an aid presented him with a map of the world overlaid with Internet traffic volume lines. Naturally, as Chapter 2 notes, most global traffic transited through the United States (because of the unique manner in which packet routing occurs through hierarchical servers). The national security implications of this were immense. To read what a terrorist in Yemen was saying via email to an associate in Afghanistan, one did not even need to leave the United States. Rather, the information would come to the United States on its way to Afghanistan. Thus, so long as the NSA could figure out how to capture relevant data, signals intelligence work was poised to benefit immensely from an artifact of the Internet's technological development.

There were a number of implications of such a move. In fact, McConnell's notion matched the ideas held by many in the intelligence community from at least the late 1980s onward. Those voices, future head of the NSA Keith Alexander among them, had envisioned a sophisticated analytic apparatus in which massive amounts of data were sifted and dissected to provide the agency with unparalleled predictive capacity. The challenge in setting up such a thing would be twofold. First, the NSA would require said massive amounts of data. And since it would be impossible to collect Internet communications after the fact, the implication of a future need to find terroristic needles in a haystack was that the NSA needed to first possess "the whole haystack." In other words, the NSA needed to collect and store all Internet data prior to actually acting on an analytic need. This was accomplished in two ways. First, the NSA worked with Internet backbone operators to collect data "upstream," meaning that traffic was captured as it transited routing nodes in the United States. Second, the NSA was able to retrieve data directly from major technology firms like Apple and Google who were compelled by the Protect America Act (PAA) and a revised Foreign Intelligence and Surveillance Act (FISA) to aid the defense community.

But the data challenge implied another, one that the PAA, the Patriot Act and revised FISA were designed to address. The NSA has no purview to spy on those living within the United States or on U.S. citizens more broadly writ without a warrant. The challenge would be in separating the data of those people from foreign traffic such that the NSA could fulfill its SIGINT mission legally. The answer was a reconceptualization of the rules around SIGINT on two main fronts. First, the NSA would not be considered to be collecting Americans' information by capturing the "whole haystack" of information. Rather, this would simply be data put into storage. "Collection" would only occur if and when the NSA took that data "off the shelf" in order to perform analysis, something that would still require a warrant if Americans' information was involved. Second, the revised legal maneuvers made by the Bush administration in the 2000s removed legacy language that impractically required validation that information to be collected was definitively from persons located outside of the United States.

The line between exploitation and attack is a fine one and one that we take up in chapters to come. Next, we discuss the shape of **offensive cyber operations** (OCO). However, it is worth here noting that the line between exploitation and defense has been legally massaged over the past 20 years to allow intelligence organizations in the West to attack foreign network systems for informational gain. Specifically, in 1997, the NSA was permitted to engage in such operations broadly writ as part of a redesignation of what kinds of actions fell under the legal requirements of Title 50 of the U.S. Code (which outlines intelligence restrictions and responsibilities). For the purposes of our discussions of cyber conflict in this book, the line between exploitation and attack is meaningful only insofar as the term "cyber conflict" has increasingly been applied to an immensely broad series of conflict phenomena that have benefited from the possibilities of computer network operations. The reality, as we note consistently from now on, is that there exist few technical distinctions in the execution of operations variably labeled CNE or CNA. Rather the differences exist in the intentions of those who hack for intrusive or disruptive purposes and the kinds of effects they are able to cause. We discuss this more in detail across the next several chapters.

Toward cyber warfighting

Since the late 1980s, the Pandora's box of cyber conflict possibilities has consistently been opened wider and wider. As we describe later, events in the 1990s brought about broad-scoped national fears of "netwar" and "cyberterrorism" that persist in public discourse to this day.[33] Since at least the mid-1990s, much of that concern has been about the security of critical infrastructure that is computerized, network-enabled and operated not by the government but by private owners. And most recently, Western nations are concerned about new avenues for the conduct of information warfare in the form of propaganda and political manipulation enabled by cyberattack.

In many ways, a surprisingly small cross section of major cyber conflict topics fall under the umbrella of what we might think of as the traditional SIGINT domain. Certainly, the NSA has come under fire – much as it did in the 1970s – for apparent heavy-handedness related to the massive collection of Internet signals from within the United States since the mid-2000s. Those programs rest on a twofold logic; first, that the logical nature of the Internet as pushing most global packet traffic through U.S. servers allows for collection of foreign signals from within the country and, second, that future investigations to find needles in the haystack (i.e. terrorists, foreign agents) clearly require that the NSA collect "the whole haystack" by constantly recording what happens online. Likewise, the NSA and the broader defense community have faced several major threats in the form of cyber-enabled economic espionage, such as Titan Rain, which we describe in later chapters.

But it's important to note here that even traditional intelligence functions, because of the way that the information revolution has run its course, are about more than just spying. After all, today's Internet-oriented SIGINT is not simply about the computerized analysis or storage of data; it is about the use of computers and networks to attack opponents' information infrastructure as a prerequisite to conducting core intelligence tasks. Cyber exploitation *is* cyberattack, and Western espionage communities have at every turn been both the pivot upon which critical design decisions were made and the force that has shaped many of the operational realities we face today in problematizing cyber conflict as a broad-scoped, dynamic phenomenon.

Notes

1 For perhaps the best book on the subject, see Corera, Gordon. *Cyberspies: The Secret History of Surveillance, Hacking, and Digital Espionage*. New York: Pegasus Books, 2016. Other notable sources include Kozaczuk, Wladyslaw. *Enigma: How the German Machine Cipher Was Broken, and How It Was Read by the Allies in World War Two* (Foreign Intelligence Book Series). Lanham, MD: University Publications of America, 1984; Ratcliff, Rebecca Ann. *Delusions of Intelligence: Enigma, Ultra, and the End of Secure Ciphers*. Cambridge, UK: Cambridge University Press, 2006; Bloch, Gilbert, and Cipher A. Deavours. "Enigma before Ultra Polish Work and the French Contribution." *Cryptologia*, Vol. 11, No. 3 (1987), 142–155; and Hinsley, Francis Harry, and Alan Stripp, eds. *Codebreakers: The Inside Story of Bletchley Park*. New York: Oxford University Press, 2001.
2 For a description of the economic and industrial extent of the Empire prior to the World Wars, see Ferguson, Niall. *Empire: The Rise and Demise of the British World Order and the Lessons for Global Power*. New York: Basic Books, 2008.
3 Lovelace, Colin. "British press censorship during the first world war." *Newspaper History from the Seventeenth Century to the Present Day* (1978), 306–319; Messinger, Gary S. *British Propaganda and the State in the First World War*. Manchester, UK: Manchester University Press, 1992.
4 Corera (2016).
5 Benson, Robert Louis. *The Venona Story*. Fort George G. Meade, MD: National Security Agency, Center for Cryptologic History, 2001.
6 Corera (2016). See also Kozaczuk, Wladyslaw (1984).
7 Churchill, Winston. *The World Crisis Volume IV: 1918–1928: The Aftermath*. New York: Bloomsbury Publishing, 2015.
8 Yardley, Herbert O. *The American Black Chamber*. Annapolis, MD: Naval Institute Press, 2013.
9 Rejewski, Marian. "How Polish Mathematicians Broke the Enigma Cipher." *Annals of the History of Computing*, Vol. 3, No. 3 (1981), 213–234; Christensen, Chris. "Polish Mathematicians Finding Patterns in Enigma Messages." *Mathematics Magazine*, Vol. 80, No. 4 (2007), 247–273.
10 Stengers, Jean. "Enigma, the French, the Poles and the British 1931–1940." *Revue Belge de Philologie et D'histoire*, Vol. 82, No. 1 (2004), 449–466.
11 Corera (2016).
12 For perhaps the best layman's description of the principles of secret communications, which we developed in more detail in Chapter 2, see Schneier, Bruce. *Secrets and Lies: Digital Security in a Networked World*. New York: John Wiley & Sons, 2011.
13 Corera (2016).
14 Ibid.
15 See Welchman, Gordon. "From Polish Bomba to British Bombe: The Birth of Ultra." *Intelligence and National Security*, Vol. 1, No. 1 (1986), 71–110; Wesolkowski, Slawo. "The Invention of Enigma and How the Polish Broke It Before the Start of WWII." In *IEEE Conference on the History of Telecommunications (Canada), University of Waterloo*, 2001; and Gaj, Kris, and Arkadiusz Orłowski. "Facts and Myths of Enigma: Breaking Stereotypes." In *International Conference on the Theory and Applications of Cryptographic Techniques*. Berlin: Springer, 2003, 106–122.
16 Turing, Alan Mathison. "On Computable Numbers, with an Application to the Entscheidungsproblem." *Proceedings of the London Mathematical Society*, Vol. 2, No. 1 (1937), 230–265. See also Turing, Alan Mathison. "On Computable Numbers, with an Application to the Entscheidungsproblem. A Correction." *Proceedings of the London Mathematical Society*, Vol. 2, No. 1 (1938), 544–546.
17 Turing, Alan Mathison. *The Essential Turing*. Oxford, UK: Oxford University Press, 2004.
18 Minsky, Marvin L. *Computation: Finite and Infinite Machines*. Indianapolis, IN: Prentice-Hall, Inc., 1967.
19 Corera (2016).
20 For further work describing the Tunny, see Copeland, B. Jack. "Colossus: Its Origins and Originators." *IEEE Annals of the History of Computing*, Vol. 26, No. 4 (2004), 38–45; Copeland, B. Jack, ed. *Colossus: The Secrets of Bletchley Park's Code-Breaking Computers*. Oxford, UK: Oxford University Press, 2010; and Copeland, B. Jack. "Tunny and Colossus: Breaking the Lorenz Schlüsselzusatz Traffic." In K. de Leeuw, and J. Bergstra, eds. *The History of Information Security*. Oxford, UK: Elsevier, 2007, 447–477.
21 Corera (2016).
22 Ibid.

23 Ibid.
24 Ibid.
25 See Haynes, John Earl, and Harvey Klehr. *Venona: Decoding Soviet Espionage in America*. New Haven, CT: Yale University Press, 2000.
26 See Leiner, Barry M., Vinton G. Cerf, David D. Clark, Robert E. Kahn, Leonard Kleinrock, Daniel C. Lynch, Jon Postel, Larry G. Roberts, and Stephen Wolff. "A Brief History of the Internet." *ACM SIGCOMM Computer Communication Review*, Vol. 39, No. 5 (2009), 22–31.
27 See Leiner, Barry M., Vinton G. Cerf, David D. Clark, Robert E. Kahn, Leonard Kleinrock, Daniel C. Lynch, Jon Postel, Lawrence G. Roberts, and Stephen S. Wolff. "The Past and Future History of the Internet." *Communications of the ACM*, Vol. 40, No. 2 (1997), 102–108.
28 Kaplan, Fred. *Dark Territory: The Secret History of Cyber War*. New York: Simon & Schuster, 2016.
29 Corera (2016).
30 Ibid.
31 Ibid.
32 Kaplan (2016).
33 See Arquilla, John, and David Ronfeldt. *Networks and Netwars: The Future of Terror, Crime, and Militancy*. Santa Monica, CA: RAND Corporation, 2001; and Arquilla, John. *The Advent of Netwar*. Santa Monica, CA: RAND Corporation, 1996.

Further readings

Brantly, Aaron F. "A brief history of fake: surveying Russian disinformation from the Russian Empire through the Cold War and to the present." *Information Warfare in the Age of Cyber Conflict* (2020): 27–41.

Corera, Gordon. *Cyberspies: The Secret History of Surveillance, Hacking, and Digital Espionage*. New York: Pegasus Books, 2016.

Gilli, Andrea, and Mauro Gilli. "Why China has not caught up yet: military-technological superiority and the limits of imitation, reverse engineering, and cyber espionage." *International Security* 43, no. 3 (2018): 141–189.

Hodges, Andrew. *Alan Turing: The Enigma*. London: Random House, 2012.

Kozaczuk, Wladyslaw. *Enigma: How the German Machine Cipher Was Broken, and How It Was Read by the Allies in World War Two* (Foreign Intelligence Book Series). Lanham, MD: University Publications of America, 1984.

Nakayama, Bryan James. "Information vs the cyberspace domain." *Journal of Cyber Policy* (2022): 1–17.

Yardley, Herbert O. *The American Black Chamber*. Annapolis, MD: Naval Institute Press, 2013.

5 Attack

From exploitation to offensive cyber operations

In this chapter, we shift gears from the historical introduction of Chapter 4 and provide a descriptive introduction to offensive cyber operations (OCOs). OCOs are comprised of computer network attack (CNA) and computer network exploitation (CNE). OCOs vary in breadth and scale and are part of the overall "offense-persistent strategic environment" of cyberspace.[1] An offense-persistent environment is one in which there is a constant and continual range of OCOs and anything beyond defense in the moment is difficult.[2] The Biden administration's 2021 *Interim National Security Strategic Guidance* highlights its response to this environment when it says:

> We will expand our investments in the infrastructure and people we need to effectively defend the nation against malicious cyber activity. . . . And we will hold actors accountable for destructive, disruptive, or otherwise destabilizing malicious cyber activity, and respond swiftly and proportionately to cyberattacks by imposing substantial costs through cyber and noncyber means.[3]

United States Cyber Command's 2018 "command vision," entitled "Achieve and Maintain Cyberspace Superiority," seeks to help the United States address OCOs and "achieve and maintain superiority in cyberspace." Here, we describe the broad context of OCOs in order to help understand how and in what ways the United States and other actors are engaging in conflict online. First, we provide an introduction to the concept and to different types of OCO. Then, we discuss the argument that OCOs constitute a revolution in military affairs and overview the special characteristics associated with cyber weapons. We conclude with a review of some notable examples of OCO, which are expanded on in Chapter 7's brief history of several major cyber conflict episodes and exploration of the broader empirical record of incidents.

OCOs and the digital domain today

As earlier chapters indicate, the digital domain has been conceptualized and defined in numerous ways and has only recently emerged as a strategic security concern. In the United States, the domain was originally defined by the U.S. Department of Defense (DoD) in 2000 as the "notional environment in which digitized information is communicated over computer networks."[4] This computer-centric definition was significantly modified in 2006 when the U.S. Air Force constituted a broader definition that was subsequently adopted by the Joint Chiefs of Staff in late 2006 and ultimately codified for all

DOI: 10.4324/9781003246398-5

of DoD.[5] The new military definition of cyberspace – which applies to the military and non-military sectors – is as:

> [a] global domain within the information environment consisting of the interdependent network of information technology infrastructures, including the Internet, telecommunications networks, computer systems, and embedded processes and controllers.[6]

This definition encompasses the **Internet**, the **World Wide Web**, smartphones, computer servers, iPads and other common elements of our digital lives. The U.S. government's 2003 *National Strategy to Secure Cyberspace* went even further and highlighted the virtually all-encompassing list of societal sectors that are particularly reliant on cyberspace. The document broadly discusses the agriculture, food, water, public health, emergency services, government, defense industrial base, information and telecommunications, energy, transportation, banking and finance, chemicals and hazardous materials and postal/shipping sectors.[7] Given the breadth of functions of daily life reflected in this list, cyberspace is unmistakably central to the United States and global economy. Further, the United States is utterly dependent on cyberspace with over 239 million regular Internet users (a 77.3% penetration rate).[8] Cyberspace is also a key supporting element of U.S. military power. The DoD relies heavily on information technology networks for Command, Control, Communications, Computer, Intelligence, Surveillance and Reconnaissance and the planning and execution of day-to-day military operations. This reliance on cyberspace, while particularly relevant for the United States, also applies to the rest of the international community. As the Obama administration's *International Strategy for Cyberspace* points out:

> [t]he last two decades have seen the swift and unprecedented growth of the Internet as a social medium; the growing reliance of societies on networked information systems to control critical infrastructures and communications systems essential to modern life; and increasing evidence that governments are seeking to exercise traditional national power through cyberspace.[9]

The ITU, the UN agency for information and communication technologies, reported that approximately five billion of the world's nearly eight billion people were online at the end of 2021, a fivefold increase since 2005.[10] Multilateral security organizations such as NATO are still grappling with how to approach cyber threats and develop consensus on regulative norms and approaches for collective defense.[11] Further, the cyber domain is largely owned and controlled by private industry and, thus, many actions in cyberspace require a public–private partnership.[12] This raises a host of ethical and legal questions associated with conducting warfare through a domain largely privately owned and controlled. For example, what are the responsibilities of ISPs to detect, report and block malicious traffic intended to harm their host nations? This legal question and many others arising from this rather unique aspect of the domain have yet to be resolved.

While it can be challenging to reach agreement on what constitutes cyberspace as a domain, hostile action in cyberspace is even more difficult to define – yet it is even more pivotal to understand the dynamics of cyber warfare. OCOs are the employment of cyber capabilities where the primary purpose is to achieve objectives in or through cyberspace and cyber warfare and generally understood to be CNE- and CNA-style attacks. As the previous chapter alludes to, CNE and CNA often go hand in hand as CNE is conducted to collect information and conduct pre-attack reconnaissance prior to a CNA. In a very real sense, using unauthorized cyber access to steal information allows the option to destroy information and progress into a cyberattack. Tom Gjelten described this phenomenon when he said that:

> [t]he difference between cyber crime, cyber-espionage and cyber war is a couple of keystrokes. The same technique that gets you in to steal money, patented blueprint information, or chemical formulas is the same technique that a nation-state would use to get in and destroy things.[13]

Box 5.1: Cyberspace as the fifth domain . . . too actuarial?

It is worth, at this early stage, noting that the U.S.-centric view of cyberspace as a domain in which human interaction – such as warfighting – can occur is challenged by many as being so actuarial as to preclude certain notions of what cyber conflict actually ends up being. In short, Western defense communities tend to think of cyberspace as a domain much like land, sea, space and air. It has a certain terrain and a unique mode of interaction, much as exists in the other domains. The strong implication is that we can problematize conflict in the cyber domain in much the same way that we do in the others by understanding the characteristics of the landscape and tools involved.

The issue with such a view, though it remains popular and has clear appeal in thinking about conflict online, is that it encourages categorization of cyber techniques for engagement in line with the dynamics of the domain itself. Cyber actions might be analyzed as meaning or accomplishing particular things and not others simply because much theory about how actors fight online emerges from an understanding of *only* domain-specific characteristics. For instance, if there is an inability to differentiate between intrusion for aggressive purposes or for reconnaissance purposes, then we might be inclined to think that limited cyberattacks are more indicative of the latter than the former. And yet, such attacks may be designed to enable conflict in other domains, such as a kinetic attack or a HUMINT-style intelligence operation.

This might seem like a minor point and, indeed, we do not argue that the notion of cyberspace as a unique domain of warfighting is unhelpful. But it is certainly the case that non-Western countries have conceptualized Internet-based technologies and their effects on global conflict without resorting to the sort of domain-specific conceptualizations that parallel the organization of Western defense establishments so neatly. The Chinese, for instance (and as we discuss later on), conceptualize information age conflict as occurring more broadly in the context of an "informationization" trend in global society akin to industrialization at the end of the 19th century. Others have envisioned the phenomena not as an entirely new feature of human interaction, but rather as a sort of new source of background pollution that adds complexity to international conflict and contention. Ultimately, of course, the question is one of the utility of different frameworks for thinking about cyber conflict. The "fifth domain" mindset is useful and certainly the prevailing framework present in Western cyber conflict discourse. But it is easy to see how such an approach to conceptualizing cyberspace could be limited. Indeed, as we discuss at length in later chapters, Western defense communities have regularly undergone "realization" episodes wherein previously under-problematized cyber threats have manifested and forced planners to consider a broader scope to cyber issues than they had in the past.

As a result, many today refer to cyber espionage as "cyber warfare" or "cyberattacks" when in actuality no damage (other than secondary damage caused by the relative advantage the stolen information provides) occurs. Security scholar John Arquilla has pointed this out by highlighting the fact that international law defines an attack as "violence against the adversary" and that such a term does not necessarily apply to all cyber operations (namely, here, to CNE-style operations).[14] A good example of this blurred line between CNE and CNA would be the 2014 cyberattack that occurred during the political crisis in Ukraine involving a weapon known as "Snake" (or Ouroboros). Snake is of suspected Russian origin, but positive attribution has not been achieved.[15] It is a CNE, possibly CNA, tool kit that in 2010 began infecting Ukrainian computer systems.[16] Since 2010, researchers have identified 56 incidents of Snake, 32 of which were found in Ukraine, and believe it was used not only for CNE but also to conduct highly sophisticated CNA-style attacks.[17]

This imprecise lexicon when it comes to the term "cyber warfare" and "cyberattacker" complicates the environment in which perspectives on cyber warfare emerge and in which cyber conflict is itself conducted. Moreover, Security and Defence Agenda, in collaboration with computer security company McAfee, published a report in February 2012, which identified the lack of agreement over key terms such as cyber war and cyberattack as a major impediment to norms and regulating cyber conflict.[18]

Offensive cyber operations as a revolution in military affairs

It is commonly thought that cyber warfare and OCOs represent a major revolution in military affairs, a technological shift so dramatic that the character of conflict and the paradigms of strategic thought that govern how militaries prepare for (and engage in) war are fundamentally transformed. Some have gone so far as to predict that it will "soon be revealed to be the biggest revolution in warfare, more than gunpowder and the utilization of air power in the last century."[19] Further, in all likelihood, the threat of emerging-technology cyber weapons will only increase. As far back as 2011, CSIS identified more than 30 countries that are taking steps to incorporate cyber warfare capabilities into their military planning and organizations and today it is considered a nearly ubiquitous aspect of modern warfare.[20] Adam Liff has argued that the use of cyber warfare as a "brute force" weapon is likely to increase in frequency.[21] Adversaries such as China have increasingly focused on developing "informationized" warfighting strategies (discussed in Chapter 8) that are heavily reliant on computers and information systems and focus on attacking such systems possessed by their adversaries.[22] Increased international interest in cyber warfare is also based on the recognition that information networks in cyberspace are becoming operational centers of gravity in armed conflict.[23] This was reflected in DoD's final and most recent Quadrennial Defense Review in 2014 (as it was in previous reviews), which said:

> [t]he United States has come to depend on cyberspace to communicate in new ways, to make and store wealth, to deliver essential services, and to perform national security functions. The importance of cyberspace to the American way of life – and to the Nation's security – makes cyberspace an attractive target for those seeking to challenge our security and economic order. Cyberspace will continue to feature increasing opportunities but also constant conflict and competition – with vulnerabilities continually being created with changes in hardware, software, network configurations, and patterns of human use.[24]

Cyber warfare plays a role at the tactical, operational and strategic levels of war: from impacting engagement systems at the tactical level, to the adversary's ability to mass and synchronize forces at the operational level, to the ability of senior leadership to maintain clear situational awareness of the national security environment at the strategic level.[25] Additionally, many of IR's most well-known perspectives, such as Michael Horowitz's theory on the diffusion of new military capabilities (Adoption Capacity Theory), indicate that cyber weapons are likely to spread quickly. Adoption Capacity Theory, for instance, argues that the diffusion of military innovations depends on two intervening variables: the financial intensity involved in adopting the capability and the internal organizational capacity to accommodate any necessary changes in recruiting, training or operations to adopt the capability.[26] The low financial and organizational barriers to developing cyber warfare capabilities indicate that the adoption of OCOs will likely be widespread.

The special characteristics of cyber "weapons"

Cyber warfare involves many special characteristics that often do not apply to other forms of conflict, especially conventional military conflict. These include the challenges of actor attribution, multiuse nature of the associated technologies, target and weapon unpredictability, potential for major collateral damage or unintended consequences due to cyberspace's "borderless" domain, questionable deterrence value, the use of covert programs for development, attractiveness to weaker powers and non-state actors as an asymmetric weapon, and the use as a force multiplier for conventional military operations.[27]

Challenge of attribution

The first major characteristic of most cyber "weapons" – this term is extremely common in scholarship and practitioner work on cybersecurity, but we place it in quotation marks because of the broadly acknowledged dispute over how accurate it is to call code a weapon – is the challenge of attribution following their use. This is a result of the tremendous difficulty in conclusively determining the origin, identity and intent of an actor/attacker operating in this domain if the actor wishes to remain anonymous, and defenders generally lack the tools needed to reliably trace an attack back to the actual attacker. Thomas Rid argues that all cyberattacks to date have been examples of a sophisticated form of sabotage, espionage or subversion and are reliant on this attribution difficulty.[28] Cyberspace is truly global, and nearly all action passes through networks and ISPs in multiple countries. Additionally, the hardware used to conduct cyber warfare can be owned by innocent noncombatants, illicitly harnessed for malicious use through the use of computer viruses. In 2007, some computer experts estimate that between 10 –and 25% of computers connected to the Internet (approximately 100–150 million devices) were compromised and used illicitly as part of various networks of compromised computers – known as "botnets" – utilized to conduct attacks.[29] That percentage estimate is likely still accurate today.

The use of these types of proxies provides plausible deniability to state-sponsored activity. The Conficker worm, first detected in November 2008, is another example of the challenge of attribution in cyberspace. It is suspected that it is of Ukrainian origin, largely because it did not target Ukrainian IP addresses or computers using Ukrainian-configured keyboards; however, a savvy adversary could have deliberately programmed it that way as part of a deception strategy.[30] Another attack, this one on DoD computer systems and known as "SOLAR SUNRISE," was initially traced back to Israel and the United Arab Emirates. U.S. officials suspected that the attack was orchestrated by operatives in Iraq. However, later investigations determined it was conducted by two teenagers in California.[31] Yet another cyberattack, known as "Night Dragon," targeted five multinational oil companies and stole gigabytes of highly sensitive commercial information regarding Western energy development activities. Investigators traced the attack to IP addresses in China, confirmed that the tools used in the attack were largely of Chinese origin and that the attacks were conducted between nine and five Beijing time (indicating the likelihood that government or government-affiliated personnel conducted the attack). However, in spite of this significant evidence indicating Chinese involvement, probably even official Chinese government involvement, it was not possible to conclusively attribute the attack and Chinese officials claimed innocence.[32] These attribution challenges make it very difficult to conclusively link hostile action in cyberspace to a particular individual, organization or nation state. This reality makes cyber warfare particularly appealing for an adversary seeking to achieve certain effects anonymously or at least with reasonable deniability.

Offense as emerging from defensive considerations

Something to consider about OCOs up front is that not all offensive action taken in cyberspace is designed to be aggressive or even about espionage. Rather, states and non-state actors often hack in order to better their defensive situation. In general, we think of cyber defense as being constituted of both computer network defense (CND) *and* proactive measures taken to scope out the nature of threats on the horizon. CND is a relatively simple set of operational activities to understand. In short, CND typically involves passive defensive measures designed to help secure the perimeter of an organization and that organization's information systems. Actors might seek to improve cyber hygiene among employees (i.e. by training the workforce in better data protection practices and enforcing better standards for password usage). Likewise, they might employ firewalls, intrusion detection systems (IDS) and antivirus software in their systems. **Firewalls** are programs that sit on information exchange points (i.e. a router) and either allow or prohibit Internet traffic from entering a network based on a set of rules (e.g. does this packet of information come from an IP address in a blacklisted country?). IDS programs are similar but look at traffic internal to a network as opposed to just at the exchange point. IDS applications, though the line between these and firewalls has blurred a great deal in recent years as applications have become more sophisticated, look for anomalous behavior and report findings to a network administrator. Antivirus software does something similar, either comparing the content of data packets to a database of known malicious signatures or (in the more sophisticated instances) looking for code that could be linked to malware.

By contrast, any actor concerned with cybersecurity has significant incentives to take *active defense* measures. The logic here is pretty similar. If you are trying to defend a vault in a building containing valuable commodities (gold, diamonds, bearer bonds, etc.), you are not only going to invest in security guards, ID systems, bullet-proof glass and so on. You are also going to hire individuals to actively investigate who might want to steal your goods and who might be capable of doing so. This lets you better prepare your perimeter defenses. In some instances, this might let you disrupt the preparation of someone you find that is interested in attacking you or at least signal to them that you're onto them, thus hopefully changing their operational calculus.

The logic is identical with cyber operations. More often than not, a good defense means a good offense. Reconnaissance of those threat actors you suspect of future transgressions is useful for your own defensive preparations. Moreover, doing something like "burning a vulnerability" (i.e. drawing attention to your own ability to intrude on an enemy's systems) can signal your awareness of the situation to an opponent and hopefully deter their future attack. This dynamic with offensive cyber operations is particularly significant for our later discussions of the strategic dynamic that exists between states in cyberspace. In short, if you are unable to tell the difference between what is an attack, an intrusion that is a prelude to an attack, and a mere reconnaissance exercise, how can you plan for different conflict eventualities?

Box 5.2: "Hack-back"

An additional element to the discussion of CNA as multipurposed in these sections is the practice of "hack-back." Hack-back is quite simply the practice of victimized actors – often via some specialized service provider hired to help mitigate the costs of an attack – undertaking their own offensive cyber operations designed to find and delete stolen information, monitor an attacker after the fact and potentially minimize the chance of a future assault. In practice, hack-back often occurs in several stages, which are illustrated in Figure 5.1.

The first stage of a hack-back operation is to track the culprit and assess their approach to your systems. This stage is risky, largely because tracking a culprit often means entering innocent third-party systems that were compromised for purposes of supporting the initial attack. We return to this point later. Second, the victim gains access to the hacker's system, often the command and control (C2) server used to launch the attack and which now (ideally) is the location upon which stolen data resides. At this juncture, the idea is not to be seen to be counterattacking, and so efforts are taken to hide the presence of the victim in the attacker's system. Once able to do so, the victim can then affect a means of control over the system and monitor attacker actions. Then, when ready, the victim attacks with the goals of mitigating the costs of the original assault and disabling the attacker's ability to further antagonize.

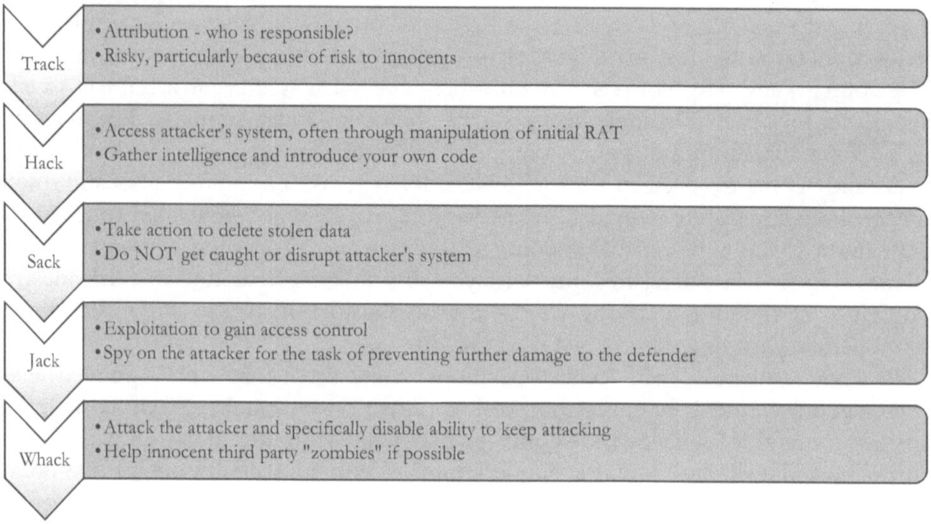

Figure 5.1 Operational stages of "hack-back"

Hack-back is not, at the point of writing this, legal in the United States without a special court warrant. Such a warrant has rarely been issued. Without such authorization, hack-back attacks violate the Computer Fraud and Abuse Act. In Europe, it violates the Council of Europe Cybercrime Convention. In particular, hack-back clearly contains risks to innocents that have been compromised and manipulated in the initial attack. Nevertheless, a niche industry exists around the practice that makes use of proxy servers in foreign countries and personnel based overseas to bypass legal restrictions on the practice. More significantly, a bill is at the time of writing this book in Congress that aims to legalize certain forms of hack-back.

Multiuse nature of Internet technologies

Another characteristic of cyber weapons is that their underlying technologies are multiuse. This means that cyber IT systems can have defensive and civilian applications and purposes in addition to any offensive cyber warfare application. In fact, many IT and hardware and software components usable for cyber warfare are ubiquitous, commercial, off-the-shelf technologies with many peaceful applications. According to the NRC, advances in IT are driven primarily by commercial needs and are widely available across the globe to nearly all groups and individuals.[33] Additionally, Forrester Research projected that the number of computers worldwide – and therefore the number of individuals with access to these tools – would grow from one billion in 2008 to probably two billion by 2015, although this would be difficult to quantify accurately.[34] The military and intelligence community IT required to conduct cyber warfare is drawn from these globally distributed and commercially developed resources. In some cyber operations, such as those utilizing distributed denial of service (DDOS) attacks, private and commercial computers themselves may deliberately and surreptitiously be utilized as part of the attack. A DDOS attack uses multiple compromised systems (collectively known as botnets), usually infected with a Trojan virus that can be developed by simple criminals or state actors, to target a single system. Victims of a DDOS attack consist of both the end-targeted system and all systems maliciously used and controlled by the attacker (also known as a "botherder" or "botmaster") in the distributed attack. The most common form of DDOS attack is simply to send more traffic to a network address than it is equipped to handle. This multiuse nature of cyber warfare technology has obvious implications for the ability to address cyber threats by restricting access to the hardware or software involved – namely that doing so would likely not be particularly effective or practical.

Unpredictability and potential for collateral damage

Another characteristic of cyber weapons is the unpredictability and potential for collateral damage associated with their use. Due to the ever-changing innovations in enterprise architecture and network operations, as well as any IT interdependencies, predicting the precise effects of an attack is very difficult. As in other warfighting domains, an actor may have conducted intelligence, surveillance and reconnaissance (ISR) operations and mapped out vulnerabilities in an adversary's cyber network as would be done to plan for a conventional ground attack with tanks and troops. However, unlike in the conventional realm, the targeted actor is capable of flipping a switch and instantly changing the network (i.e. the target set) or even unplugging it altogether. This factor is a destabilizing force as it rewards immediate hostile action to prevent network modification if cyber ISR intrusions are later detected. It is in effect the opposite of deterrence, incentivizing early offensive strike when an advantage is present. Defenders may also have unknown automated countermeasures that negate the desired effects of cyberattacks (such as instantaneous network reconfiguration or firewalls). For example, the Stuxnet attack is probably no longer able to continue to attack Iranian nuclear facilities as the zero-day exploits it utilized have been plugged by Iranian officials. In addition to network/target evolution, cyber weapons themselves can also be unpredictable since many can evolve. A cyber weapon can adapt – as has been seen with the Conficker virus. Conficker includes a mechanism that utilizes a randomizing function to generate a new list of 250 domain names (used as command and control rendezvous points) on a daily basis – remaining adaptable and staying one step ahead of those seeking to shut down or hijack the illicit Conficker-enabled network.[35]

Network interdependencies also contribute to the potential for collateral damage that is characteristic of cyber weapons. The Internet is made up of hundreds of millions of computers connected through an elaborate and organic interwoven network and, as previously discussed, it is the backbone of much of the global economy; any major attack could pose significant unintended and collateral impacts if it spurred a ripple effect through the network. For example, if an attack on a particular Internet node resulted in the blackout of an entire regional ISP, not only would the intended target be affected but also all other users of the ISP and other individuals who relied on the services of those users directly impacted. The second and third degree impacts of some forms of cyber warfare are nearly impossible to predict. These effects are not limited to the theoretical: cyberattacks have already led to real-world collateral damage. Israel's suspected cyberattack on Syrian air defense radars in advance of their 2007 attack on a Syrian nuclear reactor under construction may have also inadvertently caused damage to Israel's own cyber networks.[36]

Questionable deterrent value

Another characteristic of many cyber weapons is their questionable value in achieving deterrent effects. Deterrence theory and OCOs are discussed in more detail in subsequent chapters, but their relevance as a characteristic of cyber weapons is briefly discussed here. The uncertain effects of cyber weapons coupled with the availability of defenses and the need for secrecy and surprise reduce their ability to serve as a strategic deterrent. Available defenses and the potential for network evolution to mitigate the effects of an attack given early warning require cyber attackers to rely on surprise for much of their effectiveness. To achieve surprise, secrecy is required, reducing the ability of a state to make credible threats without compromising their cyber warfare capabilities. Credible threats regarding specific means of attack or targets invite the threatened state to take protective actions which could blunt the deterrent value of a threat. Although cyber weapons have the potential to inflict unacceptable damage against an adversary, they are probably unable to offer states an "assured" capability for doing so. This deficiency significantly undermines their suitability as a deterrent tool and instead they are more likely to be used preemptively or as force multipliers. Additionally, because of the attribution challenge discussed previously, there is often limited public discussion regarding cyber warfare capabilities and intent.

Importance of secrecy and surprise

A feature of cyber conflict discussed more fully in later chapters, another important characteristic of cyber weapons is the frequent use of covert programs to develop them and the related prospect for unexpected technological breakthroughs of tremendous significance. Due to the sensitivities of cyber weapons and the uncertain international response, their development is rarely publicly acknowledged or demonstrated. Further, because of the multiuse nature of IT, the development of offensive cyber capabilities is similar in many ways to the development of defensive capabilities or even civilian and commercial activities. Thus, it can be very difficult to gauge the intentions of the adversary based solely on their public indicators. Cyber warfare does not require large facilities with distinctive signatures and easily detected emissions as would a nuclear weapons program. This makes national technical means such as intelligence collection satellites fairly ineffective for understanding adversary cyber activities. Intelligence on foreign capabilities and intentions in both areas is likely to be poor barring well-placed human sources who pose challenges of their own.[37] Further, the utility of cyber CNE-type espionage activity incentivizes keeping secret efforts to develop such cyber tools and countermeasures against them.

Another distinct aspect to the importance of secrecy and surprise in cyberspace is the potential emergence of revolutionary technology such as a quantum computer, which would render all forms of encryption obsolete. Any category of weapon is always subject to an advance in technology that gives someone an edge, but with cyber the risk of a breakout is much more pronounced. A quantum computer would utilize the principles of quantum mechanical phenomena to process data at spectacular speeds. The first nation to develop and field a full-blown quantum computer could be able to utterly dominate cyberspace for a period of time. Looking beyond the acute example of quantum computing, smaller technological advances could also have a dramatic effect on the balance of power in cyberspace. The life cycle of advanced computer technology is much more accelerated than other weapon systems. Moore's Law, developed by Intel co-founder Gordon Moore in the 1960s, rather accurately predicted that computer technology would advance dramatically. He predicted that "the number of transistors which can be manufactured on a single die will double every 18 months."[38] Moore's Law arguably continues to apply today. If a nation fails to keep up with these advances, their ability to defend against or wage cyber warfare will be dramatically reduced.

Asymmetric warfare

Another distinguishing aspect to cyber weapons, which is fully the focus of Chapters 10 and 11, is their attractiveness to weaker powers and non-state actors as asymmetric weapons. This attractiveness is based on the potential for anonymity and associated plausible deniability, as discussed previously, as well as the relatively low cost of developing cyber weapons and the global power projection they can provide. Cyber weapons are attractive to relatively weaker actors (state and non-state) due to their low cost compared to other weapons. The most successful known cyber weapon – Stuxnet – likely cost in the low double-digit millions of dollars to produce.[39] Alternative weapons for achieving similar effects against the Iranian nuclear program – Stuxnet's target – would have necessitated weapons costing billions of dollars (e.g. producing a single B2 bomber costs over $2 billion).[40] Cyber expert Adam Liff has contested the financial ease of acquiring potent cyber weapons and argued that obtaining advanced cyber weapons such as Stuxnet would in most cases exceed the reach of weaker states.[41] However, Liff's argument does not take into account the ease with which computer code, once developed, can be replicated and modified. Antivirus software company founder Eugene Kaspersky has said that given Stuxnet's code is now publicly available, it would be "quite easy to disassemble the code to discover how it works, to extract the components and to redesign the same idea in a different way."[42] As a result, the cost of cyber weapons will likely decrease as they (and their associated code) proliferate and are increasingly deployed. Dorothy Denning further described this appeal to weaker actors when she highlighted that the cost of launching cyber warfare operations could be "negligible" while the cost to the attack victims could be "immeasurable."[43]

In addition to relative low cost, cyber weapons also provide global power projection capability to almost any adversary due to the global nature of cyberspace. This characteristic is particularly appealing to states with very limited expeditionary capabilities but with global aspirations, such as China. "Thanks to computers," one Chinese strategist writes, "long-distance surveillance and accurate, powerful and long-distance attacks are now available to our military."[44] Operations in cyberspace, unlike those in other domains (with the possible exception of space), immediately give a state global power projection capability. The National Research Council (NRC) has highlighted this prospect of "remote-access" attacks where computers are attacked through the Internet or connection nodes present in wireless networks or dial-up modems.[45] By tapping into global ISPs and other IT-based networks, attackers are able to effectively conduct expeditionary warfare in an area distant from their own territory. Prior to the advent of cyber warfare, very few nations had the resources to develop the sizable and robust military assets required to overcome global logistical challenges and project power far outside of their neighborhood. Through pre-existing global computer networks, a cyberattack with global reach can be conducted as rapidly as electrons can traverse the electromagnetic spectrum. While IT networks and advanced technologies have enhanced the command and control required for traditional power projection, cyberattacks can now be conducted from completely within cyberspace itself. This effectively removes the high entry costs required in conventional warfare to develop aircraft carrier battle groups, strategic bombers, intercontinental ballistic missiles and so forth, associated with power projection. Denning describes this characteristic, stating cyber warfare operations:

> [c]an take place in an instant and come from anywhere in the world. They can be orchestrated and conducted from the comfort of a home or office, without the risks of spies and undercover operations, physical break-ins, and the handling of explosives. The number of targets that potentially could be reaching is staggering.[46]

Cyber warfare has clear limitations as compared to traditional expeditionary capabilities, but it is understandably attractive to less-developed or advanced states, such as rising peer competitors to the United States – such as China – who are seeking to exert global influence. The 2007 DDOS attacks against Estonia, discussed in greater detail in Chapter 7, provide a good example of this power projection capability. During a two-week period, attackers were able to successfully disrupt the Estonian government, media outlets, banking, ISPs and telecommunications websites by launching attacks from approximately 100 million computers distributed to more than 50 nations.[47] Due to the asymmetric nature of cyber warfare, it is likely to be a favored form of warfare by adversaries unwilling to directly challenge conventional military capabilities with similar conventional capabilities (particularly China, which as discussed, has demonstrated a heightened interest in cyber warfare).[48]

Force multiplier

The ability to use cyber weapons as a force multiplier for conventional military operations is another significant characteristic of cyber warfare. Cyber weapons are well suited for attacks on logistical networks, reinforcements and command and control facilities in order to "induce operational paralysis, which reduces the enemy's ability to move and coordinate forces in the theater."[49] While cyber weapons may not have a direct kinetic effect on an adversary's tanks and aircraft, it is still possible for cyberattacks to render these weapons useless. Additionally, because cyber weapons can achieve such effects without kinetic destruction, they can be employed in ways similar to those intended for the infamous neutron bomb (which killed troops with a blast of lethal neutron radiation but did not cause physical damage) by degrading or destroying enemy military capabilities while preserving transportation infrastructure, etc. Thus, cyber weapons can provide an attacker with the capability of seizing valuable natural resources or industrial facilities without risking their destruction.

Similarly, cyber weapons, particularly those allegedly being developed by China to exploit the U.S. military's logistics IT network, would complement conventional military operations. A RAND Corporation report on Chinese anti-access strategies explained that Chinese military strategists believe cyberattacks are likely to be effective in disrupting U.S. military operations because military IT systems are connected to commercial networks. The report indicated that one Chinese official said that in the United States, "95 percent of military networks pass through civilian lines and that 150,000 military computers pass through normal computer networks. This characteristic of computer networks makes it easy to conduct a virus attack."[50] Additionally, the U.S. Cyberspace Solarium Commission reported in 2020 that "Chinese operators constantly scan U.S. government and private-sector networks to identify vulnerabilities they can later exploit in a crisis."[51] Despite their general lack of transparency on defense issues, Chinese strategists have had a handful of open discussions about how they would exploit this weakness as a force multiplier for a conventional conflict. Early Chinese thinking in this area, from a report in 2000, said that the goal of Chinese cyber warfare was to "cut off the enemy's ability to obtain, control, and use information, to influence, reduce, and even destroy the enemy's capabilities of observing, decision-making, and commanding and controlling troops, while we maintain our own ability to command and control."[52] This still holds true today.

What do offensive cyber operations actually look like?

Though Chapter 2 does strive to provide a technical foundation upon which students using this book can expand their understanding of what dynamics cyber conflict emerges from, it is not an in-depth forensics accounting of how cyberattacks take place. Likewise, the earlier sections in this chapter are written so as to offer a strategic perspective on what offensive cyber operations look like. Simply put, it is not the purpose of this book to offer a set of operational training resources for those interested in OCOs. Rather, we aim to offer context and content for those interested in understanding the theoretical, doctrinal and policy implications of cyber conflict, all of which we turn to over the next few chapters.

That said, it is worth briefly describing the series of actions typically involved in cyberattacks. One way to do this is via something called the **Kill Chain** (or "Cyber Kill Chain"). In essence, the Kill Chain is a model for deconstructing cyber operations via an understanding of the operational phases involved for the attacker. The assumption with a Kill Chain is that some attacker is going to achieve some objective and that they'll need to go through various operational steps to get there. Given that, we can start to make assumptions about what an attacker might need to be successful, how they'll have to plan their attack, what resources they'll need, and so on. In doing so, it is possible to construct a pretty good generic outline of what attackers are likely to do in detail at different stages of planning and execution. Thus, the Kill Chain is immensely useful for those interested in forensically breaking apart cyber incidents to understand the relevance and significance of different particular actions taken by aggressors online.

It is important to note here that there are a number of other models similar in function to the Kill Chain but different in the details. Most, however, deal with similar kinds of assumptions. There is an argument out there among experts that the Kill Chain is outdated, but a common position is that it has immense analytic value.

Figure 5.2 visually outlines the Kill Chain. The Kill Chain envisions cyberattacks as occurring in four stages, one of which is actually constituted of another four sub-stages. First, malicious cyber actors will take initial action to preliminarily compromise a target. In all of this, it's important to remember what we read and talked about in Chapter 2 regarding access control. Simply put, all cyber insecurities – insofar as they emerge from the *design* of information systems – have to do with loss of access control (i.e. with the loss of the ability to prevent an unauthorized access of information or data controls). The implication of this is that attackers will rarely be able to directly attack – i.e. attempt to gain access to – their primary target. They'll need to figure out a way in toward their objective. This involves reconnaissance and the preparation of resources geared toward compromising some part of a target system. This is the phase where traditional intelligence resources

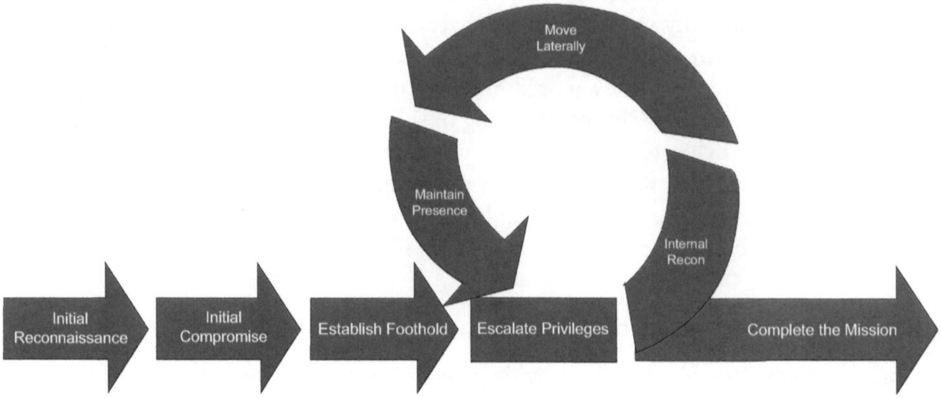

Figure 5.2 The cyber Kill Chain

are often employed for the purpose of, say, subverting employees of a target organization, stealing credentials or inserting malware. The initial compromise then allows basic access to a system wherein an attacker eventually aims to move around and find the right target.

The second stage of the Kill Chain is the establishment of a beachhead or a foothold. Once the initial compromise has been made, an attacker will act to ensure their continued initial access to a system. This means that an attacker might do something like change logs (records) to make their use of stolen credentials look innocuous or create themselves additional low-level access credentials.

After this, the Kill Chain describes cyberattacks as going into a multi-phase stage wherein the attacker attempts to move toward their ultimate target. This entails four different activities. The fourth activity is that, now inside, the attacker will look for new opportunities to escalate their privileges. This means exploring the environment they have access to and looking for ways to affect new compromises that will let them into other areas of a network or computer system that they currently can't access.

Three activities aid this effort to upgrade access privileges. First, an attacker will undertake reconnaissance of their new environment. Second, they will move laterally within the system in order to achieve the right conditions for further compromise. And third, they will continue to take actions to maintain their presence, including masking their actions and providing for future abilities to re-compromise the system. The final part of the Kill Chain is then simple. At some point, the attacker will complete their mission, having escalated their access to such a point that they are able to achieve the disruption they intend or are able to steal information, etc.

An important point to be made here is that not all cyber operations mirror the linear attack profile outlined by the Kill Chain model all the way to a clear end point. Indeed, there is a robust criticism of the Kill Chain as being more useful for conceptualizing the tasks involved in planning and executing computer network attacks than for actually mapping out specific operations. For one thing, the Kill Chain is particularly useful for describing efforts to insert malware and then use infected systems to affect greater control of a target's platform. By contrast, extremely sophisticated intelligence and warfare operations, which are largely the focus of this book, might involve an immense volume of intrusions, malware employments, and non-cyber actions as a necessary means of attacking highly secure or complex systems. Often, understanding how an APT was planned and executed precludes the use of something like the Kill Chain, simply because it oversimplifies the attack pathway taken by an attacker. In reality, perhaps the best way to think about cyber campaigns is quite simply to think of multiple Kill Chains arrayed alongside one another to represent multiple vectors of attack, both cyber and conventional. Taken together, these different intrusive and manipulative actions constitute a campaign that provides appropriate access control exploitation potential and allows for the execution of a mission – the delivery of a payload. Nevertheless, as we mentioned earlier, it is a useful model with which to conceptualize and from which to infer about OCOs in general.

Understanding cyber warfare

In this chapter, we have discussed offensive cyber operations from a strategic perspective. The next section of the book extends what we've started here in outlining major cyber conflict episodes and engaging with questions of *why* and *how* states go about making the decision to use cyberspace for aggression, espionage and subversion.

Notes

1 Harknett, Richard and Emily Goldman. "The search for cyber fundamentals." *Journal of Information Warfare* 15, no. 2 (2016): 81–88.
2 This concept underpins recent shifts in strategic posture in cyberspace by the United States, which we discuss in a later chapter. Initially, see Smeets, Max. "US cyber strategy of persistent engagement & defend forward: implications for the alliance and intelligence collection." *Intelligence and National Security* 35, no. 3 (2020): 444–453; Fischerkeller, Michael P., and Richard J. Harknett. "Persistent engagement, agreed competition, and cyberspace interaction dynamics and escalation." *The Cyber Defense Review* (2019): 267–287; and Healey, Jason. "The implications of persistent (and permanent) engagement in cyberspace." *Journal of Cybersecurity* 5, no. 1 (2019): tyz008.
3 The White House. *Interim National Security Strategic Guidance,* 2021
4 Woolley, Pamela. "Defining cyberspace as a united states air force mission." *Air Force Institute of Technology* (June 2006): 2–3.
5 United States Department of Defense. *Joint Publication 1–02: Department of Defense Dictionary of Military and Associated Terms* (May 2011): 93.
6 United States Department of Defense. "The definition of cyberspace." *Deputy Secretary of Defense Memorandum* May 12, 2008.
7 United States Department of Homeland Security. *U.S. National Strategy to Secure Cyberspace* (2003). www.whitehouse.gov/pcipb/cyberspace_strategy.pdf.
8 International Telecommunication Union. *Measuring Digital Development: Facts and Figures 2021,* 2022.
9 United States. *International Strategy for Cyberspace: Prosperity, Security, and Openness in a Networked World.* Washington, DC, May 2011.
10 International Telecommunications Union. *The World in 2011 – ICT Facts and Figures,* December 2011.
11 Spencer Ackerman. "NATO doesn't yet know how to protect its networks." *Wired.com,* February 1, 2012.
12 United States Department of Homeland Security. *U.S. National Strategy to Secure Cyberspace,* 2003.
13 Gjelten, Tom. "Cyber insecurity: U.S. struggles to confront threat." *NPR.org,* April 6, 2012. www.npr.org/templates/story/story.php?storyId=125578576.
14 Arquilla, John. "Twenty Years of Cyberwar." *Journal of Military Ethics,* April 17, (2013), 85.
15 Sanger, David E., and Erlanger, Steven. "Suspicion falls on Russia as 'snake' cyberattacks target ukraine's government." *The New York Times,* March 8, 2014.
16 Ibid.
17 Ibid.

18 Grauman, Brigid. "Cyber-security: the vexed question of global rules." *Security Defence Agenda and McAfee*, February 2012, 6.
19 Bender, Jeremy. "Israel: cyber is a bigger revolution in warfare than gunpowder." *Business Insider*, February 4, 2014. www.businessinsider.com/the-internet-is-the-next-battlefield-2014-2.
20 Lewis, James, and Timlin, Katrina. "Cybersecurity and cyberwarfare: preliminary assessment of national doctrine and organization." *Center for Strategic and International Studies* (2011).
21 Liff, Adam P. "Cyberwar: a new 'absolute weapon'? the proliferation of cyberwarfare capabilities and interstate war." *Journal of Strategic Studies*, 35, no. 3 (2012): 401–428.
22 United States. "USCC 2007 report to Congress." *U.S.-China Economic and Security Review Commission*, 2007, 94.
23 United States Department of Defense. *DoD Information Operations Roadmap* (October 30, 2003). www.gwu.edu/~nsarchiv/NSAEBB/NSAEBB177/info_ops_roadmap.pdf.
24 United States Department of Defense. *2014 Quadrennial Defense Review* (2014). www.defense.gov/pubs/2014_Quadrennial_Defense_Review.pdf.
25 United States Department of Defense. *Joint Publication 3–13.1: Electronic Warfare* (January 2007). www.dtic.mil/doctrine/jel/new_pubs/jp3_13.1.pdf.
26 Horowitz, Michael. *The Diffusion of Military Power: Causes and Consequences for International Politics*. Princeton, NJ: Princeton University Press, 2012.
27 Koblentz, Gregory, and Brian Mazanec. "Viral Warfare: The Security Implications of Cyber and Biological Weapons." *Comparative Strategy* 32, no. 5 (2013): 418–434.
28 Rid, Thomas. "Cyber war will not take place." *Journal of Strategic Studies* 35, no. 1 (2011): 5–32.
29 Paget, Francois. "How many bot-infected machines on the internet?" *McAfee Labs*, January 29, 2007.
30 SRI International. *An Analysis of Conficker's Logic and Rendezvous Points*, March 19, 2009.
31 Lyons, Marty. *Threat Assessment of Cyber Warfare: A White Paper*. University of Washington, December 7, 2005, 17.
32 Entous, Adam, and Hodge, Nathan. "Oil firms hit by hackers from China, report says." *The Wall Street Journal*, February 10, 2011.
33 United States National Research Council. *Technology, Policy, Law, and Ethics Regarding U.S. Acquisition and Use of Cyberattack Capabilities*. Washington, DC: National Academies Press, 2009, 82.
34 Webber, Liz. "Computer use expected to top 2 billion." *Inc. Magazine*, July 2, 2007.
35 Burton, Kelly. "The Conficker worm." *SANS*, October 23, 2008.
36 Lewis, James. "The Korean cyber attacks and their implications for cyber conflict." *Center for Strategic and International Studies*, October 2009.
37 Koblentz, Gregory, and Brian Mazanec. 2013.
38 Kopp, Carlo. "Moore's law and its implications for information warfare." *3rd International AOC EW Conference*, January 6, 2002.
39 Bell, Stephen. "Cut-price Stuxnet successors possible: Kaspersky." *CSO Magazine*, March 28, 2011.
40 United States General Accounting Office, *B-2 Bomber: Status of Cost, Development, and Production*, GAO/NSIAD-95-164, August 1995.
41 Liff, Adam P. 2012.
42 Bell, Stephen. 2011.
43 Denning, Dorothy E. *Information Warfare and Security*. Boston, MA: Addison-Wesley, 1999, 17.
44 Mulvenon, James C. "China's proliferation practices and the development of its cyber and space warfare capabilities." *U.S.-China Economic and Security Review Commission Hearing*, May 20, 2008, 70.
45 United States National Research Council. *Technology, Policy, Law, and Ethics Regarding U.S. Acquisition and Use of Cyberattack Capabilities*. Washington, DC: National Academies Press, 2009, 87.
46 Denning, Dorothy E. 1999.
47 Cyber Conflict Studies Association. "Implication for an estonia-like cyber conflict for the government and the private sector." *Cyber Conflict Studies Association's Annual Symposium at Georgetown University*, February 26, 2008.
48 Liang, Qiao and Wang Xiangsui. *Unrestricted Warfare*. Beijing: PLA Literature and Arts Publishing House, 1999.
49 Pape, Robert. *Bombing to Win: Air Power and Coercion in War*. Ithaca, NY: Cornell University Press, 1996, 72.
50 Cliff, Roger et al. *Entering the Dragon's Lair: The Implications of Chinese Antiaccess Strategies*. Santa Monica, CA: RAND Corporation, 2007, 55.

51 U.S. Cyberspace Solarium Commission, *Final Report*. March 9, 2020.
52 Houqing, Wang, and Xingye Zhang, eds. *Science of Campaigns*. Beijing: Beijing National Defense University Press, 2000.

Further readings

Borghard, Erica D., and Shawn W. Lonergan. "Cyber operations as imperfect tools of escalation." *Strategic Studies Quarterly* 13, no. 3 (2019): 122–145.

Buchanan, Ben. *The Cybersecurity Dilemma: Hacking, Trust, and Fear Between Nations*. Oxford, UK: Oxford University Press, 2017.

Fischerkeller, Michael P., and Richard J. Harknett. "Persistent engagement, agreed competition, and cyberspace interaction dynamics and escalation." *The Cyber Defense Review* (2019): 267–287.

Healey, Jason. "The implications of persistent (and permanent) engagement in cyberspace." *Journal of Cybersecurity* 5, no. 1 (2019): tyz008.

Libicki, Martin. *Cyberspace in Peace and War*. Annapolis, MD: Naval Institute Press, 2016.

National Research Council. *Technology, Policy, Law, and Ethics Regarding US Acquisition and Use of Cyberattack Capabilities*. Washington, DC: National Academies Press, 2009.

National Security Strategy 2017. Washington, DC: The White House.

Rid, Thomas, and Ben Buchanan. "Attributing cyber attacks." *Journal of Strategic Studies* 38, no. 1–2 (2015): 4–37.

U.S. Cyberspace Solarium Commission. *Final Report*. Washington, DC, March 2020.

6 Shape

Subvert, deceive, poison

At the start of the 1980s, the Soviet Union found itself substantially behind the West – particularly the United States and the United Kingdom – in the development of computers. What had begun as a race in the late 1940s and 1950s had gotten away from those in Moscow responsible for ensuring technological parity with the capitalist nations of Western Europe and North America.[1] At the start of this race, the Soviet Union had barely been behind at all. In 1948, the country developed its own digital computer replete with pre-stored programming.[2] More than that, Soviet capabilities for cryptanalysis and mathematics-based technology development were arguably more advanced than those of the Western Allies. After all, the experience of total national mobilization just a few years earlier had not only strengthened the pool of available intellectual talent but also organized them under the central auspices of the state. In some cases, of course, this capacity was far from organic. Much computer-oriented work in the early years of the Cold War took place at Marfino Sharashka, a secret laboratory within gulag hard labor camps that marshalled imprisoned intellectuals to develop exceptional single-purpose machines.[3] More broadly, central economic planning worked to produce task-specific devices quite quickly but prevented innovative thinking and investment in pure scientific discovery. As such, general-purpose computers were underemphasized and ultimately, by the 1970s, made by reverse engineering what designs could be bought or stolen from Western companies like IBM and Sperry.[4]

This deficiency in technological innovation was one of several shortcomings of the centrally planned Soviet economy that led Russian security services to prioritize espionage against Western industry and defense production entities. Computers, in particular, were a core focus of KGB attempts to harvest information from the West given their role in several key industrial functions – nuclear power generation, for instance, and space-based infrastructure – that sat at the heart of the science-driven utopian vision espoused by Party leaders going back to Khrushchev.[5] If the Soviet Union claimed that it was driving toward an ideal society underwritten by the best that human innovation in science and technology had to offer, then any kind of perceived deficiency in those areas had to be remedied as a matter of national reputation and power in world affairs.

The KGB proved reasonably proficient between 1950 and the early 1970s in stealing and repurposing information from the West. In one case, a senior executive of IBM based out of Paris was in the employ of Moscow for more than two decades.[6] This agent – codenamed ALVAR – was key to the development of basic computer networking techniques and hardware by the Soviet military establishment, passing secrets through his retirement toward the end of the 1970s.[7] Elsewhere, West German companies were particularly targeted for their association with American companies like IBM and the relative

DOI: 10.4324/9781003246398-6

ease of using East German agents as infiltrators. Indeed, from the 1960s onward, East German technology industries became so heavily dependent on the transfer of information from their West German counterparts that research and development wings of major companies often lacked personnel in key expertise categories.[8] After all, imitation was the key to keeping up with the West, not innovation or even improvement on what was being stolen.

Soviet and Eastern bloc efforts to transfer as much know-how about advancing scientific and technological capabilities from the West as possible had a major Achilles' heel, however. In the early 1980s, the President of France – François Mitterand – surprised his American counterpart by reporting that French intelligence had managed the unthinkable, the placement of a mole inside the KGB.[9] The agent's name was Vladimir Vetrov, and he had successfully made copies of thousands of classified Soviet documents detailing the attempts of one major wing of the spy organization – called "Line X" – to steal Western technology.[10] From this infiltration, it became clear that the Soviet attempt to steal technology was not just opportunistic; it was also systematic and extremely well-resourced. Vetrov's exfiltrated documents constituted a shopping list of target systems, products and scientific know-how linked to everything from satellites to computers and jet fighters.[11] French intelligence estimated that KGB success had already saved years of time in the Soviet development of things like fourth-generation fighter aircraft. And the emphasis placed on computer systems in KGB activities was noticeable. Clearly, additional effort was being made to catch up to the West by the time Mitterrand briefed Ronald Reagan on what France had uncovered.[12]

The Western mole in Soviet intelligence was too good an opportunity to pass up. Vetrov's position in Soviet intelligence was fairly unique. He was one of several individuals tasked with being a liaison between the KGB and Soviet industry, which quite naturally had a shifting spate of demands and desires that informed the spy organization's "shopping list."[13] Vetrov was also quite singular in his disaffection with the Soviet Union.[14] His recruitment by French intelligence had been opportunistic, and his handling was a constant source of uncertainty in Paris, as he womanized, drank and generally grumbled about every part of his job (both of them). This led to Paris turning to MI6, the British foreign intelligence agency, for advice about how to effectively use Vetrov. London brought in the CIA, and the suggestion came in from Washington – indeed, from an adviser close to Reagan – that newfound, evolving knowledge of KGB efforts could be used to feed the Soviet Union bad apples. That is to say that certain technologies could be allowed to fall into Soviet hands. This would be planned and controlled by the CIA, however, so that technologies reaching the Soviet Union would be special "enhanced" versions that would pass authenticity checks but either fail or incorporate major flaws as the Soviets attempted to repurpose them. Even if Moscow caught wind of the plot, the West would still win, as the Soviets would suddenly doubt the quality and functionality of all repurposed technology.

The plan proceeded. Misleading information produced odd, deviant and downright dangerous Soviet technology production through the 1970s and 1980s. The famed Soviet space shuttle Buran was abandoned after countless design faults pushed program costs through the roof. For years, Moscow had apparently no idea that this debacle stemmed from the counterintelligence effort centered on Vetrov's discoveries years earlier. With computers, in particular, Western sources were able to introduce various faulty pieces of hardware components that became standard in certain military-industrial efforts. Production plants and facilities were plagued by constant breakdowns and inefficiencies, issues not faced by their Western equivalents. The gas pipeline explosion referenced in the last chapter – never confirmed and considered by some to be apocryphal – was ostensibly the result of a backdoor introduced in (intentionally) faulty hardware leaked to the KGB in the 1980s. And equipment issues inevitably filtered out into Soviet consumer products, leading even to an inability of Russian firms following the collapse of the Soviet bloc to compete against Western competitors in quality and reliability. At the end of the day, the impact on Soviet industry was enormous, a testament to the scale at which well-planned and well-resourced clandestine activities can impact the conditions under which more conventional strategic and economic competition takes place.

Shaping, not signaling

In international relations, states approach competition with other states in one of two distinct fashions. First, states visibly maneuver about one another, sending signals about intention, capability and resolve via one of a diverse set of mechanisms of state power. States might develop and deploy military force, make statements, sign treaties and more to signal in such a manner as to achieve national interests. Second, states can attempt to shape their environment – the international system – toward favorable conditions such that they are more likely *to be able* to maneuver and signal effectively, thus securing national interests, in the future. Ben Buchanan, in his book on the relationship between hackers and states, describes these competing approaches to international relations in the context of a poker game.[15] If one is playing a high-stakes game, it is possible to signal intent and capability (i.e. the cards you possess) in a multitude of different ways. A player might make bets (or withhold a raise) designed to credibly communicate the strength of the hand being held, for instance. Players can also shape the game itself, however, by cheating in a number of different ways, for example keeping extra cards up one's sleeve or distracting another player. Shaping is about changing the state of play, the ruleset by which the game gets played. Science fiction fans may recognize another famous example of shaping in the Kobayashi Maru, a simulation game played by officers in Starfleet in the eponymous television show *Star Trek*.[16] The game is intentionally designed to be unwinnable with the idea being to teach officers lessons about command under difficult circumstances. And yet, the game is beaten by a central character, James Kirk, when he reprograms the game to make winning possible. Kirk changes the game rules, shaping favorable conditions to grasp victory where previously none would have been possible.

The past two chapters have described the roots of the technological revolutions leading to the digital world of today as firmly entrenched in the experience of intelligence communities, clandestine security circles and the context of major geopolitical conflicts (both hot and cold). They then turned to the idea that machines and networks became more than just the tools of spying, but rather the targets and the terrain of exploitation, a world-spanning assemblage of human engagement, socioeconomic value and real societal functionality that has opened the door to unprecedented insecurity and opportunities to engage strategic competitors. As the last chapter suggests, of course, so much thinking about cyberattack, cyber operations and cyberspace itself has been couched in the nomenclature and the trappings of warfighting institutions.[17] Particularly in the West, the need to grapple with new threats of a digital nature and to coordinate effective defensive postures has, for more than two decades, demanded appropriate vehicles (i.e. institutions) within which cyber capacity can be housed and cultivated. This has led to the perpetuation of core assumptions about cyber operations as an adjunct of warfighting – i.e.

something that is sometimes useful for coercion, more often useful for signaling, and certainly responsive to the many deterrence and conflict logics that dominated the Cold War era – as well as popular representations of cyberspace as a distinct domain of engagement between international competitors.[18]

As the rest of this book illustrates along various lines, this mode of thinking about cyber operations is limiting. Even in Chapter 5's description of the characteristics of cyber instruments and methods, we see a number of constraints that equate to poor abilities to clearly signal national positions and intentions. As the next chapter will initially show, very few cyber operations or campaigns look anything close to conventional warfighting activity. Cyber instruments are immensely versatile multipurpose tools via which states (and other actors) might shape favorable conditions in international affairs. Indeed, as we hope the case of Vladimir Vetrov and the Achilles' heel of Soviet technology development planning illustrated earlier, cyber capabilities have the potential to dramatically reshape and rewire the landscape of international engagement to the benefit of some countries, some companies and some individuals over others. Indeed, they have already played such a role.

The idea that signaling is a less significant perspective for understanding the shape of so much digital activity in the world might be, for some, a hard pill to swallow. Despite the immense scope of intelligence efforts and clandestine attempts to shape international relations during the years of the Cold War, the focus of so many practitioners, policymakers and scholars of international relations (IR) rests firmly on those elements of statecraft that equate to signaling. The deployment of military forces to communicate resolve, capability and intention was a constant feature of Cold War geopoliticking, made permanent by the stationing of armed forces along substantial frontiers in Europe that separated the Eastern bloc from the West. This emphasis on understanding Soviet approaches to *realpolitick* and crisis signaling can be seen even in the institutional setup of Western defense organizations during the Cold War. As Chapter 4 described, for example, it was not until after the fall of the Soviet Union that the National Security Agency (NSA) transitioned resources and personnel toward the technology programs that today define the organization's mission and chosen methods. For decades, the linguists, analysts and Kremlinologists reigned supreme at Fort Meade and elsewhere in the American intelligence community, a reflection of the significance placed by successive executive administrations on understanding the moves in the geopolitical game being played.

But to ignore the significance of shaping activities as a method of statecraft amenable to altering the fundamentals of geostrategic competition is to ignore two distinct realities.[19] First, the actions of numerous governments to shape the international environment in eras past lie at the heart of sea change and important inflection points in history. Going back hundreds of years, rulers like Louis XIV, Ivan the Terrible, Peter the Great and Frederick Wilhelm I of Prussia quite famously sponsored years-long campaigns to alter public opinion, buy off elites and subvert both the political and economic foundations of territories that might one day be the target of military action. During the Cold War, both sides fought clandestine battles to subtly disempower their opponents. For the Soviet Union, in particular, the *maskirovka* ("little masquerade") formed a key formal pillar of national efforts to hamstring Western political institutions and leadership by deceiving individuals, intelligence organizations, political campaigns and more.[20] These efforts, like those of Vetrov, were naturally clandestine, which remains a key reason that students of strategic studies and international affairs focus less on shaping vis-à-vis signaling. Some such activities were discovered and exposed, of course, and provided each side key insight into the thinking of the other. But it is important to note that information gleaned from these exposed clandestine operations was rarely incorporated into strategic analyses as a distinct form of signaling. Rather, the *maskirovka* and other national equivalent practices were dirty tricks and tactical gambits, never intended to provoke a strategic reaction by some adversary but simply to shape favorable conditions.

Second, to ignore the significance of shaping activities in the history of statecraft is to ignore the clear characteristics of cyber instruments as poor analogs to weapons of war. The use of cyber "weaponry" may be harmful. But there have only rarely been circumstances where the employment of cyber operations has genuinely appeared analogous to the use of military force. As the next chapter discusses in more detail, cyber instruments simply don't allow for the securing of conventional military objectives, such as territory occupation.[21] They are also temporally limited in their effects and lead to physical effects only under the rarest of circumstances. Instead, cyber operations are about exploitation, espionage, destabilizing and manipulating informational conditions and more.

Cyber operations, in perhaps the simplest sense, allow one to hack an adversary. As previously discussed, hacking is an exploitation of some system that subverts the rules or norms of that system. This often hurts the system but is not always something that is explicitly forbidden; much hacking is simply something not anticipated or intended by system designers.[22] Hacking an adversary can obviously mean finding and taking advantage of exploits in networked computer systems. But hacking also takes on broader meaning in the context of international relations. Nations are complex systems of systems – political, social, economic and infrastructural. By leveraging diverse digital tools to subvert, deceive and poison the ruleset of the game of strategic competition that is the core constant of international affairs, states are able to hack one another in a much more meaningful sense, altering the state of play a manner much divorced from the arena of signal and maneuver that dominates the attention of so many policymakers, scholars and national security practitioners.[23]

The inescapable appeal of shaping

In 1948, a man called Claude Shannon proposed what would become the seminal mathematical representation of communication.[24] More specifically, his model effectively described the transmission, processing, extraction and use of information via reference to a simple concept – uncertainty. Uncertainty is some amount of unknown information that it is impossible to place a value on. Some have described uncertainty as the opposite of risk. Risk is a known quantity, probabilistically speaking.[25] Those choosing to take risks are informed of the odds and potential outcomes involved in their decision-making. Taking risks is, thus a gamble, something that can be calculated and incorporated into a cost-benefit calculation. Uncertainty is not perceptible in the details, and so one cannot gamble on uncertainty.

Shannon described communication as the resolution of uncertainty. Information traveling from a point of origination – a person, say – to a destination (another individual) is affected by the environment in a few different ways.[26] First, information is coded for communication. Meaning that forms in the human brain to be communicated is formed into language, either oral or written, to be given to the other person. Often, coding will also involve some technology, like a telephone translating words into electrical pulses or a computer interpreting keyboard inputs to produce binary code. Second, information is decoded before it can be received at a destination, again, for instance, by an individual's ears or some technological device. Between these influences on the communication process, third, is noise. Telephone calls might lose quality due to interference on the line. Smoke signals or drumbeats used by pre-modern civilizations to communicate over long distances might suffer from weather interference.

One of the primary contributions of Shannon's model to numerous modern fields of science and engineering was the idea of **channel capacity**. Channel capacity is simply the idea that there is an upper limit on how much information can be communicated (which varies depending on the medium involved) reliably without much risk of losing meaning in the process. This concept is highly useful for readers of this book. After all, in the most general sense, we are thinking in this book about the interaction of information technologies and human society. In this context, channel capacity represents the established practices that allow society to function given the existence of a set of information technologies that are well-integrated into social, economic and political processes. In the 1700s, for instance, the growing global reach of colonial empires dictated that regional governors and managers be given substantial degrees of autonomy over what went on in their assigned territory. After all, with communication between colony and home country taking weeks or months by sea or land in many cases, imperial interests would best be served via the installation of strong executive authority. The telegraph changed this. With near instantaneous communication possible between home nation and outposts of empire, autonomy on the part of regional leaders – or military commanders, diplomats, etc. – was not necessary. And yet, such authority persisted in many cases, producing political tension, social tumult and strategic fumble where institutions had not yet adapted to the sea change in basic societal functionality brought about by a sufficiently dynamic information technology.

When a sufficiently dynamic information technology comes along, the channel capacity of society changes. New forms of communication, of social, political and economic function become possible that simply were not previously. But human institutions, cognition and norms of sociopolitical behavior don't instantly change to reflect this new normal of worldly function. This is why revolutionary information technologies are associated historically with immense social and political upheaval, often at the same time as they help produce economic transformations.

For this chapter's discussion of cyber operations in the context of shaping, this process helps contextualize the rapid proliferation and the diversification of so much digital engagement in the world today. Simply put, the information revolution around computers and the Internet "hacked" global society, changing countless key features of the way in which global commerce, finance, politics, social process and security practice are constituted.[27] The actions of states and a great diversity of non-state actors to hack one another (in the more conventional sense) and to take advantage of new technology systems – like social media platforms, cloud-based consumer products and more – are occurring in the context of that fundamental and ongoing shift in the underlying function of the international system.

This reality forms the basis of a logic of digital security issues as fundamentally centered on shaping beyond just the technical characteristics of code. Shaping is an unavoidably appealing activity for political actors of every stripe because some new tools seem particularly well suited to manipulation and subversion, but not destruction. Rather, shaping is the prevailing conventional logic of security behavior and strategic competition in the age of the Internet because the systems and practices of humankind across all facets of global society are still yet to compensate for the transformations dictated by revolutionary technology adopted at scale and in an evolving fashion since the mid-20th century.

Subvert, deceive, poison

As we move forward to think about the empirical shape of the global cyber conflict landscape and to discuss the various ways in which scholars and practitioners talk about cybersecurity in the context of international security, it's worth differentiating several concepts from those we've already introduced in the course of the last several chapters. These are operational and strategic concepts typically associated with efforts to shape favorable conditions in international affairs, whether by state actors, their proxies or non-state political actors. Naturally, in outlining how states and other actors have hacked one another over the past several decades, we will, in forthcoming chapters, talk about espionage, about efforts to degrade military capacity and about attempts to disrupt information systems for a host of purposes, from information warfare to transgressive digital activism (hactivtism). These concepts should be easy to grasp for readers of the book to this point. Others bear explication before we move on, namely **subversion**, **deception** and **poisoning**.

Subversion

Subversion is a concept that often gets invoked in the context of cyber operations or digital security more broadly. As such, definitions of subversion tend to be quite use case specific. Maschmeyer, for instance, describes subversion "as exploiting vulnerabilities to secretly infiltrate a system of rules and practices in order to control, manipulate, and use the system to produce detrimental effects against an adversary." Many cyber conflict experts, including the authors of this book in other forums, have described subversion as the fundamental characteristic of the underlying technologies that constitute cyberspace. After all, cyber operations rely on mechanisms of secret exploitation; of finding design shortcomings and faults that might be manipulated for some sociopolitical, economic or other gain. Cyber operations are the indirect methods producing strategic or criminal effects of one kind or another. And, as such, those hacking for gain are operationally constrained by the need to shape favorable conditions rather than contest, to maintain some level of secrecy or to reveal knowledge of exploits only in very specific fashion, for instance.

Most writing on subversion and cyber operations anchors the concept to the technical character of the tools involved. But **strategic subversion** exists as an approach to statecraft that requires thinking about the concept in the context of international relations, meaning the real sociopolitical conditions within which statecraft occurs in the pursuit of national interests.

As indicated earlier, the word "subversion" describes a particular kind of outcome. In the social sense, subversion is the successful manipulation of expectations and sociopolitical processes such that previously taboo issues and outcomes – or those beyond reproach in contemporary society – become legitimately considerable. Subversion is all about information, which means in so many instances that it is about hearts and minds insofar as it describes persuasion of a population to a position radically juxtaposed to what was formerly the norm. Though the definitional boundary is certainly somewhat fuzzy, the main distinction between subversion and more common forms of advocacy or persuasion is that the term describes a *wholesale* change in the way a society operates in normative terms. The traditional example offered for what successful subversion looks like is that of the Nazi Party in Germany from the time of the Beer Hall Putsch in 1923 – where the party failed to violently overthrow the German government – to Hitler's democratic ascension to the chancellorship a decade later. Over the course of that period, changing national conditions and the narrative sales pitch of Hitler's party machine succeeded in persuading a large part of the German population to a position of staunch irredentism and tolerance of discrimination that would have seemed anathema to someone in Germany circa 1920.

While social subversion and task-specific subversion have some engagement, strategic subversion remains to this day somewhat difficult to define, a relatively understudied facet of international affairs among IR scholars. Studies of subversive actors often take place as a component part of projects focused primarily on political extremism, civil militancy, terrorism and insurgency. But while it is certainly the case that there are common linkages between such phenomena and subversion, it would be inaccurate to assume that these political activities are synonymous with subversive activities. Terrorists, for example, do attempt subversion. However, subversive behavior is relatively rare and terrorists, focused as they often are on forcing policy changes on the part of national or international authorities, must often undertake activities broadly designed to alienate – rather than persuade – elements of a population.

Paul Blackstock[28] offers the definition of subversion perhaps most free of the assumption that subversives are terrorists or inherently seditious in arguing that it "is the undermining or detachment of the loyalties of significant political and social groups within the victimized state, and their transference, under ideal conditions, to the symbols and institutions of the aggressor."[29] Blackstock's definition is well articulated for a number of reasons. First, it detaches an understanding of subversion as being explicitly tied to the overthrow – violent or otherwise – of governments or sub-governmental institutions. This is important because, as noted earlier, subversion is not always seditious. Modern history is full of cases – from LGBT movements in culturally oppressive regimes to white supremacist movements in Central and Eastern Europe – in which subversion either occurs or is attempted without a stated ambition for structural transformation or violence. Subversion is about ideas and perspectives that are often, but not necessarily, reflected in structures. Second, in referencing the loyalties of individuals, Blackstock links ideational perspectives to a population's preferences. Again, this is critical because subversion takes place under conditions of contestation. New ideas that are tolerable given the progressive nature of a particular society or culture are not subversive, even if they are controversial. Subversive activities are inherently undertaken in an effort to affect a polar shift in the political and social preferences of a population. In short, there must be contest; otherwise, there is no struggle. Finally, Blackstock's definition does well to describe the transformation of ideational conditions and the transfer of normative loyalties to the "symbols and institutions" of the subversive force insofar as it describes subversive efforts as bound up in the unique sociopolitical spaces of particular cultures and nations. No subversive effort is identical to another, even when the cause and the argument is the same. Even in the over-connected world of the 21st century, attempts at subversion naturally take place across different theaters of the global public sphere that boast unique characteristics and challenges.

What does this mean for subversion as a strategic concept applicable our review of cyber conflict and digital security issues here? In simple terms, while the term often conjures images of sedition and societal sabotage, subversion is ultimately just a type of strategy for addressing prevailing power structures and relationships. In order to change society, most sociopolitical actors – states, political organizations, etc. – either directly contest the status quo or "flee" from it. Contestation involves either evolutionary or revolutionary acts taken to directly change the status quo, such as coercion, warfare, voting or a coup. "Fleeing" power involves seeking out or creating alternate sociopolitical spaces where different logics of power might be considered, often either by some form of secession or by actions taken to restrict external influence in a space. By contrast, subversion is fundamentally a strategy of what some scholars have labeled "fleeing in place."[30] It is an effort to displace prevailing norms and power structures without directly challenging them – i.e. to "escape" power by seeking its dissolution. Subversive actors are those that fundamentally seek to *dis*empower competitors, something distinct from the prevailing paradigm of coercion emphasized by so many foreign policy experts. Whereas an activity like coercion is concerned with dynamics of relational power, subversion targets power that emerges from perceptions of legitimacy, whether that means (1) the legitimacy of a social or economic system of some kind, (2) the legitimacy of established logics of operational engagement or (3) the legitimacy of those rules of the road that govern international interactions. Thus, strategic subversion is an attempt to shape international relations by remaking the ruleset of competition, to undermine the fundamentals of adversaries' statecraft and to shift the value proposition of homegrown capacities versus foreign ones.

Deception

If subversion is a core strategy emphasized by actors attempting to shape international affairs toward favorable conditions, then deception is a key tactic favored by those actors. In forthcoming chapters, we speak about information operations and disinformation in more empirically grounded terms. However, it is worth conceptually differentiating deception here from other core methods of subversion and shaping as a distinctly information set of activities. Likewise, it should be noted that deception as an approach to competition is age-old. Sun Tzu, writing sometime between the 3rd and 5th centuries B.C.E., famously stated that "all warfare is based on deception" and indeed was himself drawing on centuries of accumulated knowledge and empirical explication to make his seminal arguments about the nature of human conflict.[31]

Deception is, at its core, an exercise in manipulating informational conditions so as to either simulate or dissimulate – i.e. to make things appear to be true when they are not or to hide the true character of some feature, fact or development in human affairs.[32] Deception need not occur only in clandestine fashion, of course. Diplomacy and blatant interest-oriented propaganda are both overt methods of attempting to persuade national publics, specific elites or other entrenched social forces to change their perspective to align more closely with some strategically desirable position. And yet, so much deception is indeed the stuff of covert operations and schemes. In the information age, this is arguably especially the case.

All polities practice information manipulation in one, other or many fashions. However, as Brantly has noted, there are certain countries that have elevated the practice of deception to core pillars of their foreign and domestic policy.[33] These states are often prompted to do so by geostrategic circumstance, leaning on asymmetric methods of engaging foreign powers either when resource constraints prevent more conventional engagement or when geography makes even the most spirited defensive practices an exercise in futility. Among those states, the Russian Federation and the states it descended from – the czarist Russian Empire and the Soviet Union – stand apart as a power that has not only consistently leveraged deception to secure substantial foreign policy objectives but that has also originated many of the practices and key associated terms that are now at the heart of debates about cyber power and digital deception.

Most prominent among these is the concept of disinformation, which describes an intentional effort to spread false information – either entirely artificial or altered from some original version – for a range of purposes. Importantly, Russia's conceptualization of disinformation (dezinformatsiya or дезынформáция) differed historically from the view of deception held by many of its great power contemporaries over the centuries. Whereas nations like the United Kingdom or France have strong centuries-old traditions of deception as a key practice in attempting to disrupt the command-and-control capabilities of opposing military formations, Russia has – going back to at least the mid-19th century – utilized disinformation alongside other clandestine methods in what is often called active measures (активные мероприятия), activities that target individuals, organizations or foreign publics as a whole to achieve some favorable outcome without provoking the relevant state in a conventional fashion.[34] These activities might include bribery, blackmail or targeted propaganda. Likewise, they might constitute more broad-scoped efforts to alter the perceptions of foreign publics to some advantage. One of the most successful of this latter type of campaign, for instance, is *Operation Infektion*, an effort to convince Western publics that HIV was the intentional result of American government planning and development.[35] Indeed, the episode highlights the non-confrontational character of true efforts to shape international affairs via deception. After all, *Infektion* was a slow burn, an initial suggestion of government involvement in an obscure journal followed by other tidbits in the months and years to follow culminating in a groundswell of public suspicion and anger about the origins of HIV. Though an American commission did finally (and quite thoroughly) unravel Soviet involvement in the HIV panic of the 1980s, Russian active measures had nevertheless achieved some considerable operational success in shaping the political environment of a key competitor via the use of nothing more than clever planning and basic information manipulation.

Poisoning

If deception and its component tactics are the informational manifestation of strategic subversion, then **poisoning** might be said to be the more tangible manifestation of the phenomenon. As a disclaimer, the term "deception" is often used to describe espionage and sabotage beyond the informational by scholars of intelligence studies and IR. Nevertheless, we – the authors – feel the need to differentiate the operational underpinnings of something like *Infektion* from the kind of broad-scoped industrial and infrastructural trickery that we recounted at the opening of this chapter. We selected **poisoning** for several reasons, not the least of which is its relevance to a rapidly growing dimension of cybersecurity – the development and use of artificial intelligence (AI) – which we discuss in our penultimate chapter.[36]

In international affairs, states and other political actors make decisions based on a series of factors. In addition to those informational factors bound up in the signaling inherent in all strategic competition, decisions are made based on an understanding of the sources and current composition of both soft and hard capabilities (i.e. economic or normative influence vs. military power). Poisoning is a tactic that involves undermining those fundamentals without alerting an adversary to such a change. With AI, as we discuss later in the book, poisoning refers to the undermining of algorithms such that they perform differently in practice than their designers intended.[37] One method of doing this, for instance, would be to make changes to the data that is used to train a machine learning model. The data tells the model about some specific dimension of the world and allows the algorithm to respond to tasks it is given in a fashion that as closely as possible mimics human decision-making. If a malicious actor were to get access to that data, they might alter key elements or delete tranches of information, given the AI system an incomplete or unrepresentative understanding of the relevant issues.

The massive technology transfer trick that began with Colonel Vetrov's theft of thousands of classified KGB documents was a successful attempt to poison the ruleset of the game being played by both the Soviet Union and the United States-led Western bloc.[38] Soviet investments in science and engineering were riddled with faults, backdoor manipulations and intended obsolescence that was not recognized in the strategies of Moscow's military-intelligence services or in the decision-making of the USSR's political leaders. Another example of poisoning as a method of statecraft utilized by states attempting to shape favorable conditions was the postwar dissemination of Enigma machines to developing nations referenced in passing in Chapter 4. Having broken Nazi Germany's elite codes, the Allies worked extremely hard to make sure that knowledge stayed under lock and key. So successful were their efforts that the end of the war brought almost no recognition that the Allied counterintelligence effort had played a substantial role in ending the war. As such, American and British intelligence forces were able to make sure that key German cryptanalytic infrastructure made its way into the hands of both friendly and adversarial regimes all around the world, thus ensuring a capability to effectively spy on strategically critical – in the context of the Cold War, particularly – communications for nearly three decades.

Next steps

Over the past three chapters, we have discussed cyber operations in no small amount of historical context, walking you – the reader – through the development of such techniques and infrastructure as the topic of spymasters, warfighters and strategic planners. The concepts, history and discussions of this section of the book will be expanded on over the next few chapters as empirical context is brought in and contemporary discourse on the minutiae of cybersecurity in a complex, adversarial world is brought in. Specifically, the next chapter of the book extends what we've started here in outlining major cyber conflict episodes, looking at macro data about trends in global cyber conflict and engaging with questions of why and how states appear to go about making the decision to use cyberspace for aggression, espionage and subversion.

Notes

1 Oreskes, Naomi, and John Krige, eds. *Science and Technology in the Global Cold War*. MIT Press, 2014.
2 Corera, Gordon. *Cyberspies*. Simon and Schuster, 2016.
3 Andrey Zverev of Ruselectronics speaking on Ekho Moskvy Radio, 14 October 2013.
4 The Soviet Bloc's Unified System of Computers, available at http://old.cistp.gatech.edu/programs/inter-diff-innov-info-tech/docs/The%20Soviet%20Bloc's%20Unified%20System%20of%20Computers.pdf
5 Carera (2016).
6 Andrew, Christopher, and Vasili Mitrokhin. *The Mitrokhin Archive: The KGB in Europe and the West*. Penguin UK, 2015.
7 Ibid., 245.
8 Wolf, Markus, and Anne McElvoy. *Memoirs of a Spymaster*. Pimlico, 1997.
9 Corera (2016).
10 Ibid.

11 Weiss, Gus W. *The Farewell Dossier*. Washington DC: Central Intelligence Agency, Center for the Study of Intelligence, 1996.
12 Ibid.
13 Berghel, Hal. "A farewell to air gaps, part 1." *Computer* 48, no. 6 (2015): 64–68.
14 Corera (2016).
15 Buchanan, Ben. *The Hacker and the State: Cyber Attacks and the New Normal of Geopolitics*. Harvard University Press, 2020.
16 A good description of which can be found in Conti, Gregory, and James Caroland. "Embracing the Kobayashi Maru: Why you should teach your students to cheat." *IEEE Security & Privacy* 9, no. 4 (2011): 48–51.
17 A point that Kaplan makes abundantly clear in his book *Dark Territory*. See Kaplan, Fred. *Dark Territory: The Secret History of Cyber War*. Simon and Schuster, 2016.
18 This is in spite of the efforts of numerous scholars and researchers who have pushed back on this narrative since at least 2010. See among others Rid, Thomas. "Cyber war will not take place." *Journal of Strategic Studies* 35, no. 1 (2012): 5–32; Gartzke, Erik. "The myth of cyberwar: Bringing war in cyberspace back down to earth." *International Security* 38, no. 2 (2013): 41–73; and Valeriano, Brandon, and Ryan C. Maness. *Cyber War versus Cyber Realities: Cyber Conflict in the International System*. Oxford University Press, USA, 2015.
19 It should be noted that some international relations (IR) works have returned to the idea of shaping and the study of various aspects of such activities in world politics. See, among others, Lee, Melissa M. *Crippling Leviathan: How Foreign Subversion Weakens the State*. Cornell University Press, 2020; Maschmeyer, Lennart. "The subversive trilemma: why cyber operations fall short of expectations." *International Security* 46, no. 2 (2021): 51–90; Wohlforth, William C. "Realism and great power subversion." *International Relations* 34, no. 4 (2020): 459–481; and Poznansky, Michael. *In the Shadow of International Law: Secrecy and Regime Change in the Postwar World*. Oxford University Press, USA, 2020.
20 Brantly, Aaron F. "A brief history of fake: Surveying Russian disinformation from the Russian Empire through the Cold War and to the present." *Information Warfare in the Age of Cyber Conflict* (2020): 27–41.
21 Gartzke (2013).
22 On the nature of hacking, see Brantly, Aaron F. "Risk and uncertainty can be analyzed in cyberspace." *Journal of Cybersecurity* 7, no. 1 (2021): tyab001; Erickson, Jon. *Hacking: the art of Exploitation*. No Starch Press, 2008; and Schneier, Bruce. "Invited talk: The coming AI hackers." In *International Symposium on Cyber Security Cryptography and Machine Learning*. Cham: Springer, 2021, pp. 336–360.
23 This idea is taken up in a range of discourse about cyber conflict. See, for instance, Farrell, Henry, and Bruce Schneier. "Common-knowledge attacks on democracy." *Berkman Klein Center Research Publication* 2018–7 (2018); and Whyte, Christopher. "Cyber conflict or democracy "hacked"? How cyber operations enhance information warfare." *Journal of Cybersecurity* 6, no. 1 (2020): tyaa013.
24 Shannon, Claude Elwood. "A mathematical theory of communication." *The Bell System Technical Journal* 27, no. 3 (1948): 379–423.
25 This view of risk and uncertainty is found most prominently in Bernstein, Peter L., and Peter L. Bernstein. *Against the Gods: The Remarkable Story of Risk*. Vol. 383. New York: John Wiley & Sons, 1996.
26 Verdu, Sergio. "Fifty years of Shannon theory." *IEEE Transactions on Information Theory* 44, no. 6 (1998): 2057–2078.
27 This notion is already somewhat common among thinkers on global cyber conflict. See, for instance, Segal, Adam. *The Hacked World Order: How Nations Fight, Trade, Maneuver, and Manipulate in the Digital Age*. Hachette UK, 2016; and Demchak, Chris C., and Peter Dombrowski. "Rise of a cybered Westphalian age." *Strategic Studies Quarterly* 5, no. 1 (2011): 32–61.
28 See Blackstock, *The Strategy of Subversion: Manipulating the*, 1964.
29 Ibid., p. 56.
30 Pierce, Joseph, and Olivia R. Williams. "Against power? Distinguishing between acquisitive resistance and subversion." *Geografiska Annaler: Series B, Human Geography* 98, no. 3 (2016): 171–188.
31 Griffith, Samuel B, and Sun Tzu. *The Art of War*. New York: Oxford University Press, 1971.
32 Concept applied to cyber conflict in Gartzke, Erik, and Jon R. Lindsay. "Weaving tangled webs: Offense, defense, and deception in cyberspace." *Security Studies* 24, no. 2 (2015): 316–348.
33 Brantly (2020), pp. 27–28.

34 Ibid.
35 Bates, Stephen. "Disinforming the world: Operation INFEKTION." *The Wilson Quarterly* 34, no. 2 (2010): 13–15.
36 Whyte, Christopher. "Poison, persistence, and cascade effects." *Strategic Studies Quarterly* 14, no. 4 (2020): 18–46.
37 See Whyte, Christopher. "Problems of poison: New paradigms and" agreed" competition in the era of AI-enabled cyber operations." In *2020 12th International Conference on Cyber Conflict (CyCon)*, Vol. 1300, IEEE, 2020, pp. 215–232; Comiter, Marcus. "Attacking artificial intelligence." *Belfer Center Paper* 8 (2019).
38 As were wartime operations like Mincemeat that sought to convince the Axis powers that certain facts of Allied planning and operations were quite simply outside the realm of possibility. On Mincemeat, see Macintyre, Ben. *Operation Mincemeat: The True Spy Story that Changed the Course of World War II*. A&C Black, 2010.

Further readings

Hughes, Geraint. "War in the Grey Zone: Historical Reflections and Contemporary Implications." *Survival* 62, no. 3 (2020): 131–158.

Stevens, Tim. "Knowledge in the grey zone: AI and cybersecurity." *Digital War* 1, no. 1 (2020): 164–170.

Wigell, Mikael. "Hybrid interference as a wedge strategy: a theory of external interference in liberal democracy." *International Affairs* 95, no. 2 (2019): 255–275.

Whyte, Christopher. "Beyond tit-for-tat in cyberspace: Political warfare and lateral sources of escalation online." *European Journal of International Security* 5, no. 2 (2020): 195–214.

7 The topology and history of major cyber conflict episodes

Though the link between information technologies, cyberspace and national security seems to be timelessly cited as something governments are *increasingly* worried about, the fact is that the history of interactions between states online in international affairs is already decades old. Broadly speaking, cyber conflict has been a hallmark of international relations – particularly between advanced industrial countries – for many years. States regularly intrude upon the digital systems of their peer competitors in attempts to disrupt, to steal information and to contest the digital control of diverse national functions. Perhaps the first known incident of state-sponsored hacking of American government systems – involving the infiltration of American computer systems by an East German freelance agent connected to Soviet intelligence forces – happened almost 40 years ago. Today, states are faced with the challenge of detecting, deterring, redirecting and defending against many thousands of cyberattacks against critical state functions on a daily basis. Moreover, national militaries have increasingly adopted doctrine and assets designed to incorporate digital operations into campaign deployments.

That said, it would be disingenuous to suggest that cybersecurity has changed the entire calculus of national security and of strategizing on future conflict for states. Certainly, governments face new challenges and opportunities in the digital age. However, it is arguably as yet unclear as to the degree to which cyber interactions have actually affected state foreign policymaking beyond the confines of specific episodes. Indeed, initial evidence on the matter suggests that cyber conflict and non-digital responses to thousands of cyber "interactions" over the past couple of decades have been remarkably restrained. Cyber assaults can provide unique gains for an aggressor. However, cyberattacks rarely lead to negative foreign policy responses, with barely a few notable exceptions. Cyber conflict, it appears, tends to meaningfully manifest as the result of relatively nuanced strategic realities, like the existence of an established regional rivalry or government investment in parallel strategic initiatives.

This chapter deals with the history of interstate cyber conflict. Conflict is defined as no more or less than the incidence of conflictual interactions between state entities, such as militaries, intelligence agencies or other governmental organizations. Such interactions lie on a spectrum of severity from unauthorized access to certain computer systems all the way to actions designed to cause major digital or physical disruption, and include acts of espionage, sabotage and, in some quite rare cases, physical violence. In the following sections, we use data to outline the dynamics of cyber conflict over the past few decades. In breakout boxes, we simultaneously cover briefly major events in the history of cyber conflict to date with a particular focus on those incidents involving the United States, China, Russia, Iran, North Korea and Israel. We aim to describe the nature of state vulnerabilities

DOI: 10.4324/9781003246398-7

to different forms of cyber conflict and the challenges that stem therefrom before outlining the empirical contours of cyber interactions over the past few decades. This tees us up for the discussions included in Chapters 9 and 10, where we debate the utility of cyber methods for state security strategies and consider the applicability of traditional IR concepts to international security in the information age.

Cyber conflict terminology: cyberwar versus cyber conflict

Before discussing conflict in the information age, it is first worthwhile that we briefly revisit issues of terminology with particular regard to two terms – "cyberwar" and "cyber conflict." Though these phrases have variously been used interchangeably in this book so far – and will be used as synonyms in a great range of literature on the subject of cybersecurity alongside related terms (cyber warfare, cyber battle, cyber incursion, etc.) – there is a discrete variation in meaning in the way that academics particularly use them to describe different interstate interactions.

First, it is important to revisit what *cyberspace* actually is. One might be forgiven (because they would be partially correct!) for describing cyberspace as the set of network technologies that undergird the Internet and similar networks that do not actually link directly to the searchable Internet (like the DeepWeb or darknet overlay networks). The problem with this, as some point out, is that this definition precludes two important considerations. First, this definition technically eschews capture of network systems that don't relate to the Internet in any way. The design of malware employed during BUCKSHOT YANKEE and Olympic Games, episodes that we discuss in this chapter, demonstrates that a lack of connection to the Internet – in those cases due to the "air gap" defense procedures in place – is simply an architectural feature of cyberspace in particular areas to be overcome. Another secondary problem is that the aforementioned definition ignores the possibility that future network technologies radically different in design from current Internet-connecting platforms could be considered component parts of cyberspace. The Internet of Things, where network devices governing all manner of biological and physical infrastructure are linked, is a commonly cited example of what is meant here in that a more diverse landscape of network systems utilized for specialized functions of industry and government in the future is a more likely topological description of cyberspace than is the current cohesiveness of the Internet.

Valeriano and Maness's description of cyberspace as being "[a]ll computer, network, digital, cellular, fiber, and space-based forms of communications, interactions, and interconnections" seems appropriately broad.[1] To be fair, this imagining of cyberspace complicates (accurately so, admittedly) the landscape of what must be considered to effectively address multifaceted questions of cyber conflict. Such a definition naturally implies that cyberspace is far more than a technological domain. Indeed, much variation in the constituent parts of cyberspace – including information stored in servers dispersed around the world, members of far-flung social media networks, etc. – is defined legally, politically, socially, economically and geographically, and *not* technologically. This makes substantial sense in historical context, of course, as the idea of cyberspace as a distinct spatial "domain" of warfare – alongside the conventional land, sea, air and space – became the centerpiece of efforts by one group within the U.S. defense community to centralize perspectives on computer security issues and build relevant institutions. In this way, from the mid-1990s onward, thinkers like General Michael Hayden and Vice Admiral McConnell popularized the idea of a terrain of digital engagement that goes beyond just recognition of the physical and logical elements of network architecture. This view spread outward from this successful effort to the rest of government, to (some) other countries and to society writ large.

Scholars studying cyber conflict further assign different meanings when using the phrases "cyberwar" and "cyber conflict." The latter phrase is often used to describe most conflictual actions that take place between actors – both state and non-state – in world politics. This includes one-off cyberattacks, any directed employment of malware, the use of computer exploitation techniques to gather intelligence through unauthorized systems access and more. Cyberwar, by contrast, is used to describe a discrete episode in which two politically recognizable entities are engaged in hostile activities against one another entirely via cyberspace. Politically recognizable entities do not have to be states, but naturally often are. Moonlight Maze (discussed later) is often held up as an example of what is often meant by cyberwar, wherein a state actor engaged in a broad-scoped campaign of exploitation and intrusion designed to radically alter strategic dynamics between the countries involved.

Naturally, "cyberwar" is not as monolithic in form as is something like nuclear war. Though one might imagine nuclear exchanges as being limited or total in terms of the destruction wrought, the results are massive destruction. That is not necessarily the case with the employment of cyber weaponry. Moonlight Maze was broad-scoped but not overly disruptive. The intent was to exfiltrate information, not to engage in anything that might be construed as aggression under traditional norms of armed conflict between states. Nevertheless, massive disruption and even physical damage is possible under some potential cyberwar scenarios. Attacks against national energy grids and utilities alongside blockade activities targeting a country's ISP complex could produce inadvertent deaths and would undoubtedly produce billions of dollars of damage to infrastructure, even if much of it were to be non-physical. This and other cyberwar scenarios are further discussed in the following sections.

The history, patterns and evolutions of interstate cyber conflict

States have hacked one another countless times and for a broad range of reasons over the past several decades. And yet, interstate warfare conducted via cyberspace – meaning the universe of identifiable episodes of large-scale cyberattacks directed by one country against another – has been a relatively limited set of affairs, at least insofar as individual incidents have erupted into what we might categorize as "warfare." Though students of international security might be forgiven for assuming that low costs of entry and attribution provide incentives for states to liberally engage in belligerent behavior online, only a relative handful of episodes constitute the noteworthy topography of the timeline of what we might think of as *meaningful* digital interstate conflict – meaning sophisticated, targeted assaults and not simply low-level nuisance operations – over the past three decades. In this section, we employ data on cyber conflict developed in recent years to explore these trends alongside breakout boxes that speak about key formative episodes reflective of the evolving shape of cyber threats to nations over time.

The objectives of state hacking

Naturally, any summary of the history of interstate cyber conflict episodes will focus on the actions of great powers – those countries with immense capacity for conflictual operations and the geopolitical motivation to undertake them – and this section is no exception. There are two reasons for this. The first is that advanced nations were the first to commit significant resources to the development of cyber operations capabilities. Likewise, advanced industrial states are among the most technically vulnerable owing to the sheer complexity of the societal functions wired and rewired by information technologies. In other words, larger and more advanced societies are also more technologically fragmented in the Internet age, implying not only great value in a large spectrum of potential targets but also superior options for compromise. It is important to note, of course, that being more technically vulnerable to broad-scoped intrusion does not itself mean that small states are less susceptible to disruption from targeted cyberattacks. Indeed, there are many advantages to the fact that great powers constitute dense, complex networks of information systems connected to societal functions. Some of these will be discussed later in the chapter. Nevertheless, any student of strategic studies in the digital age must recognize that the level of integration of ICT in advanced industrial states does at the very least provide would-be belligerents a more robust set of options for attack.

The second reason that summaries of the history of cyber conflict between states focus on the great powers is pedagogical and, in many ways, more significant than the point about complexity. Simply put, interstate cyber conflict that can be identified and verified has been relatively rare in modern international affairs and only a handful of interactions exist as example of the several main archetypes of attack commonly discussed by scholars and analysts of the topic. Indeed, though the description of specific episodes that follows is arrayed chronologically, it is important to note that cyber conflict between states has essentially been conducted in four principal ways, each of which is characterized by a unique set of attacker objectives.

First among these is (1) **information exfiltration**. These attack episodes are often the longest, with coordinated cyberattacks spanning timeframes from a few minutes to multiple years. The purpose of these operations is to steal significant information from or monitor activity within government, military or related industry systems, such as defense contractors or security firms working for a particular government agency. Such information is then repurposed to some goal by a foreign entity, such as the acceleration of technology development or the execution of diplomacy from an improved informational position.

Second, cyberattacks have been employed for the (2) **direct disruption, degradation** or **destruction** of core foreign security assets.[2] Stuxnet, which will be discussed further below, is perhaps the most prominently cited example of this type of conflict episode wherein a state employed digital-only means to disrupt the physical operation of an important facility (in that case, the uranium enrichment facility at Natanz in Iran). Two things characterize this type of cyber conflict episode. First, such operations employ only malicious code to achieve an identifiable disruptive outcome (i.e. they do not rely on secondary support from traditional military assets to be effective). Second, they aim to disrupt the function of a particular security entity. This can be achieved through a number of methods, including simply deleting information through physically damaging infrastructure. Kinetic results – i.e. when a system is physically harmed by a cyberattack – are rare. Nevertheless, such outcomes are possible.

It is critical at this stage to draw a definitional line around a particular manifestation of such attacks. Though most conflict episodes in this vein have involved targeting of specific facilities or, when multiple systems are targeted, particular organizations or security entities, it is possible that an entire country can be the target of this kind of disruptive conflict action. Such an operation is called a **cyber blockade**.[3] The term describes a particular instance in which an entire national system is, in practice, disconnected from the Internet. This is done through massive denial attacks directed against a country's set of Internet Service Providers in combination with a series of complementary attacks on critical regulatory and information infrastructure targets. There are few instances of this type of attack episode taking place and blockades are usually very short-lived. The resources involved in successfully carrying out such an attack are enormous and limited recovery by the foreign state is likely within a matter of hours. Nevertheless, such strategic employment of disruptive cyberattacks differs so drastically in its potential for political messaging or large-scale warfighting scenarios to be worthy of particular note.

Likewise, it is possible – and even desirable for researchers – to think about disruption as a desired outcome. In doing so, we must recognize that not all cyberattacks aimed at technical disruption are motivated by a desire to simply wholesale deny an opponent the ability to function. Many cyberattacks are designed to limit the function of opponent's capabilities and prevent efficient employment of assets, security or otherwise. In this way, we might think about sabotage operations as differently being either about *disruption* (i.e. where the aim is broadly to disrupt core target functions) or about *degradation* (i.e. where attacks are tailored to reduce the efficiency of an opponent's processes). By contrast with broadly disruptive operations, degradation attacks are the preference of attackers with sophisticated political or economic designs.

Related to such targeted disruptive cyber efforts, the third principal way that cyber conflict is conducted is in the form of (3) **enabling attacks**, where some disruptive action is taken as an aid to more conventional forms of military operation.[4] The common example cited both in scholarship and in forecasting efforts to problematize the role of cyber capabilities in warfighting scenarios is that of a cyberattack against localized military defense systems, such as a radar station, a missile emplacement or a military checkpoint. Disruption of a small section of specific national security systems opens up new space for traditional military assets to achieve a primary objective, such as a kinetic airstrike against foreign targets or the exfiltration of intelligence assets.

Finally, cyberattacks can directly aid efforts to (4) **manipulate the information environment** within which politics, policy debates and policy construction occurs. Here, cyberattacks are construed broadly as operations composed of dozens of actions designed to obtain, redirect and modify information in tandem with more traditionally contentious actions to provoke a particular political response. Particularly in democratic states, where there are clear and direct linkages between public conversations and national deliberations on different policy approaches to a range of issues, the global adoption and integration of information technologies has altered the dynamics of information transmission and dissemination such that there exist new angles for foreign manipulation. Russia's 2016 efforts to manipulate public-facing political conversations in the United States, which at the time of writing this book are still being investigated in full, stand as a good example of how foreign powers might increasingly engage in cyber conflict designed to achieve favorable ideational outcomes.

We cover information warfare in greater detail in Chapter 10. However, for purposes of clarity, it is worth noting here that new abilities for states to engage in information warfare aimed at subverting the internal politics of peer competitors cover more than simply another mode of cyber conflict interaction. Much of what is bound up in this form of contention has to do with the way in which cyber actions increase the effectiveness of non-cyber instruments, such as the employment of traditional propaganda tools or direct human intelligence efforts. Among other things, information exfiltration attacks might be used to more effectively equip domestic subversive elements that irritate status quo forces in adversaries' political systems or to produce international efforts aimed at shaming the same. Likewise, cyber vandalism and information modification – i.e. attacks that change the content of websites or data repositories – are useful techniques for causing confusion and altering the shape of popular discourse, while the release of private information and even the mere appearance of systems being compromised by a foreign entity can be of service to any attempt to cast doubt on the integrity of domestic political procedures.

The section that follows outlines significant interstate cyber conflict episodes chronologically. In Chapter 9, the book then turns to specific questions regarding the dynamics of interstate cyber conflict. First, we build on this chapter's illustration of the scope of cyber conflict to delineate national security processes as uniquely diffuse in the digital age. Specifically, we will discuss the technical vulnerabilities and organizational challenges involved in safeguarding and building cybersecurity capabilities across four main sectors of national security apparatuses – (1) critical infrastructure, (2) military systems, (3) innovation and research (i.e. intellectual property) and (4) the national information environment. Chapter 9 then shifts gears and moves beyond history to both conceptually consider unique digital threats to interstate relations and to match such conjecture to empirical evidence on the nature of cyber conflict. This effort informs the discussion outlined in the final sections of this chapter on the applicability of traditional strategic concepts – such as power, coercion and deterrence – to the cyber domain.

The scope of cyber conflict over time

The landscape of global cyber conflict is multifaceted and has evolved in several distinct ways over even the early years and decades of the phenomenon. This section pulls data from several different sources – most notably the latest version of the Dyadic Cyber Incidents and Dispute (DCID) dataset – to illustrate many of these trends, moving from the broadest details of cyber conflict dynamics to more nuanced information intended to inform readers for the discussions of strategy and international relations theory found in the chapters to come.

Figure 7.1 reflects perhaps the simplest illustration of global cyber conflict interactions. The DCID dataset from which the data used to make this map is drawn catalogs major

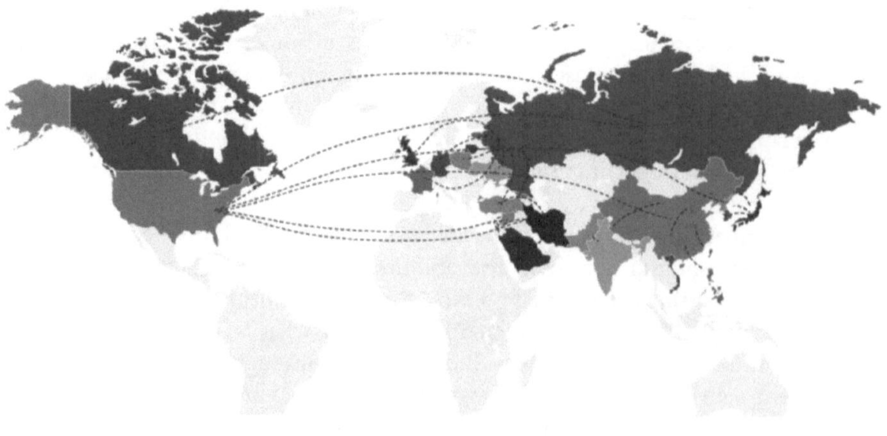

Figure 7.1 The global landscape of rivalrous cyber engagement

cyber operations incidents that have taken place between rivalrous states over a two-decade period (2000–2021).[5] The data is recorded dyadically, meaning that each incident has a national initiator and a national target. As can be seen in this figure, a substantial number of relationships already dot the landscape of global cyber conflict. Specifically, DCID identifies 429 significant cyber conflict episodes involving around two dozen countries over the two decades period. The international interactions include long-standing entrenched rivalries – such as that between India and Pakistan or that between Israel and Iran – as well as a host of lesser such relationships, for instance, between various European nations and the Russian Federation.[6]

Though dozens of nations have been observed to undertake cyber operations for purposes of national defense and espionage, just over a dozen core nations have been involved in what we might think of as the history of *significant* cyber conflict incidents since the 1990s. It will surprise no reader that these nations include those developed nations that play especially prominent roles in global affairs alongside those states that have conventionally taken a highly militarized approach to engagement with the bulk of the international community. On the former hand, countries like Russia, China, the United States and Israel lead the pack in terms of nations that have actively employed cyber instruments in efforts to secure national interests. On the latter hand, states like North Korea and Iran – and, to a lesser extent, Pakistan and even Russia – have utilized cyber capabilities to produce the full range of effects described earlier.[7] Figure 7.2 shows a simple pie chart proportionate distribution of all major cyber conflict incidents recorded in the DCID organized around the *initiators* – meaning those countries offensively responsible for cyber incidents, not victims – of cyber conflict episodes. Clearly, four countries are responsible for the vast majority of all major cyber conflict incidents – China, Russia, Iran and North Korea – with the first two themselves being tied to over half of all recorded

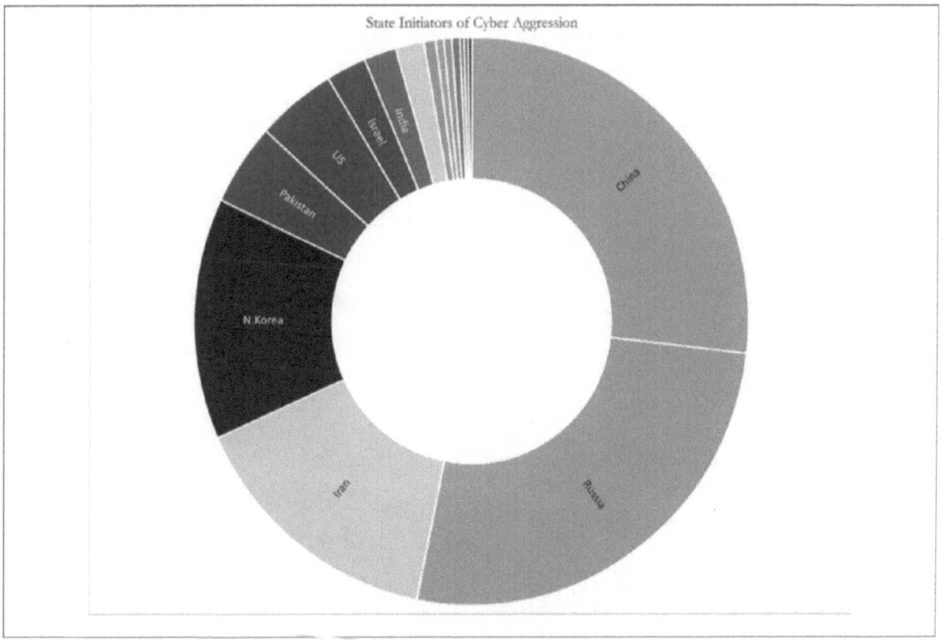

Figure 7.2 Who hacks?

events. To be clear, this does not definitively make a statement that global cyber conflict is synonymous with conflict involving primarily these states (which, naturally, might be grouped together as the core set of countries that have tended to adopt adversarial positions vis-à-vis Western interests over the past three decades). Rather, these states represent the vanguard of developing national cyber capacities in recent years.

Figure 7.3 illustrates this notion that we should not think of cyber conflict principally as a component element of foreign policy involving these specific states, but rather that these states might be ahead of the curve. Naturally, it makes no small amount of sense that Russia, China, North Korea and Iran are involved in most cyber aggression observed to date. The first two are advanced industrial nations with immense resources that present as the primary strategic competitors to American, European and other regional interests (e.g. Indian interests). The latter two are rogue nations in a number of regards – North Korea, in particular – with clear incentives to develop asymmetric means of competing with both regional peers and global powers (i.e. the United States). But, as Figure 7.3 suggests, these reasons are only sufficient to tell us why these four nations have been involved with so many cyber conflict incidents *so far*.

The story told in Figure 7.3 is of the growth of national cyber forces, including military commands and less conventional hybrid institutions, over nearly two decades between 2000 and 2018.[8] Readers should note that, until around 2009–2011, there were remarkably few formal institutions of national security oriented on a cybersecurity mission around the world, both inside the North Atlantic Treaty Organization (NATO) and outside. Indeed, in the year 2000, only seven countries that were members of the United Nations (for which membership is nearly universal) had cyber forces, including China, Russia and the United States. By 2018, that figure was up to 61 nations with such institutions.[9]

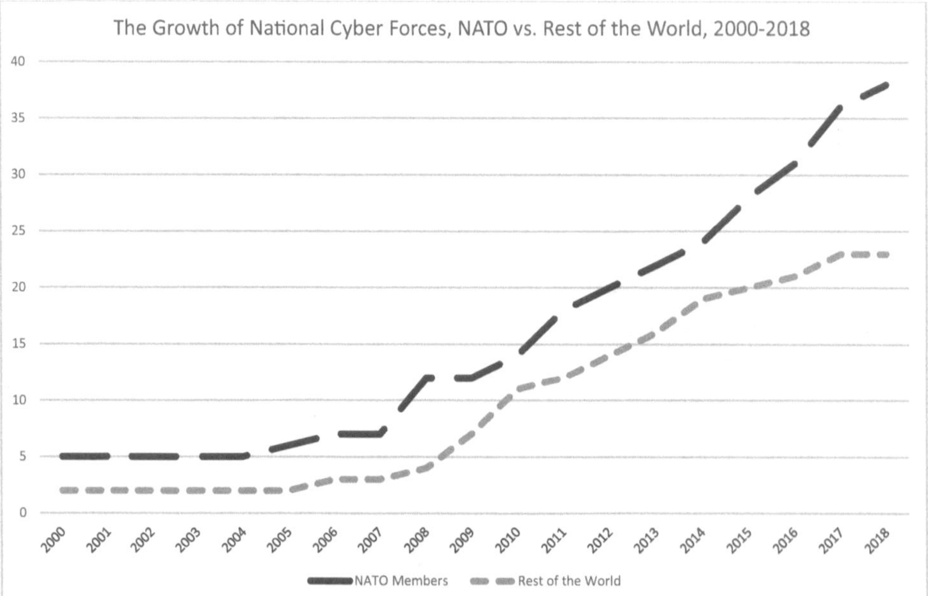

Figure 7.3 The rise of national cyber forces

The specific patterns of and impetuses for the growth of cyber forces through the 2000s and particularly in the 2010s are still being explored by researchers at this juncture. However, some initial research suggests two obvious underlying conditions for the exponential construction of cyber conflict infrastructure through the end of the last decade. One is quite simply the incidence of several major wake-up calls that had the effect of galvanizing nations' support for such institutions.[10] The growth in NATO member states' cyber forces following the Estonia episode in 2007, which we introduce later in this chapter, is an obvious likely example. Another condition identified in recent research is the development of cybersecurity-related policies within the context of existing regional and multilateral diplomatic relationships. Simply put, countries with shared interests, identities and strong diplomatic ties – such as those within NATO or another intergovernmental setup like the African Union – that talk about cyber issues often simply mimic each other's adoption of certain institutional structures and practices. This phenomenon is called **coevolution** and explains why cooperation and new capabilities seem to emerge so much more quickly and easily, as we discuss further in a few sections' time, within the trappings of beneficial, pre-established international relationships.[11]

Box 7.1: Formative episodes: early wake-up calls

Trans-Siberia pipeline attack

While cyberspace as we know it today has at most existed for only a few decades and most sophisticated cyberattacks have occurred only in the last two decades, the first purported CNA-style cyber operation dates back to 1982. This attack is largely still shrouded in uncertainty and it is possible that the attack did not actually take place, as there only is a single source for the episode.[12] In 1982, a portion of the Trans-Siberia pipeline exploded within the Soviet Union, allegedly as a result of computer malware implanted in the pirated Canadian software by the CIA which caused the SCADA system that ran the pipeline to malfunction.[13] The main source of information on this cyberattack is the Farewell Dossier.[14] Among other things, the document points out that "contrived computer chips (would make) their way into Soviet military equipment, flawed turbines were installed on a gas pipeline, and defective plans disrupted the output of chemical plants and a tractor factory."[15] While the accuracy of this attack is disputed to this day, it allegedly resulted in the "most monumental nonnuclear explosion and fire ever seen from space" and the embarrassed Soviets never accused the United States of the attack.[16] For the purposes of understanding emerging norms, this attack is significant because it involved an attack on critical infrastructure that was not explicitly military in nature. The Trans-Siberian pipeline was responsible for transporting natural gas to western Ukraine and ultimately to the broader energy market in order to generate revenue of about $8 billion a year.[17]

The Cuckoo's Egg

Though the incident did not directly involve one state employing malicious code against another per se, the earliest cyber conflict incident worthy of note that involved state actors in some way (that we verifiably know of) is what has now come to be called the Cuckoo's Egg.[18] This name was taken from a book written about the incident by Clifford Stoll, the researcher-administrator at Lawrence Berkeley National Laboratory who in 1986 led a somewhat impromptu investigation into the targeted infiltration of the organizations computer network. In the Cuckoo's Egg case, suspicion that malicious hackers were attempting to gain access to lab computers was aroused, as is often the case in the realm of information security, when a logistical discrepancy was presented to administrators that seemed out of place even beyond the usual profile of such things. In this case, Stoll was asked to account a billing discrepancy of only 75¢ for computer usage (at the time, lab computer usage was billed and monitored much as an electronic payphone might be). Though this seemed to be a minor discrepancy – the billing anomaly translated to only nine seconds of computer usage – Stoll managed to ascertain that it was linked to an unauthorized entry into lab systems. Indeed, more than simply a compromise of a lab user's account, Stoll say that an intruder had managed to obtain administrators privileges by exploiting the movemail function written into the GNU Emacs used in the lab. In a matter of days, Stoll's examination of the intruders actions led him to escalate his initial assumption that the lab was dealing with an amateur prankster of some kind. Clearly, though nothing was initially known about intentions, the intruder was patient and knowledgeable.

Despite reticence on the part of authorities to get involved early on (as will be discussed further in the next chapter), Stoll's subsequent investigation was path-breaking in that he developed for the first time a range of forensic and counter-intrusion techniques that would become mainstays of digital investigation. After identifying the port of entry for the attacker, Stoll was able to monitor the hacker's activity and make inferences about the belligerent's interests and identity. The attacker was active at particular points in the day that suggested a European work schedule. Moreover, they were clearly interested in files related to the Strategic Defense Initiative ("Star Wars") and other critical military topics. This nugget of information allowed Stoll to employ a honeypot – a technique wherein a target of value, which in the case of Cuckoo's Egg was a seemingly important "SDInet" account, is placed so as to attract the attention of the hacker – and invite the intruder to reveal himself. Markus Hess, who had for some years been selling hacked secrets to the KGB, was arrested soon thereafter in West Germany.[19]

Moonlight Maze

Cuckoo's Egg was a relatively limited case of intrusion. A decade later, the United States would stumble upon evidence of another spy campaign that made Hess's activities look inconsequential.[20] The episode began in late 1996 when the U.S. Navy assessed a digital break-in had occurred in the computer network of the Colorado School of Mines in Golden, Colorado.[21] The immediate purpose of the attack was relatively clear. Hacker's had compromised the network and hacked a specific machine for the purpose of setting up a proxy base. From this temporary access point, the hackers spent many hours exploring and examining machines belonging to the Navy, the Air Force, NASA and to the National Oceanic and

Atmospheric Administration (NOAA). Some weeks later, the hackers struck again, breaking into NAVSEA Indian Head (a Navy system commands facility) in Maryland. Here, they exploited a known vulnerability, seemingly from a computer based at the University of Toronto. This time, however, investigation revealed that the original location of the intruders was a machine in Moscow. This pattern would repeat itself throughout the next year across hundreds of intrusions and attempted intrusions aimed at elements of the government's organizational infrastructure, including the Navy's Naval Research Laboratory and the Department of Energy.

Much as was the case with Cuckoo's Egg, it was unclear if this set of incidents originated from a foreign state entity or from black hat hackers operating either independently or as a proxy. It was also unclear if there was some sort of connection between the various attempted and actual intrusions. Was this the work of some broad-scoped operation targeting the U.S. government or was this the advent of a new normal of systems vulnerability? Clearly there was some link between some hacks, as evidenced by source information extracted during investigation, but the total shape of the threat remained complex and irregular in the eyes of examiners. Indeed, despite some commonalities in the way that intrusions were executed, there was no evidence that could concretely tie one or a handful of concentrated efforts with the increasing number of intrusions across government, military and contractor entities.

This changed finally in the middle of 1998, when the Wright Patterson Air Force base was the target of a number of attempted intrusions. The Defense Information Systems Agency (DISA), an agency tasked with providing combat and logistical support of various kinds at the Pentagon, investigated and found that a machine in Cincinnati at the university was the initial point of contact in the chain. Much as occurred at the Colorado School of Mines, University of Toronto and elsewhere, the machine had been compromised for the purposes of using that computer as a beachhead from which to launch attacks and confuse investigators. Further investigation, however, found that this compromised machine was itself subverted from another compromised computer in the United Kingdom. Over the weeks that followed, federal authorities working with Scotland Yard reenacted Stoll's earlier efforts to isolate and monitor the activity of the known point of entry. The results were astonishing. The U.K. terminal was linked directly to Moscow and had been used to launch the intrusions not only at Wright Patterson but also across a massive number of government systems in the United States. Some of these were known intrusions; some were not. Regardless, the incident could now be bound as a large-scale foreign-based attack on various parts of the U.S. government. The FBI investigation, arguably now dealing with a substantially more worrying set of challenges, suddenly became the first national law enforcement and counterintelligence effort to problematize and neutralize an interstate cyber threat. Agents dubbed the ongoing episodes "Moonlight Maze," taking the investigation code name that referenced both the stunning complexity of the networked attacks and the midnight hours in which most intrusions were attempted.

The scope of the information theft enabled by Moonlight Maze was stunning. Russian-based hackers were stealing unbelievable numbers of sensitive files from across government agencies, linked research labs, universities and military units.

The classified Congressional report stated that all the files stolen would be taller than the Washington Monument if stacked one on top of the other. And the patterns of intrusion were unusually hard to predict, up to a point. Intruders would steal information selectively, systems access would be brief and the hackers would get away with sensitive material in a matter of minutes. This, naturally, enhanced the natural of investigative challenges facing the FBI and the intelligence community. Ultimately, of course, examiners managed to explain such behavior. In yet another parallel with Stoll's early incident investigation, the content of files themselves dictated the intrusion patterns of the Moscow-based hackers. One extensive strategic planning document in particular was eventually found to be the roadmap for Russian efforts to steal information. Investigators used this information to set up a honeypot – a corrupted PDF file that would prompt the intruders to re-download a version Adobe Acrobat from a government mirror that would send IP address information back to the United States. The operation worked and, though there remained uncertainty about the usability of the information, gave the Feds a location in Moscow.

We spend some additional time on the details of Moonlight Maze because, almost two decades on, the size of the operation undertaken by one state against another in this specific case has arguably not been surpassed. Moonlight Maze was unprecedented in its complexity. In many ways, it constituted a cyberwar. Eventually, information about the episode came out in public and prompted a range of government responses aimed at improving national capacity to deal with such threats. These are discussed further in the next chapter.

There was no clear end date for the Moonlight Maze spy campaign. Indeed, Russian efforts became even more persistent and sophisticated over time. Though the concentrated intrusions of 1998 and 1999 constitute the bulk of hacks that can be specifically tied to that operation, the U.S. government would receive precious little respite in the first years of the new century. In many ways, the complete retooling of strategic planning processes after 9/11, as well as rapidly changing procedures for effective cybersecurity across agencies, helped the U.S. government mitigate the effects of Moonlight Maze insofar as procurement and development priorities were reformulated across many features of military planning in their entirety. But conflictual interstate interactions via cyberspace entered a new era after Moonlight Maze, with a range of attacks characterizing the security experiences of the United States and other countries.

Solar Sunrise

One of these episodes happened simultaneously with Moonlight Maze. In 1998, at roughly the same time as the early phase of the Moonlight Maze investigation was underway, the Department of Defense and other government agencies were the target of a series of attacks now known as SOLAR SUNRISE.[22] These attacks were sophisticated in their simplicity, mostly targeting unclassified information and systems across the U.S. government. Specifically, Air Force, Navy, Knesset (in Israel) and various university systems were compromised in what was called, at the time, the "most organized and systematic attack to date" on the United States via cyberspace. Though the culprits of such attacks were ultimately demonstrated to be an Israeli hacker and two teenagers in California basing some of their intrusions from compromised machines in the Middle East, the prevailing thought in the initial stages was that SOLAR SUNRISE was being prosecuted by Iraqi infowarriors in rapid retaliation against ongoing anti-Hussein regime airstrikes being undertaken by the United States. SOLAR SUNRISE was both an organizational wake-up call (discussed in the next chapter) and a strategic revelation insofar as planners increasingly saw the value for foreign states in low-intensity cyber conflict responses to real foreign policy clashes.

So how do states hack one another? Figure 7.4 outlines an answer to this question in the most basic sense, that is in terms of the methods and techniques employed. Almost two thirds of all major cyber conflict episodes – which will invariably involve multiple methods of approach blended together – involve network intrusion. Simply put, this category of engagement generically covers attempts to exploit vulnerabilities in a system, such as a website or operating system, in order to enter without authorization. This is often done for purposes of espionage but the purpose more broadly is to affect some measure of unintended control over a foreign network space. Network infiltration more simply describes the use of one of several techniques to force access to a computer system's functions in a fashion not desired by designer or user. In the DCID dataset, most such infiltrations are computer viruses, which we previously discussed in Chapter 2, with other techniques fairly evenly spread across incidents. Finally, a small volume of major cyber conflict episodes involve basic attacks beyond network intrusion or infiltration, namely denial of service attacks – usually a distributed attack using a botnet of compromised computers – or simple acts of vandalism.

Figure 7.5 expands our understanding of how states hack one another by arraying the types of targets selected by the 16 initiators of major incidents observed in DCID.

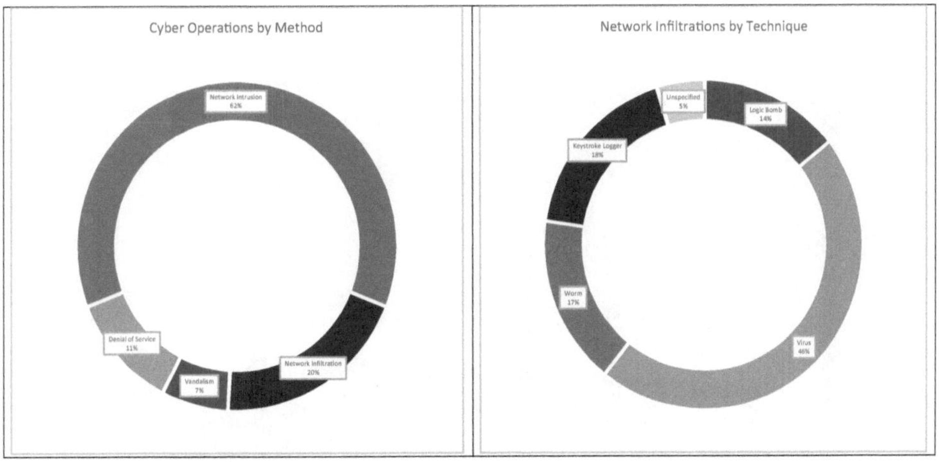

Figure 7.4 How states hack

Specifically, the dataset codes for targets among (1) private industry and infrastructure, (2) civilian governments and (3) state militaries. Apart from again illustrating the degree to which just a few nations are responsible for the preponderance of major cyber conflict incidents, Figure 7.5 also shows that those same nations also attack private industry or infrastructure targets substantially more than their strategic counterparts, particularly those democratic nations on the list. Countries like the United States, India or Israel very rarely attack private targets, instead targeting government and military networks when they employ cyber operations. Another interesting takeaway from this figure, of course, is the limited degree to which state militaries appear to be targeted for major cyber operational outcomes. This is not to say that militaries are never targeted in cyberspace. That would obviously be disingenuous and several decades of research and reporting suggests that militaries are the daily target of millions of low-intensity attempts to spy and steal useful information. However, it does suggest that few countries see cyber operations as particularly useful for military-to-military interactions. The United States is a notable exception to this, at least proportionately. While the United States has not undertaken many cyber operations against the militaries of other states, Washington clearly sees such attacks as occasionally useful in a way that Russian, Chinese or Iranian counterparts – for instance – do not. We unpack this dynamic further in the next chapter.

But what are states attempting to achieve in launching cyber operations? Figure 7.6 paints a picture similar to that of Figure 7.5. The United States appears uniquely – again, in proportional terms – interested in using cyber operations to degrade the military capabilities of foreign competitors. By contrast, countries like China, Russia and Iran use cyberspace almost exclusively for espionage, both short-term operations in aid of some broader foreign policy activity and longer-term attempts to access either sensitive data or large volumes of information. Other national entrants on the list, from Japan and India to Pakistan and Turkey, largely aim to simply disrupt their competitors.

These initial illustrations suggest a few basic conclusions for our exploration of global cyber conflict. Few countries have cause to launch major operations against competitors,

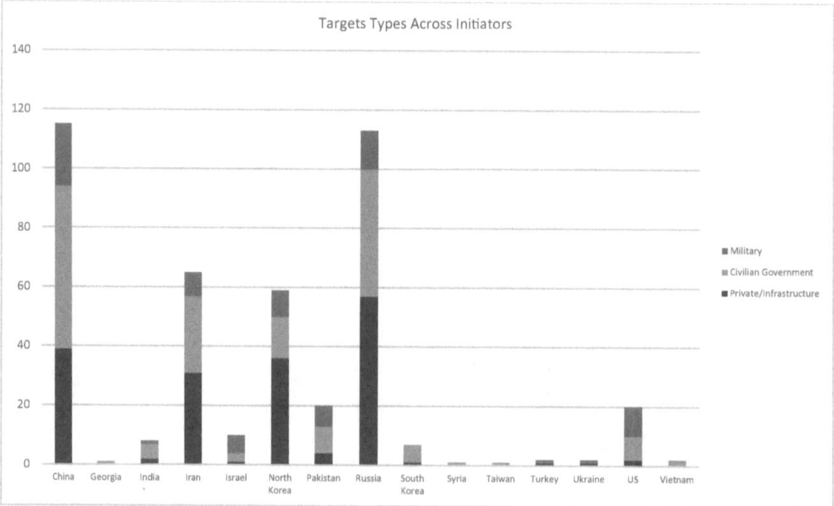

Figure 7.5 Targets of state hacking

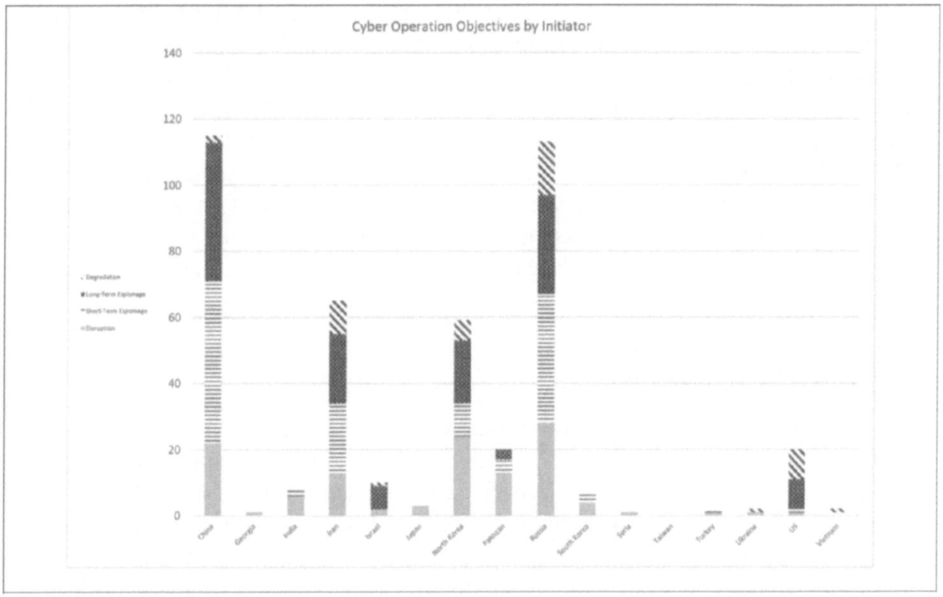

Figure 7.6 Objectives of state hacking

it would seem, for instance, at least historically. Primarily, however, we know already that different states clearly employ cyber operations for diverse purposes. Specifically, the emphasis so clearly placed on distinct types of methods and associated objectives suggests that national interests and circumstance motivate cyber behavior. But what about those interests and circumstances most strongly dictate such preferences? Geopolitical conditions, domestic politics and prevailing national security doctrine are all possible foundations upon which unique cyber strategies might be built. We explore this question further in chapters to come.

Box 7.2: Formative episodes: growing sophistication

Chinese espionage

Quite arguably the next phase of global cyber conflict, at least as it impacted the United States, kicked off just half a decade after the wake-up calls of the late 1990s. Yet another massive set of attacks ostensibly perpetrated by state actors against the United States took place starting in 2003. Originally labeled "Titan Rain" by the federal government (this codename was changed to "Byzantine Hades" after the original name leaked, and then again to an as-yet-unspecified moniker), this cyber operations campaign constituted an Advanced Persistent Threat wherein hackers based in China were able to steal a large amount of unclassified information from several computers.[23] Several possibilities exist for the identity and motivation of the hackers involved, including direct involvement of the Chinese government, low-level elements of the People's Liberation Army and independent hackers based in China. The attacks did target sensitive systems belonging to the FBI and NASA and, though they did not achieve significant success over the initial period, have been emulated in a range of major attacks over the subsequent decade (the bulk of which have been dubbed "Shady Rat" to delineate a specific set of China-based espionage cyber operations), including in one major incident in 2007 wherein (according to the Snowden leaks) more than 50 TB of information, containing blueprint information for the F-35 fifth-generation fighter aircraft and other sensitive data, was stolen.[24]

BUCKSHOT YANKEE

Though the scope of the infiltrations bound up in the Moonlight Maze, Titan Rain/ Shady Rat and SOLAR SUNRISE incidents is impressive, it is commonly argued that the most significant interstate cybersecurity incident to date was BUCKSHOT YANKEE.[25] In 2008, Department of Defense and State systems were infected by a malware application dubbed "Agent.btz." This incident, purportedly the work of Russian security services, was unique in that it bypassed the "air gap" defenses (meaning that there is no direct connection to the Internet via which hackers might gain entrance) of these secure systems. The scenario for the incident that enabled BUCKSHOT YANKEE is commonly described as one of human errors – a USB drive was purchased from a local marketplace known to be frequented by American service personnel and targeted for product infiltration by Russian security services. The unwitting carrier of the malware payload then plugged it into an otherwise secure computer on base. Agent.btz would then infect computers until it communicated with a master system, at which point it could receive targeting instructions and begin the exfiltration of sensitive data. The threat was neutralized relatively quickly. Nevertheless, BUCKSHOT YANKEE was yet another demonstration of the relative vulnerability of states to subversion and infiltration at the behest of foreign powers. Specifically, BUCKSHOT YANKEE was significant because it was the first breach of classified, protected systems by a foreign power. The incident demonstrated the various difficulties involved in protecting information from a sufficiently sophisticated effort to infiltrate systems beyond the efforts of individual hackers or teams of hackers. With appropriate design and implementation resources, the vectors for possible assault by a foreign state multiply. BUCKSHOT YANKEE, the effective response to which was almost entirely put forth by NSA personnel, also demonstrated the relative inability of military institutions to adapt to such a threat. Institutional inability to act itself, the roots of which are discussed further in the next chapter, very almost prolonged broad government exposure to malicious action.

Olympic Games

Ironically, of course, the lessons of attacks against the United States have clearly translated into new abilities on the home front. From the First Gulf War to NATO operations in Kosovo, American infowarriors have utilized ICT in conflict at the cutting edge of technological possibilities. And no history of cyber conflict, however brief, would be complete without particular mention of Operation Olympic Games, a covert series of operations perpetuated by the George W. Bush and Barack Obama administrations as an attempt to disrupt the functionality of Iran's nuclear infrastructure. To be clear, the program and operations have yet to be acknowledged in public. Nevertheless, there is anecdotal and minor direct evidence that American agents (with possible Israeli involvement) sought to both specifically disrupt Iranian nuclear operations via cyber means and more generally steal information about the function of various Middle East governmental and military organizations on an ongoing basis.

The former effort was typified in the deployment of Stuxnet.[26] An alleged joint operation of U.S. and Israeli government authorities,[27] Stuxnet was first discovered when an Iranian customer complaining of trouble with his Windows computer contacted an obscure antivirus firm in Belarus. Close inspection, first by the firm in question and later by malware experts, revealed that the affected device suffered from a complex and malicious piece of software. Indeed, the program, dubbed "Stuxnet," was one of the most sophisticated ever revealed and was targeted, according to various expert investigations, at the facilities of Iran's fledgling nuclear complex. Stuxnet reportedly went on to damage between 10% and 20% of the centrifuges in place in Iran's Natanz facility, an act that undoubtedly set any weapons' development program back some time.[28] The expert community has regularly agreed that Stuxnet's sophisticated design and deployment suggests governmental involvement[29] and a consensus has commonly emerged that the development strongly implies American agency.[30]

The latter effort, focused on a broader range of disruption and monitoring efforts, is the supposed function of Flame. Flame is malware that was discovered in 2012 that both has similarities with Stuxnet (which was derived from the much larger Flame) and presents as vastly more sophisticated than the worm that damaged Iran's centrifuges and temporarily hindered operations at Natanz. Flame is unusually large, but also uncommonly smart in its ability to evade detection. It is highly complex and is designed to record all manner of input to computer systems, including data stored, audio and video. Leaked documents from the NSA and a range of computer forensics efforts point to the development and deployment of Flame by the NSA in collaboration with the British intelligence establishment. The purpose, it seems, is broad-scoped interstate collection of private information from foreign sources and, through that process, the interdiction of foreign state-based and related threats.

Few cyber operations result in major disruption or even amount to anything that would conventionally be considered a hostile foreign policy action.[31] As Figure 7.7 outlines, no episodes have been more severe than an infiltration of a critical national security network to cause physical destructive effects against highly targeted systems. Even then, only six episodes have been this severe, including the famous Stuxnet attack on the Natanz uranium enrichment facility and the retaliatory Shamoon virus (both discussed in this chapter's breakout boxes) that wiped thousands of hard drives.[32] Very few episodes have been

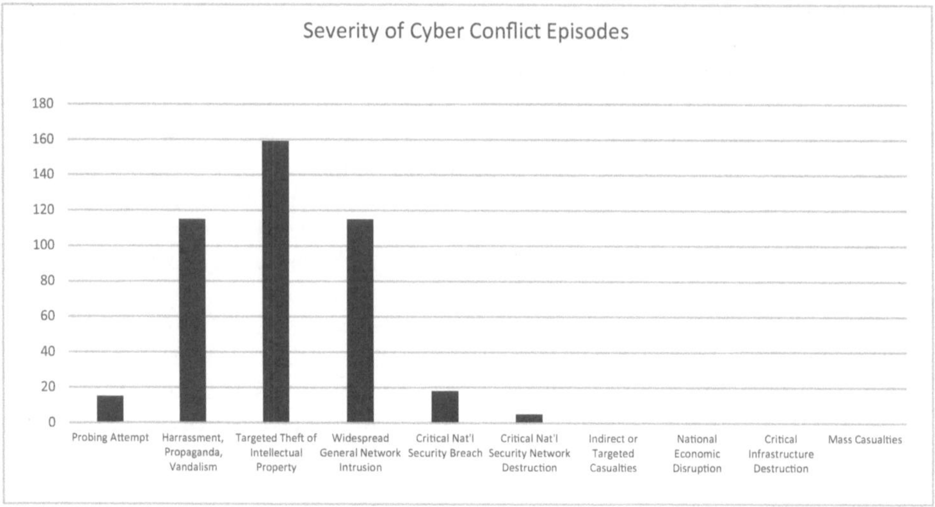

Figure 7.7 The severity of interstate hacking

unsuccessful versions of this kind of attack (i.e. infiltration occurred but no outcomes were achieved), with most episodes involving (1) vandalism or basic disruption, (2) theft of data from one critical national security network or (3) theft of data from a multi-network compromise. No cyber operation has ever led to direct or indirect deaths, caused destructive effects to national critical infrastructure or substantially disrupted a national economy entirely on its own.

Of interest, as Figure 7.8 shows, the most severe cyber operations in terms of their disruptive effects have been prosecuted by democratic nations or those more closely aligned with the United States and Europe than China or Russia. The United States, Vietnam and Israel each sport higher average severity scores for the cyber operations they are linked to than Russia, China, Iran, North Korea, Pakistan and others. This is despite the fact that these latter countries are linked to the preponderance of major cyber campaigns observed over the past two decades. This dynamic illustrates yet again that there are distinct differences in the way that countries choose to fight in cyberspace – if, indeed, it is fighting that takes place online as opposed to harassment, espionage, sabotage and subversion, that is.[33] The United States and several of her partners, particularly Israel, seem far more willing to authorize degradative cyber operations aimed at military targets. We discuss this further in the next chapter but one possible reason why is obvious enough to be mentioned now, namely that democratic nations may simply be less drawn to semi-visible dirty tricks and subterfuge than are authoritarian states. This may indicate not only a reason why countries like China and Russia are so much more engaged in long-term and short-term espionage over time but also a reason why the United States and her partners are more willing to attack harder targets for purposes of degradation. Simply put, they see value in cyber activity for signaling in the absence of as much incentive to simply harass.

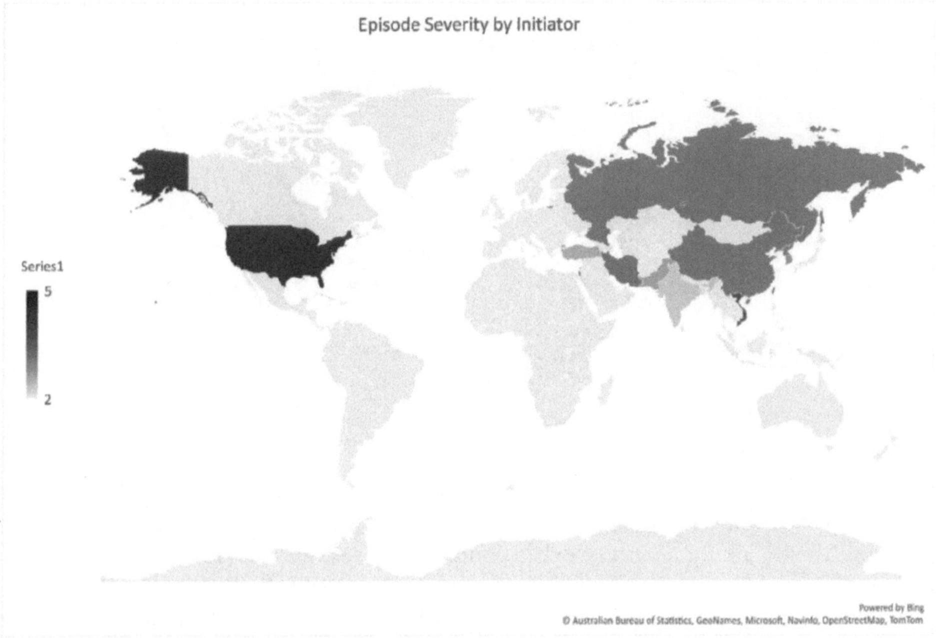

Figure 7.8 The severity of interstate hacking by initiator

This section has painted a reasonably simple picture. States hack in different ways for different reasons. We explore those reasons further in the next chapters of the book. Now, however, we turn to the temporal element of the story told by data about trends in global cyber conflict. By this, we mean that the base character of interstate cyber engagement has changed over time, something we can expect to continue out into the 2020s and beyond.

The diversification of cyber conflict

In particular, the overarching trend that can be seen in any review of cyber conflict data for the past two decades is a diversification in the types of campaigns, effects and actors involved. If the average citizen or even student of international relations were to be asked what they think has changed in that time frame, it wouldn't be unreasonable to think that most would say the cyber conflict events had increased in frequency and intensity. There is, of course, some truth in that. Consider Figure 7.9, which illustrates two trends. First, the left-hand chart shows the average length of cyber conflict episodes (in months) over time. We can clearly see that a relatively high average length of cyber campaigns in the 2000s (until about 2012) has given way to much lower average episodes lengths in recent years. This might reflect a few developments, for instance, that cyber force development has meant improved national cyber defenses and less ability for malicious actors to simply loiter in enemy networks focused on one objective (or a few related objectives) for months and years at a time. Regardless, the trend toward shorter lifespans of particular cyber operations engagements seems clear.

The other chart in Figure 7.9 shows simply that the frequency of cyber conflict episodes increases, as the average citizen above might have suggested as their answer. Specifically,

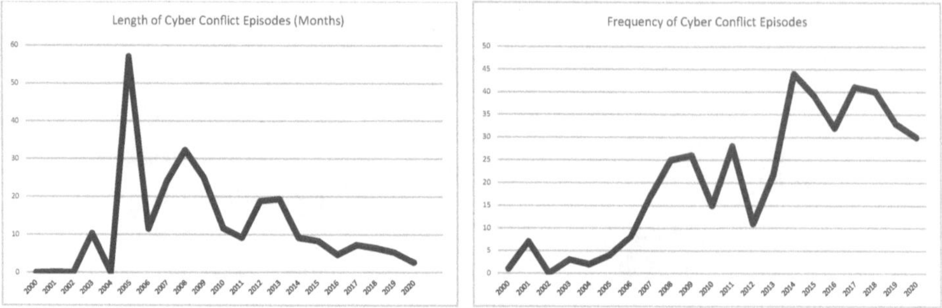

Figure 7.9 The frequency and length of engagements

there were two approximate moments in time where the frequency of episodes increased quite substantially, one following 2007 and another after 2014. Taken alongside information about the shrinking length of cyber conflict episodes, we can clearly see that there is nuance to be described in how global cyber conflict has evolved over the past 10–15 years (at time of writing). Operations have shorter half-lives, but attackers are no less devoted. The growing number of major engagements suggests that nations are more than willing to devote resources to diverse "fronts" in order to compensate for whatever dynamic is responsible for limiting the potential of long-term operations. Indeed, since the rate of change for episode frequency is greater than it is for episode length, it stands to reason that attackers are *increasingly* willing to engage in multi-pronged cyber activities.

Figure 7.10 illustrates another interesting and broadly accepted feature of the landscape of global cyber conflict, namely that there is a highly regional and a highly rivalrous nature to the thing. In fairness, this is one area where the data the charts in this chapter are produced from (DCID) must be placed in appropriate context. Specifically, the data is generated around known rivalrous relationships between nation states. This means that non-state cyber operations and a great many operations involving states – particularly small states – not against a known rival are not included. And yet, no alternative data resource – and there are several – has offered compelling evidence that the universe of interstate cyber incidents beyond what is in DCID is expansive. Certainly, almost no major incidents where attribution of a nation-state actor has been confirmed exist beyond the dataset.

Given this, the geographic illustrations of the regionalism of international cyber conflict presents an takeaway for readers of this book. Most of the time, the four major perpetrators of cyber conflict – Russia, China, North Korea and Iran – are acting against the United States or a regional rival. Russia, when not engaged in operations against the United States, has primarily acted against nearby obstacles to Russian influence in Eurasia. This includes former Soviet states like Ukraine and Georgia, as well as European members of the NATO alliance like Germany. The cases involving China exhibit a similar trend, where attacks not against the United States are localized in Asia against targets in countries like Japan, Australia or Vietnam. The Iranian and North Korean cases are similar, but are even more defined by singular rivalrous context, in those cases involving Israel and North Korea. Most Indian and Pakistani offensive actions involve one another also.

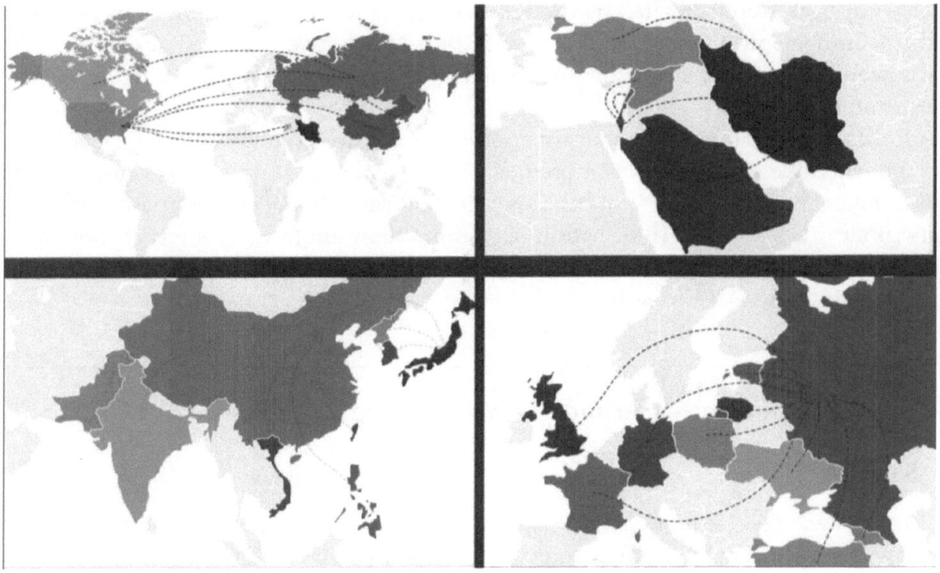

Figure 7.10 The regionalism of cyber conflict

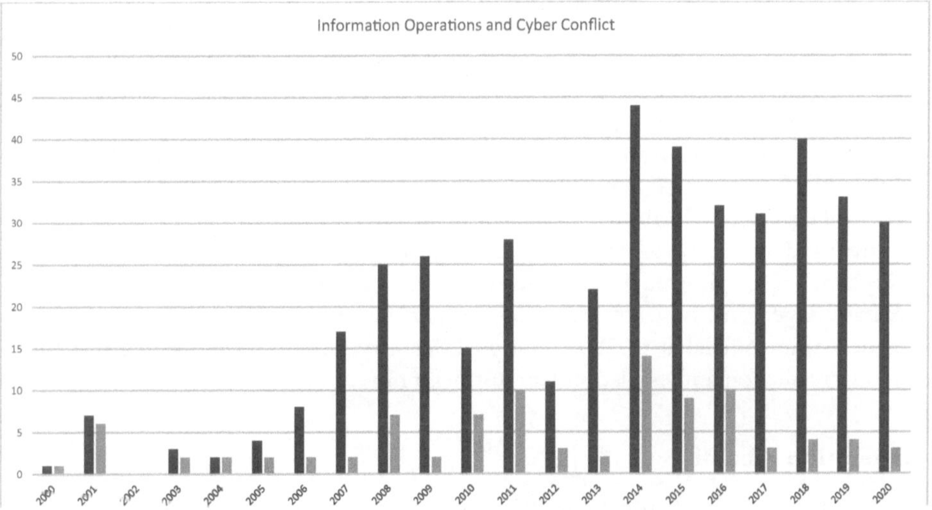

Figure 7.11 The information operations tie-in

The diversification of cyber conflict in recent years has also involved the use of cyber operations to enable other forms of interference and disruption. Information operations are one such adjacent mechanism of state power.[34] Indeed, one might argue that the 2010s were the decade of information operations targeted against democratic polities from North America and Europe to the Asia-Pacific. Though the attempted Russian interference in the American presidential election of 2016 reflects the paradigm case of information operations leveraged against democratic states, in reality over three dozen such campaigns were launched by the Russian Federation against almost as many countries in Europe and North America since 2014.[35] Nearly two dozen additional episodes have been observed by researchers launched by China, Iran, Venezuela, Saudi Arabia, Cuba and other states.

Of these information operations, of course, not all had pronounced cyber operational components. We address this in more qualitative detail later in this book. However, it's worth noting here that the rise of cyber-enabled information operations is visible in much data about cyber conflict. Information operation follow-on from attempts at short-term espionage by Russia and China, in particular, have become frequent staples of cyber conflict activities from 2014 onward.

Box 7.3: Formative episodes: rivals and regionalism

Saudi Aramco

In part as a cyber response to the damage wrought by Stuxnet, Iran is suspected to have invested heavily in offensive cyber warfare capabilities. On August 15, 2012, these investments seem to have borne fruit in an attack involving the "Shamoon" virus that was launched against the state-owned oil company Saudi Aramco (the most valuable company in the world).[36] The attack prompted U.S. Secretary of Defense Leon Panetta to describe Shamoon as a "very sophisticated" piece of malware generating "tremendous concern."[37] Over 30,000 computers were infected and in many cases data on servers as well as hard drives on individual computers were destroyed.[38] The goal of the attack was purportedly to disrupt the flow of Saudi oil by damaging SCADA control systems, but it did not succeed in achieving that effect.[39] An Iranian-linked group called "Cutting Sword of Justice" ultimately took credit for the attack, which also affected the Qatari company RasGas as well as other oil companies.[40] Ultimately the attack affected the business processes of Saudi Aramco and it is likely that some important drilling and production data were lost.[41] This attack again showed a dangerous trend of unconstrained attacks against non-military targets and was interpreted by Richard Clarke – cyber warfare expert and former senior official at the U.S. National Security Council – as a signal that this kind of retaliation and escalation was just beginning.[42]

Operation Ababil

In September 2012, not long after the Saudi Aramco attacks, further retaliation and escalation stemming from the Stuxnet attack on Iran occurred when the Iranian-affiliated hacker group Izz ad-Din al-Qassam launched "Operation Ababil" targeting the websites of financial institutions for major DDOS attacks. There institutions included the Bank of America, New York Stock Exchange, Chase Bank, Capital One, SunTrust and Regions Bank.[43] In January 2013, Izz ad-Din al-Qassam again claimed responsibility for another series of DDOS attacks again predominantly U.S. financial institutions as part of "Operation Ababil," phase two. A third phase of DDOS attacks began in March 2013.[44] U.S. officials believe that

Izz ad-Din al-Qassam is a front organization for an Iranian state-sponsored effort.[45] U.S. Senator Joseph Lieberman went so far as to state on C-SPAN that he thinks "this was done by Iran and the Quds Force, which has its own developing cyberattack capability."[46]

To this point, all interstate cyber conflict incidents discussed (bar the last two discussed as responses to Stuxnet) have involved the United States. Naturally, not all significant episodes – at least from the perspective of analysis and theorization on the nature of cyber conflict – do. Before discussing two more recent incidents that once involve again center on the experiences of the United States or her close allies – those of the 2016 Russian attacks on civil society actors and industry in the West, as well as the slightly earlier set of incidents surrounding the release of Sony's film, *The Interview* – three episodes bear particular mention.

Operation Orchard

Another incident not involving the United States, also in 2007, that bears unique mention is surprisingly limited in scope given the nature of interstate conflict interactions outlined so far. Operation Orchard was an Israeli military mission in which F-15 and F-16 aircraft assaulted a suspected nuclear reactor in Syria. This incident is of particular note in the history of interstate cyber conflicts as it is perhaps the best example of the use of network intrusion for the purposes of aiding a kinetic military action.[47] In short, Israel employed a computer program often likened to Boeing's Suter program, which is designed to attack computer networks and alter the data being provided by a sensor system to either human or linked machine observers. In this instance, Israel disrupted Syrian air defense systems and fed them false radar and other sensor information to mask the transit of Air Force jets into Syria. Alternatively, some reports hold that Israeli cyberwarriors were able to attack the computer networks controlling Syrian air defense systems and use a built-in kill switch, a piece of code incorporated during the production phase and designed to allow unauthorized control – usually highly specific – of a system.

If one could make the argument that the 2010s were the decade of information operations, at least in terms of the most novel iteration of digital security challenges facing the nations of the world, it seems fair to say that an argument could similarly be made for ransomware as the defining technique of the decade. After all, several of the world's most significant cyberattacks, at least by financial cost imposed on governments and society as a result of the attacks, occurred during the 2010s and involved ransomware.[48] Incidents like WannaCry[49] and NotPetya,[50] in particular, stand out in the memory of even casual readers of the digital dimensions of international relations because of the unprecedented spread and impact of those attacks. The costs of recovery for NotPetya alone ran into 11 digits territory and so it's no wonder that many conversations about cyber conflict involve discussion of ransomware.[51]

This said, the data on major cyber conflict episodes don't bear out the idea that ransomware is a big – and getting bigger – problem. As with many of the points made earlier, there is substantial nuance to this conclusion. Certainly, as Figure 7.12 shows, there are

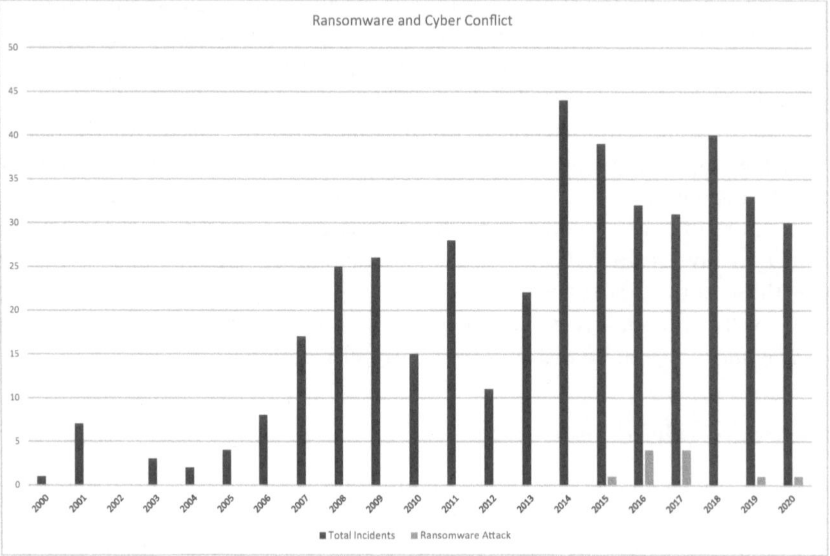

Figure 7.12 Ransomware

simply extremely few major interstate cyber conflict incidents that have a ransomware component. This makes some sense, as ransomware attacks are useful either (1) to make money or (2) as a means of permanently "destroying" data via encryption with no intention to decrypt. These are situationally valuable outcomes, such as with the case of North Korean efforts to make money as a way to get around financial sanctions, but not generally preferred methods of securing either tactical or strategic gain via cyberspace.[52] Ransomware attacks on Western nations *are* on the rise, but these attacks tend to be almost entirely criminal in nature. Even there though, they still make up a surprisingly – at least, likely surprising to the average citizen – small percentage of cybercriminal revenues, with a large handful of prolific gangs and other non-state organizations covering much of the market for ransomware as a service (RaaS).[53] All told, in sum, ransomware attacks should not be considered a significant element of the diversifying landscape of global cyber conflict.

Rising conflict, rising cooperation

How much of an impact on international relations has this groundswell of digital antagonism over the past two to three decades had? One major question that we grapple with in the next chapter is on the degree to which new cyber capabilities make conflict between states more likely, particularly in terms of kinetic warfighting. But data already gives us an idea of what this relationship might be. One major study by Valeriano, Jensen and Maness, in particular,[54] used some of the data already used in this chapter to study the link between different kinds of cyber activities – and their methods, severity, etc. – and the responses undertake by those being attacked. In particular, the study looked at whether or not victims offered concessions in response to cyberattacks – similar to what might be seen after a more conventional assertive action undertaken by an adversary – or retaliated. If they retaliated, did they do so in a proportional sense or did they escalate?

The study found that cyber operations very rarely produce concessions from victim countries in the form of, among other things, changing policy positions or clear concessionary statements.[55] Most types of cyber activity were completely un-correlated with such national responses. Only degradative attacks produced concessionary behavior among the victims of cyberattacks a noteworthy amount of the time. Here again, however, there is nuance. Degradation and a history of cyber espionage against the victim were both necessary conditions if a concession is to be produced by cyberattack alone.[56] Positive economic inducements by the attacker also helped, though it was merely a sufficient factor and not a necessary one. Very few countries have successfully secured concessions. Indeed, the study observed only the United States and Israel in this category. And target selection clearly matters. Attacks that don't involve the government in some regard were in no way linked with concessionary behavior, particularly as the researchers noted that attacks on private industry were intentionally met with tit-for-tat responses in the cases involving the United States in a clear effort to avoid escalation.[57] These results thus suggest an additional reason, at least anecdotally, why we see so much degradative emphasis by democratic nations vis-à-vis their authoritarian counterparts. Simply put, in the absence of an incentive to use cyberspace to harass as freely as their counterparts, democracies choose to be assertive because anything else stands little chance of strategic gain.

Box 7.4: Formative episodes: Russians at the gates

The attack on Estonia

To backtrack chronologically very slightly, Russia's use of cyber capabilities evolved in the mid-2000s from the espionage focus that characterized initial engagement to harassment and disruption in the context of the former sphere of influence of the Soviet Union. The most prominent such episode in this period is the 2007 cyberattacks on Estonia perpetuated by unspecified elements of the Russian government in the midst of a tense disagreement between the two countries about the relocation of a Soviet-era grave marker statue, the Bronze Soldier of Tallinn.[58] This incident is worth some additional detail as it is regularly held up as the first major example of both outright cyberwar (in the disruptive rather than the infiltrative sense) and an effective cyber blockade. The first phase of the cyber conflict between Russia and Estonia in 2007 took the form of widespread distributed denial of service (DDoS) attacks that denied service to a range of government and related organizations and prevented the public sector from conducting essential operations. The context and cause of such an attack was the anniversary of the conclusion of World War II and the recent decision by the Estonian government to move a statue – the Bronze Soldier of Tallinn – over the objections of local officials ostensibly in order to minimize tensions between ethnic Estonians and ethnic Russians in the long term.[59]

After two nights of violent clashes in riots in Tallinn, the first cyberattacks began. Based from Russia, hackers vandalized numerous websites and set about disrupting network usage on a national scale. DDoS attacks and ping flood scripts were employed to flood systems with access requests and deny both government and private corporate entities the ability to access their systems. Targets were quickly expanded to include political party websites, daily newspapers and critical service providers, including banks. These attacks were amateurish but highly effective. Estonia's response, in partnership with a range of international ISPs, was only partly effective. On May 9, a second wave of cyberattacks was unleashed against Estonia. This wave was more sophisticated. Over a million "zombie computers" around the world – machines that were compromised and directed to take part in denial of service attacks – were used to throw more than 1,000 times the normal operating volume of data at Estonian servers. This continued in third and fourth waves for several days. The effect was impressive. Estonia became largely cut off from the outside world. In addition to the other social and economic effects of the crisis, the Estonian government, Estonian industry and Estonian citizens could not access most network services to communicate beyond national borders. Quite naturally, Russia was blamed and a range of evidence points to Russian government involvement, not least because the most effective defensive effort during the first day of attacks included the blocking of Web addresses ending the Russian ".ru" identifier. Nevertheless, much as has also been the case with cyberattacks launched against Georgia and Ukraine in subsequent crises, the centralization of control of Russian cyber conflict efforts is unclear. With Estonia, many outlets claim that botnets were provided to a range of Russian-based hackers who assaulted their smaller next-door neighbor for purposes of national pride. In both Estonia and Ukraine, other sources suggest that botnets were released by organized criminal elements in Russia, perhaps at the behest of elements of Russian government officials.

Following the attack, NATO, of which Estonia is a member, established the NATO Cooperative Cyber Defence Centre of Excellence (CCD COE) on May 14, 2008.[60] This center, located in Tallinn, Estonia, seeks to enhance NATO's ability to respond to cyberattacks and as of late has been acting as an organizational platform for norm entrepreneurs, as will be discussed in more detail later in this chapter. The Estonia cyberattacks were aimed directly at disrupting and degrading civilian services and thus demonstrated the lack of a constraining cyber norm for non-combatant immunity or discrimination. However, the attack did not result in permanent damage and did not destroy any critical infrastructure – although this was likely due to the limits of the DDOS mechanism available and not to any normative constraint.

Georgia attack

Compared to Estonia, the Russian attack on Georgia in July 2008 presents a slightly more recent example of cyber warfare conducted against a former Soviet state in order to achieve tangible disruption and effects beyond CNE-style espionage. This attack began on July 20, 2008, prior to the military invasion of Georgia by Russian forces, with a large-scale DDOS attack shutting down Georgian servers. It is the best example to date of cyber weapons being used as a force multiplier for conventional military operations. As the invasion began, the attacks increased and spread to other targets.[61] This ultimately forced the Georgian government to move critical communication services to commercial U.S. sites as their own services were shut down.[62] The attack was likely organized by the Russian government to support its broader political and military objectives in the crisis, but executed by loosely-affiliated "independent" hackers that strengthen the government's plausible deniability.[63] Like the Estonian attacks, this attack demonstrated no normative constraint prohibiting targeting civilian resources. However, also like the attacks on Estonia, critical infrastructure was not attacked and permanent damage did not occur. Both of these attacks on former Soviet states – likely originating from the same source – show that the only constraint on the attacks is not a norm, rather it is the limits of what is technologically possible and effective.

Black Energy

Another major cyber conflict incident involving Russian-backed actors and European neighbors occurred much more relatively recently, in late 2015. The incident, broadly known by the name that was given to the threat group responsible by global cybersecurity firms ("Black Energy"), was a sophisticated attempt to interfere with the Supervisory Control and Data Acquisition (SCADA) systems that controlled electrical grid functions in Ukraine. Ukraine has a long set of experiences in dealing with Russia-based digital attacks against critical infrastructure and civil society (indeed, the Black Energy threat profile has been active since 2007). In December 2015, however, these efforts culminated in a successful disruption of the functions of electricity distribution companies across the country, most notable Prykarpattya Oblenergo.[64] The result was that about 1.4 million people were left without power for a few hours. While this may not seem particularly significant, the incident was the first time that a foreign actor had successfully achieved a real-kinetic effect in a cyber-only attack on critical infrastructure. Black Energy thus highlights the increasing relevance of infrastructure vulnerabilities to national security considerations.

Another fascinating – and arguably quite worrying – development in the landscape of global cyber conflict is the rise in the use of state-sponsored proxy actors. Proxies can take a number of forms, the most common of which are intelligence outfits functionally held apart from the formal auspices of a state's defense establishment and cybercriminal gangs hired by or indebted to a government. These actors hack on behalf of a state's interests in one of a number of fashions. They can do so with specific mission orders and parameters, they can do so with broad campaign objectives and no operational oversight or they can do so at their discretion with only the broadest mandate from authorities. As Figure 7.13 shows,[65] state-sponsored proxies have become extremely common in recent years even as operations directly initiated by government forces have occurred at roughly the same intensity. This lends further credence to the idea that malicious actors are increasingly willing to use cyberspace to compete, contest and harass. Indeed, use of proxies has grown to be more common than government operations by an order of some magnitude.

All this naturally poses an issue for cyber defense and for deterrence strategies because proxies are less easily tied to the political machinations of an international competitor. Plausible deniability and the condition of proxies operating under broad authorization

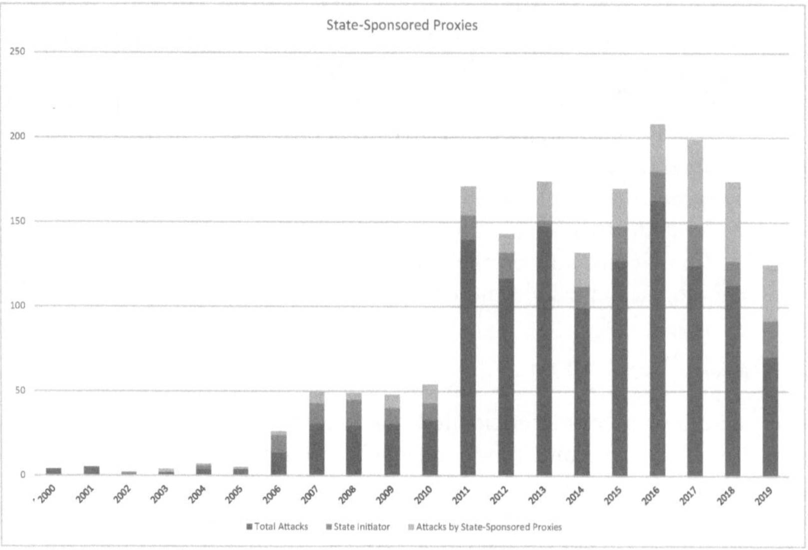

Figure 7.13 New actors weighing in

more than mission-specific permission makes sociopolitical attribution hard. Moreover, the characteristics of operating in cyberspace that make signaling difficult even between two actors that are generally aware of who they are engaging (sociopolitically speaking) are amplified when third parties are thrown into the mix. How can a country effectively launch a counterattack to draw a red line and deter further aggression when the strategic and tactic vehicles of a competitor's toolkit are not matched up? We discuss this further in later chapters, sufficed to say that the rise of proxy warfare as the norm of interstate conflict introduces numerous challenges.

Box 7.5: Formative episodes: growing reach and sophistication

Russian APTs

As the body of this chapter suggests, the 2010s might be thought of – in terms of digital security, at least – as the decade of diversifying cyber threat types, particularly the general form of information warfare and the state-sponsored use of ransomware. On the former front, one episode – actually several incidents – of Russian-sponsored campaigns to interfere with the internal politics of countries across the Western world is worth describing in some detail. Broadly put, this macro case involves several Advanced Persistent Threats (labeled "Cozy Bear," "Fancy Bear," and more by the diverse cyber threat assessment community in the West) that utilized spearphishing and sophisticated infiltration malware to attempt the theft of politically useful information from entities in the United States, Denmark, Norway,

France and the United Kingdom. Arguably most significantly, the actions of these Russian-based threat actors (potentially the same actor) has been linked with efforts to manipulate political discourse during the 2016 American election cycle.[66] It should be noted that, at the time of writing this book, investigations into the full extent of foreign cyberattacks related to the election campaigns are ongoing. Nevertheless, the broader episode is worthy of note as one of the few cyber conflict instances where the purpose of attacks was ostensibly to undergird an information warfare effort to influence the information environment of foreign politics. While much of what concerns national security administrators in the United States includes traditional propaganda efforts to influence foreign political processes, it has been clear since early 2016 that Russian efforts to hack into entities such as the Pentagon and the Democratic National Committee provided a range of actors with private information that, when released, appeared to influence the trajectory of popular discourse during the American election season. From this case, it is obviously of concern that cyberattacks enhance the ability of foreign actors to interfere in the internal political machinations of other countries. Indeed, the case of interference in the American election in 2016 is simply the paradigm case of such activity, with more than three dozen other campaigns being waged against democracies by Russia since at least 2014. These cases also highlight an important lesson about aggression and conflictual interactions in the digital age that will be discussed in more depth in the following sections – that the horizon of utility of states employing cyber techniques is measured at least partially by sociopolitical context, not simply by technical possibility.

North Korea, Sony and ransomware

The notion that sociopolitical context determines the utility of particular cyber capabilities is equally apparent in what might defensibly be called the biggest incident involving the interaction of the United States with North Korea in cyberspace (though it is worthwhile to note that North Korea has itself attempted to disrupt national services in South Korea in much the same way that Russia did in Estonia, Georgia and Ukraine in the past decade). This episode involved an obscure hacking organization and Sony Pictures Entertainment.[67] In essence, the episode was an attempt at coercion to prevent the release of a comedy embarrassing the Dear Leader of North Korea. In late November, Sony executives received an anonymous note demanding money in exchange for a commitment not to engage in cyberattacks. Several days later, cyberattacks wiped information from data storage units and infected some computers with a simple onscreen message from the group, who were called the "Guardians of Peace."

What followed was a series of events intended to gradually build pressure on Sony and coerce behavioral changes. Proprietary Sony data was leaked to various online sites and file-sharing sites were updated with company information, such as the email of Sony officials and strategic planning documents. This release lasted several weeks and resulted in some success for the attackers – Sony Pictures Entertainment said little and was forced to delay releases in order to combat the threat. The "Guardians of Peace," however, were not content to wait. Company employees received anonymous notes threatening harm if they didn't denounce Sony and threats were made regarding the movie *The Interview*. The demand was that Sony should not release the film to theaters else theaters showing it would themselves be subject to attack. Eventually, with apparently little choice, Sony announced that it would hold its release of the film and continue its own investigations.

In the weeks that followed this announcement, Sony's decision would be the basis of a broad-scoped discussion about the interaction of cybersecurity, terrorism and terrorist threats to civil society and a variety of civil liberties issues. Eventually, citing evidence from both Sony and FBI investigators,[68] President Obama confirmed that the attempt to censor the movie through coercive means was a North Korean one. Though some experts and analysts suggested, at the time, that North Korean lacked the necessary infrastructure and technical expertise to carry out such sophisticated attacks, the government line and the consensus of many private industry commentators has been that Pyongyang was the perpetrator of the attempt to compel Sony to alter its behavior. The retaliatory result was unique in Western involvement in cyber conflict episodes. For a period of ten hours, North Korean access to the Internet was entirely shut off. While the relatively primitive natural of the country's network setup makes this somewhat less impressive of an accomplishment for the presumed attacker – the U.S. government – than was the blockade of Estonia by Russia, the significance remains in that Washington clearly considers such an action to be feasibly and defensibly part of the toolkit of conflict instruments in the digital age.

And indeed, Pyongyang has clearly bought into the idea that cyberspace can be used coercively for at least some gain. Experiences with Sony aside, North Korean involvement in several major ransomware attacks through the late 2010s and into the 2020s – for instance, the WannaCry 2.0 attack in 2016 – ties the state to an interest in leveraging digital costs to either (1) rake in real revenues otherwise inaccessible due to Western sanctions or (2) degrade foreign foes by "destroying" their data.

While there are various data points that point to rising conflict trends and potentialities, so too are there developments that speak to growing constellations of cybersecurity cooperation in world politics. Perhaps the most significant development is the one referenced in the section above in this chapter, namely the tendency toward coevolution in cybersecurity policy discussions and development. As Figure 7.14 visually illustrates, a significant volume of cybersecurity calculations have taken place within the context of existing multilateral arrangements. We discuss this further in a later chapter, sufficed to say that organizations from ASEAN to the European Union and the African Union have increasingly successfully presented as vehicles via which countries have been able to start discussing and acting on a host of cybersecurity challenges. Even intergovernmental organizations whose members are primarily authoritarian nations, such as the Shanghai Cooperation Organization, have a track record through the 2010s of discussing common principles and standards for cyberspace operation on a number of fronts.

Not all cooperation is multilateral, of course. One study published in 2018 found at least 376 bilateral or trilateral compacts between states.[69] While there was substantial crossover

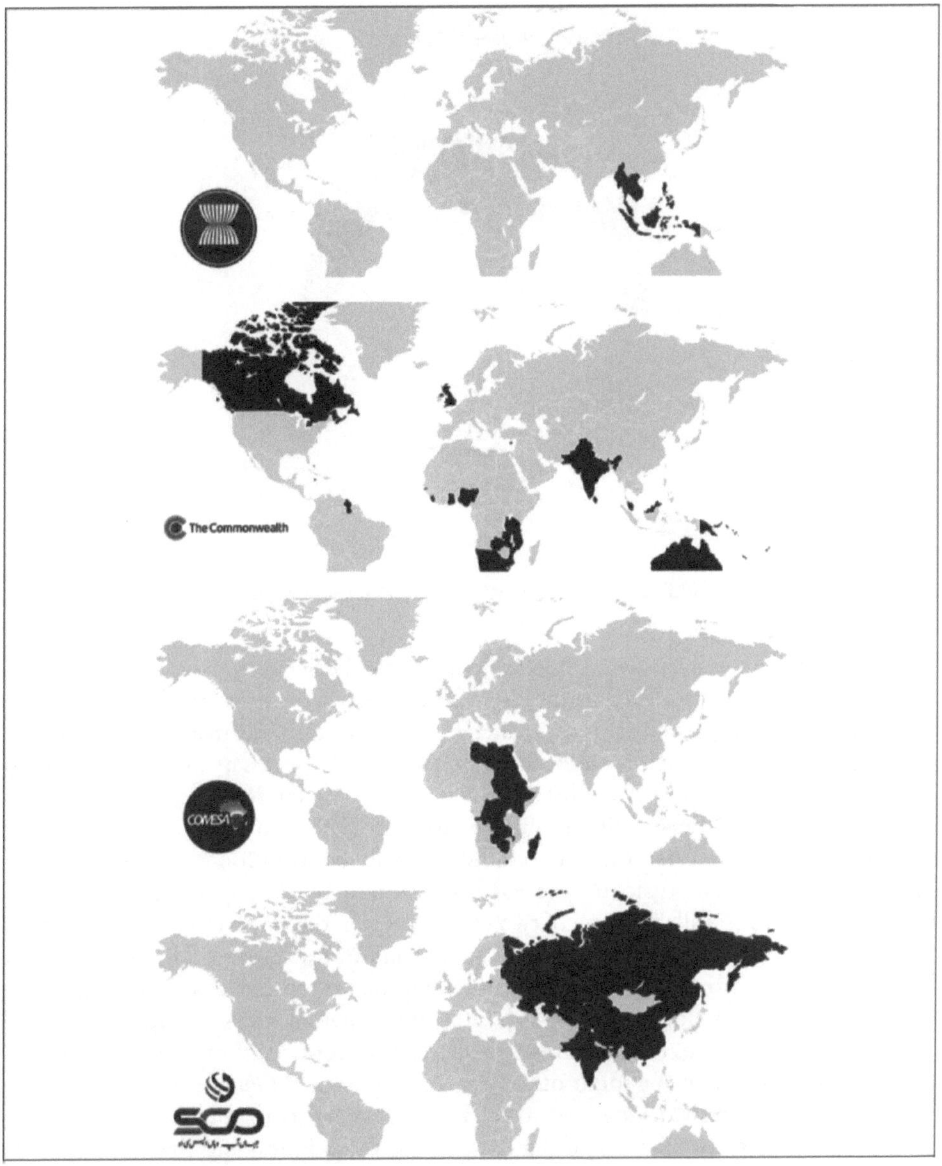

Figure 7.14 The landscape of international cyber cooperation

with the membership of multilateral organizations – the United States entering a bilateral arrangement with other members of NATO, for instance, or ASEAN members agreeing to formally cooperate on some point – there is evidence of promising ties between countries outside of these trappings. As Figure 7.15 shows,[70] by far the most common kind of such an arrangement, almost two thirds of the total, might be best described as for the purpose of dialogue and discussion.[71] These arrangements are an agreement to build toward common language and process for future cooperation on either a specific issue or

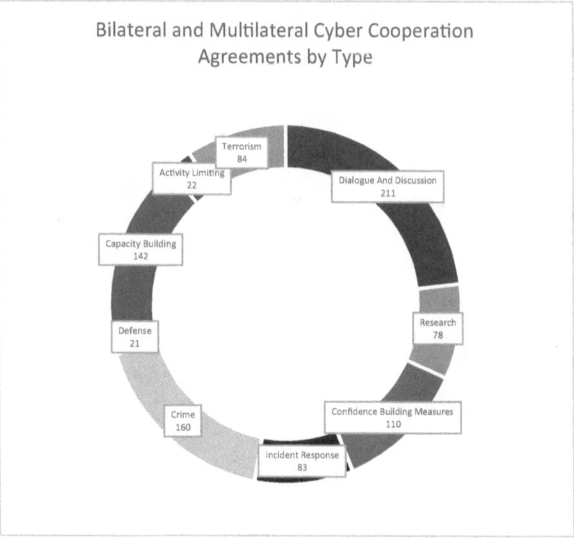

Figure 7.15 Bilateral and multilateral cyber cooperation agreements by type

a cross section of different issues. Often these include cybersecurity issues alongside more terrestrial policy concerns, such as the use of the Internet for terrorist financing as part of a broader focus on particular terrorist threats. These agreements may appear limited to the casual observer, but such arrangements are particularly necessary with novel areas of policymaking like cybersecurity where the interlocutors often do not share basic terminology, concepts or understanding of process from the start. Thus, these basic discussion and dialogue setups are often the key to more elaborate cooperation.

Beyond discussion and dialogue, the most common focus of cooperative arrangements directly between states falls into one of two categories. On the one hand, states are increasingly willing to cooperate on issues of cybercrime and terrorist uses of the Internet. This category of cooperation is arguably quite unique in that it represents an issue area where there are numerous instances of democratic and authoritarian states opting to cooperate. On the other hand, states – albeit those already closely connected by intergovernmental organization membership or some other pre-existing arrangements – have increasingly since about 2010 been working toward formal arrangements for building mutual confidence in the integrity of national digital spaces and for producing the capacity to be productive partners in cyberspace. Generally, this means the construction of capabilities for information sharing, emergency response, effective multi-level communication on developing cybersecurity issues and other similar activities.

Unfortunately, there yet remains very limited evidence of cooperation on issues of mutual national cyber defense or, particularly, on mutually constraining measures for limiting unnecessary cyber conflict behavior. Indeed, this final point underlines a key takeaway for most researchers and policy analysts on the condition of international cooperation on cybersecurity issues, namely that there is some substantial activity but that the most meaningful forms of coordination – i.e. on issues of mutual non-aggression, verification and constraint – remain sorely lacking.

Next steps

This introduction to major trends in and episodes of interstate cyber conflict was designed to be brief and is naturally incomplete. Nevertheless, these illustrations and episode descriptions demonstrate the scale and enormity of the threat faced by states in cyber interactions with competitors. As mentioned earlier, the next chapters now move to discuss the technical vulnerabilities and organizational challenges involved in safeguarding and building cybersecurity capabilities before shifting gears to consider empirical trends in cyber conflict – and the generalizations we might make from such data – in the context of strategic theory on the phenomenon.

Notes

1 Valeriano, Brandon, and Ryan C. Maness. *Cyber war Versus Cyber Realities: Cyber Conflict in the International System*. Oxford University Press, USA, 2015.

2 Valeriano, Jensen and Maness incorporate disruption, degradation and espionage (information exfiltration and monitoring via cyber means) into their study of different forms of cyber actions. They settle on these three categories from a broad review of other scholars' treatment of cyber warfare and cover actions. See Valeriano, Brandon, Benjamin Jensen, and Ryan C. Maness. *Cyber Strategy: The Evolving Character of Power and Coercion*. Oxford University Press, 2018.

3 A phrase popularized following the 2007 cyberattack on Estonia and discussed centrally in Russell, Alison Lawlor. *Cyber Blockades*. Georgetown University Press, 2014.

4 We discuss this format of attack later in describing Operation Orchard, a kinetic assault on Syrian nuclear infrastructure enabled principally via a cyber action. More broadly, however, this categorization of cyber actions fits broadly with Gartzke's now-well-known refrain that cyber conflict actions are best thought of adjunct modifiers of traditional warfare. See Gartzke, Erik. "The myth of cyberwar: Bringing war in cyberspace back down to earth." *International Security* 38, no. 2 (2013): 41–73.

5 First introduced in Valeriano, Brandon, and Ryan C. Maness. "The dynamics of cyber conflict between rival antagonists, 2001–11." *Journal of Peace Research* 51, no. 3 (2014): 347–360. The most recent version can be found at https://drryanmaness.wixsite.com/cyberconflict/cyber-conflict-dataset.

6 For a full exploration of this dataset, see Valeriano, Brandon, and Ryan C. Maness. *Cyber War Versus Cyber Realities: Cyber Conflict in the International System*. Oxford University Press, USA, 2015.

7 For more in-depth discussions of Iranian and North Korean cyber capabilities, see Jun, Jenny, Scott LaFoy, and Ethan Sohn. *North Korea's Cyber Operations: Strategy and Responses*. Rowman & Littlefield, 2016; and Baezner, Marie. *Iranian Cyber-Activities in the Context of Regional Rivalries and International Tensions*. ETH Zurich, 2019.

8 This data is a reconstruction of the trends reported in Blessing, Jason. "The global spread of cyber forces, 2000–2018." In *2021 13th International Conference on Cyber Conflict (CyCon)*. IEEE, 2021, pp. 233–255.

9 Ibid., 233.

10 An idea discussed in Healey, Jason, ed. *A Fierce Domain: Conflict in Cyberspace, 1986 to 2012*. Cyber Conflict Studies Association, 2013.

11 Choucri, Nazli, and David D. Clark. "International relations in the cyber age: the co-evolution dilemma." *Information Policy*, 2018. Chapter 2.

12 Miller, Bill, and Dale Rowe. "A survey SCADA of and critical infrastructure incidents." In *Proceedings of the 1st Annual Conference on Research in Information Technology*. ACM, 2012, pp. 51–56.

13 Brown, Gary, and Keira Poellet. "The customary international law of cyberspace." *Strategic Studies Quarterly* (Fall 2012): 130.

14 Weiss, Gus W. "The farewell dossier: Duping the soviets," *The Central Intelligence Agency*, June 27, 2008. <www.cia.gov/library/center-for-the-study-of-intelligence/csi-publications/csi-studies/studies/96unclass/farewell.htm>.

15 Ibid.

16 For disputes over the veracity of the reports regarding the attack, see Carr, Jeffrey. "The myth of the CIA and the trans-Siberian pipeline explosion." June 7, 2012, <http://jeffreycarr.blogspot.com/2012/06/myth-of-cia-and-trans-siberian-pipeline.html>; for information on the alleged

effects of the attack, see Brown, Gary, and Poellet, Keira, "The customary international law of cyberspace." *Strategic Studies Quarterly* (Fall 2012): 130.

17 Melito, Steve. "Cyber war and the Siberian pipeline explosion." *CBRN Resource Network*, November 2, 2013. <http://news.cbrnresourcenetwork.com/newsDetail.cfm?id=109>.

18 See Stoll, Clifford. "The Cuckoo's Egg: tracking a spy through the maze of computer espionage." (1989); and Stoll, Clifford. *The Cuckoo's Egg: Tracking a Spy Through the Maze of Computer Espionage.* Simon and Schuster, 2005.

19 For perhaps the fullest account of Cuckoo's Egg beyond Stoll's own in the context of cyber conflict, see Healey, Jason, and Karl Grindal, eds. *A Fierce Domain: Conflict in Cyberspace, 1986 to 2012.* Washington, DC: Cyber Conflict Studies Association, 2013.

20 For early discussions of the case, see Adams, James. "Virtual defense." *Foreign Affairs* (2001): 98–112; Terry, James P. "The lawfulness of attacking computer networks in armed conflict and in self-defense in periods short of armed conflict: What are the targeting constraints." *Military Law Review.* 169 (2001): 70; and Brooks, Richard R. *Disruptive Security Technologies with Mobile Code and Peer-To-Peer Networks.* CRC Press, 2004.

21 For perhaps the most extensive breakdown of the episode in the public record, see Rid, Thomas. *Rise of the Machines: A Cybernetic History.* WW Norton & Company, 2016.

22 Healey and Grindal, eds. *A Fierce Domain: Conflict in Cyberspace, 1986 to 2012.* 2013.

23 See Thornburgh, Nathan. "Inside the Chinese hack attack." *Times,* August 25 (2005); and Lewis, James A. "Computer espionage, Titan Rain and China." *Center for Strategic and International Studies-Technology and Public Policy Program* (2005): 1.

24 See Alperovitch, Dmitri. *Revealed: Operation shady RAT.* Vol. 3. McAfee, 2011; Li, Frankie, Anthony Lai, and Ddl Ddl. "Evidence of advanced persistent threat: A case study of malware for political espionage." In *Malicious and Unwanted Software (MALWARE), 2011 6th International Conference on.* IEEE, 2011, pp. 102–109; and Gross, Michael Joseph. "Exclusive: Operation shady RAT – Unprecedented cyber-espionage campaign and intellectual-property bonanza." *Vanity Fair* 2 (2011).

25 See Healey and Grindal (2013) for further details. Additionally, see Lynn, William J. "Defending a new domain: The Pentagon's cyberstrategy." *Foreign Affairs* 89, no. 5 (2010): 97–108; and Farwell, James P., and Rafal Rohozinski. "The new reality of cyber war." *Survival* 54, no. 4 (2012): 107–120.

26 For perhaps the most complete overview of the discovery and analysis of Stuxnet in both strategic and technical context, see Lindsay, Jon R. "Stuxnet and the limits of cyber warfare." *Security Studies* 22, no. 3 (2013): 365–404. Also see Aleksandr Matrosov, Eugene Rodionov, David Harley, and Juraj Malcho, "Stuxnet under the microscope." eset, *White Paper,* January 20, 2011; and Zetter, Kim. "How digital detectives deciphered Stuxnet, the most menacing malware in history." *Wired Threat Level Blog,* July 11, 2011, www.wired.com/threatlevel/2011/07/how-digital-detectives-deciphered-stuxnet.

27 See Sanger, David E., "Obama order sped up wave of cyberattacks against Iran." *New York Times,* June 1, 2012; and Broad, William J., John Markoff, and David E. Sanger. "Israel tests on worm called crucial in Iran nuclear delay." *New York Times,* January 15, 2011.

28 Albright, David, Paul Brannan, and Christina Walrond, "Did stuxnet take out 1,000 centrifuges at the Natanz enrichment plant?." *Institute for Science and International Security,* December 22, 2010, 3–4. See also Langner, Ralph. "To kill a centrifuge: A technical analysis of what Stuxnet's creators tried to achieve." *The Langner Group,* November 2013

29 Sanger, "Obama Order . . .;" Albright David, Paul Brannan, Andrea Stricker, Christina Walrond, and Houston Wood, "Preventing Iran from getting nuclear weapons: Constraining its future nuclear options." *Institute for Science and International Security,* March 5, 2012.

30 Sanger, "Obama Order . . ."

31 Valeriano, Brandon, Benjamin M. Jensen, and Ryan C. Maness. *Cyber Strategy: The Evolving Character of Power and Coercion.* Oxford University Press, 2018.

32 Ibid., 108–109.

33 This being the oft-cited argument of an early work by Thomas Rid, which we unpack in the next chapter. See Rid, Thomas. "Cyber war will not take place." *Journal of Strategic Studies* 35, no. 1 (2012): 5–32.

34 Foote, Colin, Ryan C. Maness, Benjamin Jensen, and Brandon Valeriano. "Cyber conflict at the intersection of information operations: Cyber-enabled information operations, 2000–2016." In *Information Warfare in the Age of Cyber Conflict.* Routledge, 2020, pp. 54–69.

35 Way, Lucan Ahmad, and Adam E. Casey. "How can we know if Russia is a threat to western democracy? Understanding the impact of Russia's second wave of election interference." In *Conference Paper. Retrieved April,* Vol. 28, 2019, p. 2020.

36 Perlroth, Nicole. "In cyberattack on Saudi firm, U.S. Sees Iran firing back." *The New York Times*, October 23, 2012. www.nytimes.com/2012/10/24/business/global/cyberattack-on-saudi-oil-firm-disquiets-us.html.

37 Bronk, Christopher, and Tikk-Ringas, Eneken. "The cyber attack on Saudi Aramco." *Survival: Global Politics and Strategy* 55 (April – May 2013): 81–96.

38 Mahdi, Wael. "Saudi Arabia Says Aramco cyberattack came from foreign states." *Bloomberg News*, December 9, 2012. www.bloomberg.com/news/2012-12-09/saudi-arabia-says-aramco-cyberattack-came-from-foreign-states.html.

39 Ibid.

40 Lewis, James. "Significant cyber events since 2006." *Center for Strategic and International Studies*, July 11, 2013. http://csis.org/publication/cyber-events-2006. 12.

41 Bronk, Christopher, and Eneken Tikk-Ringas. "The cyber attack on Saudi Aramco." *Survival: Global Politics and Strategy* 55 (April – May 2013): 81–96.

42 Perlroth, Nicole. "In cyberattack on Saudi Firm, U.S. Sees Iran firing back." *The New York Times*, October 23, 2012. www.nytimes.com/2012/10/24/business/global/cyberattack-on-saudi-oil-firm-disquiets-us.html.

43 Lewis, James. "Significant cyber events since 2006." *Center for Strategic and International Studies*, July 11, 2013. http://csis.org/publication/cyber-events-2006., 12–13.

44 Schwartz, Mathew J. "Bank attackers restart operation Ababil DDoS disruptions." *InformationWeek Security*, March 6, 2013. www.informationweek.com/security/attacks/bank-attackers-restart-operation-ababil/240150175.

45 Lewis, James. "Significant cyber events since 2006." *Center for Strategic and International Studies*, July 11, 2013. http://csis.org/publication/cyber-events-2006. 12–13.

46 Nakashima, Ellen. "Iran blamed for cyberattacks on U.S. banks and companies." *The Washington Post*, September 21, 2012. http://articles.washingtonpost.com/2012-09-21/world/35497878_1_web-sites-quds-force-cyberattacks.

47 Kaplan, Caren. "Air power's visual legacy: Operation Orchard and aerial reconnaissance imagery as ruses de guerre." *Critical Military Studies* 1, no. 1 (2015): 61–78.

48 A number of scholarly studies have rested on this assumption, for instance, Chesti, Ikra Afzal, Mamoona Humayun, Najm Us Sama, and N. Z. Jhanjhi. "Evolution, mitigation, and prevention of ransomware." In *2020 2nd International Conference on Computer and Information Sciences (ICCIS)*. IEEE, 2020, pp. 1–6; and Ryan, Matthew. *Ransomware Revolution: The Rise of a Prodigious Cyber Threat*. Springer, 2021.

49 Mohurle, Savita, and Manisha Patil. "A brief study of WannaCry threat: Ransomware attack 2017." *International Journal of Advanced Research in Computer Science* 8, no. 5 (2017): 1938–1940.

50 Greenberg, Andy. "The untold story of NotPetya, the most devastating cyberattack in history." *Wired*, August 22, 2018.

51 Ibid.

52 Mitchell, Joshua. "Ransomware characteristics by country." 2018.

53 Meland, Per Håkon, Yara Fareed Fahmy Bayoumy, and Guttorm Sindre. "The Ransomware-as-a-service economy within the darknet." *Computers & Security* 92 (2020): 101762.

54 Valeriano et al. (2018).

55 Ibid., 106–109.

56 Ibid., 108.

57 Ibid.

58 Greenemeier, Larry. "Estonian 'Cyber Riot' was planned, but mastermind still a mystery." *Information Week*, August 3, 2007. www.informationweek.com/estonian-cyber-riot-was-planned-but-mast/201202784.

59 This incident is perhaps described best in context in Russell, *Cyber Blockades*, 2013.

60 North Atlantic Treaty Organization Cooperative Cyber Defence Center of Excellence website. www.ccdcoe.org/. Accessed January 14, 2014.

61 Markoff, John. "Before the gunfire, cyberattacks." *The New York Times*, August 13, 2008. www.nytimes.com/2008/08/13/technology/13cyber.html?_r=0.

62 Ibid.; Hollis, David. "Cyberwar case study: Georgia 2008." *Small Wars Journal*, January 6, 2011. http://smallwarsjournal.com/jrnl/art/cyberwar-case-study-georgia-2008.

63 Keizer, Gregg. "Georgian cyberattacks suggest Russian involvement," *Computer World*, October 17, 2008. www.computerworld.com/s/article/9117439/Georgian_cyberattacks_suggest_Russian_involvement_say_researchers.

64 Case, Defense Use. "Analysis of the cyber attack on the Ukrainian power grid." *Electricity Information Sharing and Analysis Center (E-ISAC),* 2016.

65 This figure utilizes a different data resource than previous illustrations, namely a more recent cyber conflict incident dataset that does not code dyadically, and so aims to capture more of the secondary noise of global cyber conflict. In doing so, the dataset provides some unique additional insights into trends of diversification in global cyber conflict. See Harnisch, Sebastian; Zettl, Kerstin; Steiger, Stefan, 2021, "Heidelberg Cyber Conflict Dataset (HD-CY.CON)." https://doi.org/10.11588/data/KDSFRB, heiDATA, V1.

66 A broad range of sources offer preliminary commentary on this and related episodes, including Buratowski, Michael. "The DNC server breach: Who did it and what does it mean?." *Network Security* 2016, no. 10 (2016): 5–7; Connell, Michael, and Sarah Vogler. *Russia's Approach to Cyber Warfare.* Center for Naval Analyses Arlington United States, 2017; Ring, Tim. "The Russians are coming! Are security firms over-hyping the hacker threat?." *Network Security,* 2017, no. 3 (2017): 15–18; Pope, Amy E. "Cyber-securing our elections." *Journal of Cyber Policy* 3, no. 1 (2018): 24–38.

67 For a full summary of the timeline of the Sony episode, see Risk Based Security, "A Breakdown and Analysis of the December 2014 Sony Hack" at www.riskbasedsecurity.com/2014/12/a-breakdown-and-analysis-of -the-december-2014-sony-hack/.

68 FBI National Press Office. "Update on Sony Investigation." *FBI,* December 19, 2014. at www.fbi.gov/news/pressrel/press-releases/update-on-sony-investigation.

69 Ashbaugh, Lucas. *An Analysis of International Agreements Over Cybersecurity.* The University of Maine, 2018.

70 This figure reconstructed from the earlier analysis.

71 It is important to note in Figure 7.15 that the "Dialogue and Discussion" code is used alongside other codes, such as both "Dialogue and Discussion" and "Terrorism," to denote an initial agreement for discussion about issues of cyberspace and terror threats.

Further readings

Buchanan, Ben. *The Hacker and the State: Cyber Attacks and the New Normal of Geopolitics.* Harvard University Press, 2020.

Center, Mandiant Intelligence. "APT1: Exposing one of China's cyber espionage units." *Mandiant.com* (2013).

Greenberg, Andy. *Sandworm: A New Era of Cyberwar and the Hunt for the Kremlin's Most Dangerous Hackers.* Anchor, 2019.

Healey, Jason, and Karl Grindal, eds. *A Fierce Domain: Conflict in Cyberspace, 1986 to 2012.* Washington, DC: Cyber Conflict Studies Association, 2013.

Kaplan, Fred. *Dark Territory: The Secret History of Cyber War.* Simon and Schuster, 2016.

Langner, Ralph. "Stuxnet: Dissecting a cyberwarfare weapon." *IEEE Security & Privacy* 9, no. 3 (2011): 49–51.

Lewis, James A. "Computer espionage, Titan Rain and China." *Center for Strategic and International Studies-Technology and Public Policy Program* (2005): 1.

Rid, Thomas. "Cyber war will not take place." *Journal of Strategic Studies* 35, no. 1 (2012): 5–32.

Rid, Thomas. *Rise of the Machines: A Cybernetic History.* WW Norton & Company, 2016.

Valeriano, Brandon, Benjamin M. Jensen, and Ryan C. Maness. *Cyber strategy: The Evolving Character of Power and Coercion.* Oxford University Press, 2018.

8 National experiences with cybersecurity

Realization and institutional development

Over the past three decades, countries around the world have experienced the effects of the information revolution in radically different ways. National responses to cyber threats and to transformations of both economy and infrastructure brought about by the information revolution have been uniquely affected by different developmental episodes over time. In the United States, periods of relative calm punctuated by unique cyber threat crises have driven cyber policy development in an incremental fashion. Often, whole-of-government solutions to cybersecurity issues became the focus of the federal government only after intra-government agencies and service branches had configured themselves to deal with cyber threats in line with parochial interests. Elsewhere, where the experience of policy development has been only minimally affected by national crises of different kinds, whole-of-government approaches have resulted in alternative conceptualizations of policy and doctrine.

As a whole, however, it is probably most reasonable to say that most countries around the world are still caught up in the throes of responding to the various challenges that the digitization of national infrastructure pose and scrambling to develop appropriate institutional, legal and regulatory responses for the broad range of issues brought about by the crosscutting intrusion of information technologies. As such, no chapter can outline the shape of national cyber developments in a way that allows for nuanced generalization or even commentary on the broad range of national circumstances that countries face with cybersecurity. However, it is possible to outline the shape of what has appeared to be common trajectories of development with regard to cybersecurity policy and, specifically, national responses to the potential for different forms of cyber conflict. This chapter does just that, first by outlining the broad shape of cyber policy efforts and then by describing the national experiences of countries often at the heart of discourse on international security and the digital domain.

The shape of cybersecurity policymaking

What does it mean to develop policy and doctrine regarding security threats in cyberspace? This question has no easy answer. In many ways, an understanding of what national cybersecurity policy and approach looks like has to come from a deconstruction of the nature of cyber policy and a subsequent attempt to reconcile said policy with traditional government approaches to security issues. This section briefly attempts to do just that in order to give context to subsequent sections' narrative outline of different country's experiences with cybersecurity policymaking.

Before asking what cyber policy is, it is perhaps more prudent to try and imagine who and what cyber policy affects. Network technologies are crosscutting insofar as the

DOI: 10.4324/9781003246398-8

information revolution has radically transformed most societal functions and the Internet, broadly construed, functions as the interconnecting medium for logistical communications across any given nation's industrial sectors. Thus, policy, doctrine and practice across relatively broad segments of government functions and civil society might be said to qualify as related to cybersecurity efforts.

We might think of cyber policy as potential impact across five categories.[1] First, and foremost, it goes almost without question that cyber policy constitutes any regulatory effort to shape the governing mechanisms of cyberspace. This might refer to either the technical processes or the institutions with authority over those processes that determine the actual shape of the web. This category of things is what some have called the logical layer of cyberspace insofar as it includes the informational infrastructure for assigning domain names (the Domain Name System and the entities that are responsible for assigning IP addresses and registering URLs), the routing procedures for packets of information and the systems that attempt to ensure the integrity of those packets. We might think of this category of cyber policy as **functional** policy.

Box 8.1: The point of policy

What is policy? Policies are an enumerated set of principles and directions that are put in place to guide decision-making. Contrary to what some may believe, policymaking is not the sole domain of government. Though we focus largely on government policymaking in this chapter, the truth of the matter is that any organization can and does issue policy designed to guide the actions of those it is responsible for.

It is often best to think of policy as the generic term that we might use to categorize all constraining frameworks, from laws and constitutional documents to the guidelines of, say, a private company's human resources division. In this way, we naturally are led to recognize that policy environments are actually multilayered and hierarchical (i.e. policies at one level must reference, even if tacitly, the framework of policy at higher levels). In all countries, there is a highest level of policy formulation, a source of policymaking that frames what policies might be constructed by entities that fall under that source's jurisdiction. In the United States, that fundamental source is the Constitution, a set of articles and amendments that define the role of the government in relation to the people and states of the nation. In other countries, other documents provide this overarching framework. In authoritarian states, such documents often take the form of religious scripture or the writings of key leadership.

Policy naturally leads to the practice of governance, regardless of whether the policy involved is all-encompassing or extremely limited in scope. Specifically, policy leads to the formation of enforcement entities or to changes in the way such entities behave. The Constitution of the United States outlines the responsibilities of government and the broad guidelines within which the rule of law must function. In order that the document has meaning, it established the various wings of the federal government, which in turn established sub-organizations to undertake tasks implied but not directly listed in high-level policy (again, meaning the Constitution).

Unlike many other security areas, cyber policy is arguably most often formulated by non-government actors. The exception, and the main focus of this chapter's description of various national experiences, is with policy focused on the cybersecurity of states' national security apparatuses – i.e. of military systems and core national assets. Nevertheless, it is worth remembering up front that much of what constitutes cyber policy is the purview of private firms, non-profit governing forums and, often, ad hoc groups of technology professionals. Governance (functional) policy is, outside of authoritarian states at least, largely the domain of companies that own the logical infrastructure of Internet data routing and groups like the Internet Architecture Board that have a say in standards development. Likewise, industrial policy – or what is often labeled "enterprise management" – is the domain of technology operators who must construct decision-making regimes and best practices from an assessment of legal requirements, ethical responsibilities and business interests. This last point is worth bearing in mind below where we discuss American efforts to protect critical infrastructure – most of which is not government owned – from major cyber threats.

Second, cyber policy is naturally constituted of those regulations and protections that affect how individual and incorporated users (i.e. both citizens and organizations, from interest groups to private companies) use the web. From protection of the identity of Internet users to laws designed to ensure full compliance with civil rights to privacy and speech, user-oriented cyber policy is a broad, catch-all phrase that denotes a complex ecosystem of approaches to civil society shared by law enforcement, national executives, the intelligence community and more. In other words, cyber policy is any regulation or law that seeks to protect the rights of consumers and citizens. We might this of this category of cyber policy as **constitutional** or **user** policy.

Third, the development of doctrine regarding conflict in cyberspace and the policies that underwrite national efforts to cope with new threats in the digital age might be referred to as cyber policy. Cyber policy in this setting is nothing less than the diplomacy and strategic approach of states to international interactions. It includes the practices of intelligence communities in conducting espionage operations, military efforts to safeguard government networks, efforts to combat irregular cyber threats beyond the domestic context and the relationship between governments and non-state proxies. Here, cyber policy is doctrine and practice formed in response to pressing digital threats to the function of a state's national security apparatus. From ensuring an ability to employ cyber weapons effectively in conflict to enabling the defense of government networks, cyber policy might quite clearly be considered synonymous with the traditional functions of military and civilian arms of government devoted to addressing warfare. We might think of this category of cyber policy, even though it also covers regulation affecting quasi-military elements of intelligence and police communities, as **military** policy.

Fourth, cyber policy might be said to be any governmental approach or explicit legal action that impacts upon the management (different from the *function*) of industry and society online. Here, cyber policy affects the information infrastructure of web usage beyond the logical layer. Instead of focusing on the infrastructural function of the web (e.g. domain name assignment processes, packet security assurance, etc.), policy here is concerned with the actual security of network systems and platforms. In other words, cyber policy is the regulation and practice of enterprise procedures regarding information security. Cyber policy in this vein is about risk management, procedure and complicity with rules of production among industrial users of the web. We might think of this category of cyber policy as **law enforcement** policy.

And finally, since information technology is critical to function of all sectors of national critical infrastructure (CI), cyber policy might be any regulation that pertains to the function of different infrastructural sectors, from agriculture to energy, as well as the function of the national economy as a whole. This involves policies that, for instance, pertain to the requirements for industrial control systems or the procedures for encrypting access to medical data just as it does to efforts to safeguard intellectual property (IP) across national industrial sectors. Specifically, cyber policy is the business of regulating telecommunications infrastructure. This is quite simply because telecommunications infrastructure is first and foremost the information technology basis for broad-scoped inter-connectedness that underwrites the functions of all other critical infrastructural sectors. We might think of this category of cyber policy as **industrial** policy.

Naturally, these different areas of focus and effect are in no way indicative of a common role for cyber policy in governing the world around us. Indeed, policy implications and recommendations that emerge from scholarship on issues within any of these categories will inevitably differ in shape and the categories described earlier simply proxy for more common policy issue area labels, from enterprise management to technology configuration. Scholars of international relations interested in cyber conflict issues might be most interested, for instance, in patterns of usage of cyber "weapons" employed during interstate crises or in the shape of espionage operations conducted via the web by foreign powers. By contrast, legal scholars interested in the national security implications of web technology developments might focus their efforts on the right of private firms to deny law enforcement oversight of user encryption (arguably desirable for counterterrorism operations) in new products. Naturally, both research programs speak to governance institutions and mechanisms that play radically different goals in bolstering national security efforts (military doctrine vs. constitutional law).

Thus, though cyber policy is in the broadest sense intended to provide the means by which we might effectively govern the cyber domain, it cannot do so in a concise manner. Effective cyber policy ideally means a set of approaches that make for efficient governance of the logical functionality of cyberspace, facilitates user security online, anticipates cyber conflict issues, steers the management of IT activities toward productive ends and drives infrastructural development. And, to be fair, cyber policy in any of these domains is intended to aid in effective regulation of the others. Cyber governance policy supports robust user protectionism, which engenders greater or lesser propensity of conflict in the international system. Propensity for conflict then determines the degree to which management processes – and, by extension, infrastructural functions – are prone to disruption. But cyber policy is inherently a complex and highly diffuse concept insofar as it must inevitably be constituted of a large ecosystem of regulations and practices that reflect the crosscutting nature of information technologies.

This book is focused on the contours of cyber conflict. This chapter is an effort to outline the experiences different states have had in developing cyber policy regimes. As such, both here and in other chapters, we focus closely on issues linked with the third domain of policy impact outlined earlier – that of military policy. However, it should be clear from the start that policy relevant to national security can manifest in any of the areas outlined earlier. Indeed, this basic fact highlights an integral truth of cyber policy development, namely that efforts to "secure cyberspace" for by national security establishments have enduringly involved policy creep into non-traditional areas of government regulation. This has, at times, stymied the abilities of states to effectively adapt to the implications of the information revolution. It has, in other instances, served as the catalyst for broad-scoped changes to the infrastructure of government regulation. And it enduringly serves as a barrier to effective international cooperation on cybersecurity issues, as the complexity of regulatory efforts across countries has failed to interface well with the nuanced design of the contemporary international community.

Realization, fragmentation and militarization

The long and short of the earlier section's outline of the shape of "cyber policy" is that the efforts of national security establishments to confront a broad range of cyber threats are invariably prone to fragmentation and policy creep. From product design to enterprise management, cybersecurity issues that relate to national security imperatives manifest across a large number of governance areas *not* traditionally linked to such processes. Thus, as a whole, cybersecurity policy on conflict and state security cannot help but be characterized by growing pains associated with governments' relatively sudden need to regulate the web.

But what dictates the experience a given country will have in developing the institutions, policies and relationships to address cybersecurity problems? Naturally, the answer is different across cases. Generically, however, Jason Healey – the Director for Cyber Infrastructure Protection at the White House under George W. Bush – and his colleagues suggest an interesting framework for understanding national experiences with cybersecurity challenges.[2] In an ideal world, strategic planners and policymakers might undertake broad exploratory work to outline the shape of potential challenges that might face a country, both contemporarily and out into the future. From that assessment, national institutions and policymakers can set about adapting existing procedures such that effective, streamlined responses to different potential crises are possible.

In reality, however, the impact of the information revolution on different countries has rarely been limited or felt equally across the various domains outlined in the section above. Healey outlines one potential trajectory for national cyber policy development in discussing the case of the United States.[3] Experiences with cyberterrorism, espionage or consumer insecurity, for instance, might dictate the development of responsive policy and practices only among a country's law enforcement or domestic intelligence institutions. Over time, many "realization" episodes spread out among different arms of the state and of civil society produce a fragmented policy ecosystem wherein interactions between institutions of governance across different domains emerge from procedures designed reactively (i.e. without the benefit of ideal-world strategic planning). Of course, cohesive national cyber policy can be constructed from this fragmentation. But such a fragmented antecedent to efforts to coordinate in a holistic fashion can introduce risks to the subsequent cybersecurity regime. Foremost among these, a fragmented national cyber policy environment virtually ensures that some stakeholders are more interested and better equipped than others to shoulder authoritative roles in a whole-of-government/society regime. As the Department of Defense has played such a role in the United States Healey labels this final phase that of "militarization," where the defense establishment arguably dominates the emerging policy regime.[4]

Again, this general trajectory emerges from Healey's discussion of the United States' developmental experiences and is not a representation of what all countries experience.[5]

But this story *does* tell us much about what factors dictate the experience a given country will have in developing the institutions, policies and relationships to address cybersecurity problems. Specifically, a country's eventual policy approach to cybersecurity challenges seem to come from the interaction of two variables: (1) the diversity and magnitude of different incipient cyber threat crises and (2) the shape of existing mechanisms of security governance. How a country's institutions respond to challenges and codify the ability to act in the future largely obviously comes from the shape of the problem. But it also comes from established procedures and the bureaucratic perspective of the stakeholders affected. In an attempt to justify their relevance and maximize access to resources, government sub-agencies may seek to develop new capabilities in-house instead of collaborating with existing capacity elsewhere. Likewise, the context of a given crisis at a time when national leaders are developing holistic approaches might skew developments such that one set of stakeholders are seen as more relevant than others. A country facing legal battles over industry unwillingness to unlock a terrorist's smartphone (for fear of giving the government unconstitutional power for search and seizure) might be far more likely to let a justice ministry lead development of initial language for national cybersecurity standards than might one reeling from state-based attacks against military networks. In either case, responses to realization episodes undoubtedly have implications for the shape of subsequent public–private relationships, international diplomacy and more.

Naturally, there is no deterministic framework being described here. And yet, more so than has been the case with any other developing arena of national security issues – such as nuclear weapons development, for instance – it seems clear that the shape of different crises drives how countries develop national portfolios for cybersecurity policy beyond what we might expect to emerge from a given state's pre-existing policy development bureaucracy. The next sections pick up this flexible narrative broadly in describing the shape of national experiences in the United States and several international counterparts that are of particular relevance to the United States – partners like the United Kingdom and the intergovernmental North Atlantic Treaty Organization (NATO), and adversaries like Russia, Iran and China.

America's cyber experience

The history of the United States' effort to confront cyber threats to national security is, as noted earlier, one of fragmentation and stuttering coordination between stakeholders both inside and outside government. Today, the approach of the national security establishment toward cybersecurity is largely captured by efforts in two veins. Emerging cyber strategy articulated by the Department of Defense (DoD) eschews the long-standing position held by stakeholders in the U.S. government that cyberspace is a domain to be "dominated."[6] Rather, the most recent DoD strategy statements emphasize a limited portfolio of objectives for the military and corresponding support agencies, including the defense of DoD networks, the need to counter "meaningful" cyber threats to the country and the need to support conventional military operations.[7] In doing so, the military claims only a limited role in overseeing and protected non-traditional elements of the national security apparatus, such as critical infrastructure. Cybersecurity policy and governance efforts that *do* address non-military functions are bound up in several executive orders and the National Institute for Standards and Technology (NIST) cybersecurity framework.[8] These outline intended government responses to large-scale cyber incidents, suggest standards for public–private interactions and address affected entities (like individual citizens). Of particular note, actual lawmaking on cyber issues remains minimal at the time of writing this book. To some degree, this is a function of the fact that American approaches to cyber challenges are extensive but, at least insofar as early experiences produced a fragmented coordinative environment, not yet truly comprehensive.

Early experiences

Described previously in more detail in Chapter 7, incidents like the Cuckoo's Egg, Morris Worm and SOLAR SUNRISE forced the United States over the course of a decade to develop institutions and procedures for dealing with cyber threats that crosscut traditional areas of jurisdiction. In the Cuckoo's Egg in 1986, where German hackers intruded on computer systems at a range of government institutions looking for information about Star Wars (the Strategic Defense Initiative), law enforcement and intelligence agencies were consistently at odds about resource allocation and jurisdiction in the investigation of possible break-ins. Clifford Stoll, whose name is now famous in the cybersecurity field, was forced to conduct his own investigation of the break-ins with the aid of only a few sympathetic Air Force personnel.[9] Only after enough evidence was uncovered were government agencies moved to action. In particular, intelligence agencies were uninterested in the case, encouraging Stoll and his partners to think of the attacks as larceny despite the fact that highly sensitive national security information was in danger of compromise. Only after investigators liaised with partners in Europe to track the hackers' connection did involvement go beyond federal law enforcement to involve the military and intelligence community. Even then, as Stoll's own account notes, disagreements between the National Security Agency (NSA) and the Office of the Secretary of Defense (OSD) about resource allocation prevented a military presence at the hackers' trial, an outcome that possibly resulted in a significantly reduced (and less symbolic) verdict.[10]

With the Morris Worm's takedown of almost a tenth of the computers connected to the Internet in 1988 (then the ARPANET), the United States government faced a far more insidious problem than had been the case with Cuckoo's Egg. The worm demonstrated how quickly cyberattacks could take out large sections of national information technology functionality.[11] In this case, however, rapid response and success in quickly mitigating the threat came down to two factors not related to the abilities of government institutions. Large-scale assaults on cyber infrastructure are inherently difficult to maintain over long periods of time and, though the worm propagated quickly, the specific nature of the problem was not difficult to diagnose over a matter of a few days. At the same time, private sector responders were uniquely able to identify threats like the Morris Worm such that, even without pre-established coordination procedures, they were able to trace the infection and mitigate its effects. In part seeing this reaction, the government took a number of initial steps to bolster effectiveness for future crisis scenarios. Carnegie Mellon University was funded by DoD to become the first Computer Emergency Response Team (CERT) in an effort to improve information sharing capabilities and several departments opened centers to diagnose security problems.[12] These responses were intended to make sure future crises would not rely on *ad hoc* solutions, and Congress began to pass law designed to extend legal authority into the cyber domain (including the Computer Fraud and Abuse Act of 1986, the Computer Security Act of 1987 and the Electronic Communications Privacy Act of 1986).[13] However, government responses and the development of security institutions were not inherently multi-stakeholder from the start. Hacking collectives and conferences that emerged in the early 1990s often eschewed government interest in developing national resiliency and focused on industry practices. Likewise, the greatest resources were allocated to elements of the government – like Air Force Electronic Warfare Center and the join-agency Information Operations Technology Center – focused on cyberspace as it intersects with traditional security concerns. The U.S. government, in short, entered the final decade of the 20th century concerned with the notion, but not necessarily the full scope of cyber threats to national security.

Initial responses

This dynamic would not last forever. With Presidential Decision Directive 63 (PDD-63) in 1998, the Clinton administration acted to implement a range of suggestions by intra-government stakeholders based off of arguments about the degree to which national security increasingly depending upon information infrastructure.[14] Foremost among these suggestions was an upgraded institutional infrastructure for coordinating the means of defense of industry systems across the financial, medical, energy and other sectors. Specifically, PDD-63 aimed to actualize a sophisticated hub that could coordinate analysis, investigation and responses to threats across the gamut by 2003. Centerpieces of this effort were the Information Sharing and Analysis Centers (ISACs) and the National Infrastructure Protection Center (NIPC) at the FBI. Both through these centers and in other ways, government departments could offer expertise and information to the private sector such that the country could efficiently mitigate risks associated with different cyber threats without intrusive regulation.

Though efforts resulting from PDD-63 faltered somewhat, American efforts continued to evolve off the back of increasing awareness that government networks – and those of the military, in particular – were uniquely vulnerable to cyber disruption. In 1997, the NSA collaborated with the DoD to run a red teaming exercise called ELIGIBLE RECIEVER (ER97) – i.e. an exercise designed to simulate the conditions of an actual attack for learning purposes – that was focused on DoD network security.[15] Different teams from the NSA stationed in the United States and abroad systematically deconstructed Pentagon security measures and accessed classified systems. The exercise gave DoD firsthand experience in the kind of attack they might expect to face from a well-"armed" foreign force and prompted new planning for cyber incident resilience. Unfortunately, few such measures were in place before the U.S. government faced an actual incident of large-scale foreign-based cyberattack. SOLAR SUNRISE in 1998, described previously in greater detail in Chapter 7, seemed to be the work of a foreign power that broadly broke into government systems over a matter of days. Popular strategic thought was that Iraqi infowarriors operating under orders from Saddam Hussein's were responsible for conducting attacks in response to U.S. bombing. That turned out not to be the case, but the sudden crisis – in which Pentagon departments found themselves operating without a clear chain of command – demonstrated that the lessons of ER97 needed to move policy and organizational developments with some urgency. This led to the creation of the Joint Task Force for Computer Network Defense (JTF-CND) at the end of 1998 as a means of ensuring centralized authority for coordinating network defense across DoD.

Box 8.2: U.S. CERT

The U.S. Computer Emergency Readiness Team (CERT) is an organization that operates under DHS that is mandated to protect the Internet infrastructure of the country from disruption and exploitation. The job of U.S. CERT is fairly varied. Much of what U.S. CERT, which looks much like other CERTs do across different regions and sectors of international society, does is preventive work that includes providing advice to private (and government) entities, maintaining a cyberattack alert system, constructing resources for such entities and more. On the back end, U.S. CERT provides much the forensic support for analyzing malware instances and deconstructing cyber incidents. The organization also represents the United States in interactions with other CERTs from other countries.

And yet, America's rapid institutional pivot to cyber threats in the late 1990s would remain remarkably focused on warfighting and the direct threat to government information systems posed by foreign adversaries. To be sure, American experiences through the 1990s led to the creation of the first joint cyber command in the world. But the purview of JTF-CND (later JTF-CNO, with the "Defense" changed to "Operations," housed at the Defense Information Systems Agency (DISA), was extremely broad and there was a clear emphasis on purely defensive measures to be taken by the Pentagon. As such, interest in the possibilities for cyber offense was not realized in institutional actions, perhaps except for at the Central Intelligence Agency, until the early 2000s. Instead, JTF-CND and the NIPC at the FBI (which was later disbanded) coordinated in the late 1990s to primarily address two potential crises on the horizon – one that could emerge from a large-scale assault on American information infrastructure and another that might see cyber espionage bear strategically meaningful fruits for a foreign adversary.

Inter-agency efforts to prepare for threats in either category were quickly tested with Moonlight Maze (discussed in Chapter 7). The episode, which lasted around two years, was an unprecedented intrusion into American government networks and computers. Much like the case of Cuckoo's Egg, foreign-based hackers maliciously broke into hundreds of computers across government and government-funded organizations looking for strategically valuable information. Attacks, which were difficult to group together as part of a single concerted campaign targeted against the U.S. government, were persistent, organized in their selection strategy for intrusions and technical sophisticated. An outcome of the episode, and of subsequent experiences with computer worms and the actions of patriot hackers that targeted U.S. government systems from China around the time of the Kosovo conflict, was a shift in the way that unclassified information could be stored and accessed within government networks. Whereas classified information was somewhat easy to track given the required procedures for access, unclassified data could exist on any computer or server without security cataloguing. Thus, in the late 1990s, the government mandated that transfer of unclassified information had to be routed through one of eight core gateways that could be easily monitored by incident first responders and investigators. Unfortunately, progress through the early 2000s was limited to a growing cohesion among defense establishment practitioners and procedures. Private sector outreach and relationships continued to include the gradual standup of ISACs, but few other developments are notable in the United States with regard to public-private cybersecurity initiatives until much later.

One-sided development

The post-9/11 period in the United States has particularly been characterized by the degree to which military organizations began to dominate the landscape of cyber threat response processes in the government. The newly formed Department of Homeland Security (DHS) was guided by two major policy developments in 2003 (the Homeland Security Policy Directive 7 and the White House's National Strategy to Secure Cyberspace)[16] to coordinate cyber policy holistically for the civilian elements of the U.S. government. It largely did not succeed. Certainly, DHS took a number of important steps to shore up critical infrastructure coordination capabilities and launched EINSTEIN, the intrusion detection network that acts to protect federal systems from attack. But much of the government's mission to construct the apparatus of effective cyber operability for conflict scenarios was siloed away from DHS's own under-supported effort to coordinate with foreign partners and reconcile whole-of-government policy intentions with the complexity of the cyber threat horizon. In essence, DHS – a government department mashed together after 9/11 via the act of seconding any existing unit with any responsibility for counterterrorism to one, new banner – simply didn't have the resources or procedural abilities needed to address the full spectrum of threat and governance challenges. Likewise, by making the spectrum of responsibility for DHS so massive, the department became *not* the obvious choice of vehicle for responding to the most critical challenges. That role remained the province of America's military and intelligence personnel, most specifically the National Security Agency.

Over the few years following 9/11, the military and the intelligence communities began to flesh out their portfolio of capabilities with regard to cyber defense and offense. The NSA had, for many years, invested in new cyber capacity to the point that the organization was the sole vendor of effective Internet-age information gathering (i.e. the agency's traditional signals intelligence role) and exploitative offensive capabilities for U.S. government security services. But events in the years immediately following 9/11, such as BUCKSHOT YANKEE, added to the pressures of supporting two wars in Southwest Asia and prompted several rethinks of the Department of Defense's approach to operations in what was increasingly thought of as a new warfighting domain. Initially, in order to effectively develop and direct cyber efforts, the DoD decided to split missions among existing institutions and to house defensive and offensive units in different locations. JTF-CND, which became JTF-CNO and later the Joint Task Force for Global Network Operations (JTF-GNO), was originally the hub of all cyber operations planning and execution at the Pentagon. However, it simply was not well suited to cover the range of actions required of its resources. Therefore, responsibility for maintaining and conducting offensive cyber operations in line with strategic and conflict objectives (particularly related to the Iraq War) fell in 2004 to the Joint Functional Component Command – Network Warfare (JFCC-NW), which was under the control of U.S. Strategic Command (USSTRATCOM) but directly ran in large part by the NSA. This left the newly renamed JTF-GNO free to focus on network maintenance and defense operations.

America's experiences in siloing different parts of the national cybersecurity mission, first under military/intelligence auspices and then further within different command units for offense and defense, *was* eventually adapted to compensate for a lack of oversight on the need to defend a broad range of non-military and non-government systems in the name of national security. However, action was not taken in this vein to reverse the one-sided development of cyber institutions until almost 2008, quite possibly because of the general distraction to other security arenas caused by the wars in Iraq and Afghanistan. Of interest, though inspired by a range of emergent cyber incidents around the world, when the Bush administration returned in 2007 and 2008 to the shape of the government's cybersecurity institutions it still largely did so via reference to the events of

Moonlight Maze and scares endured with Chinese hacks between 1999 and 2004. Noting the potential for broad-scoped intrusions undertaken with strategic persistence by a dedicated state opponent, military/intelligence establishment planners and Bush officials sought to address *both* threats to government networks (civilian *and* military) and the private sector in a centralized fashion.

Much as was the case in years past, however, the Bush administration ultimately found that it was easier to address information security problems linked with government function. Thus, the result of renewed debate in 2005–2008 was the Comprehensive National Cybersecurity Initiative (CNCI). CNCI focused *only* on government network operations and protection, and saw several tens of billions of dollars funneled to the DoD. In actual fact, whereas the effort to make headway on cybersecurity issues was driven by enduring concerns about sophisticated foreign threats to the country, the specific decision to issue CNCI instead of a more holistic strategic act that considered the private sector resulted from four incidents that drove fear into the minds of strategic planners. Operation Orchard in 2007, where Israeli hackers successfully took down part of Syria's air defense network in order to facilitate a kinetic strike against an under-construction nuclear reactor,[17] worried military officers that were suddenly given an example of how similar cyberattacks could be employed as force multipliers in conflict. The other, actually a series of incidents (Titan Rain né Byzantine Hades, Ghostnet and Night Dragon), took the form of intrusions by Chinese cyber spies against a large number of DoD, Department of Energy, Department of State, Air Force and defense contractor targets.[18] Though the scale of attacks and their coordination was less pronounced than Moonlight Maze had been, the threat of Chinese espionage signaled to operators at the Pentagon that further steps to coordinate the defense of government networks were absolutely critical. Similar conclusions were reached in observing the 2007 cyber campaign against Estonian websites by Russian hackers. Moreover, Russian actions in Estonia and later in Georgia emphasized the need for better American leadership in international interactions, particularly in coordinating CERT interactions and countering non-state cyber threats. And BUCKSHOT YANKEE, where a malicious piece of software spread across a large number of military systems after being delivered via USB stick, demonstrated to military planners that operational separation of capabilities only made sense if better coordination between units during crises was achievable. Thus, CNCI was pushed out to direct massive financial resources to the defense and intelligence community and the Pentagon took further steps to centralize command and control by establishing U.S. Cyber Command (USCYBERCOM).

Box 8.3: Cyber Command and the Cyber Mission Force

Cyber Command is one of the United States' unified combatant commands and has full purview over cyber operations aimed at aiding conventional military operations, protecting DoD networks and acting to disrupt foreign-based efforts to harm American national security. As of 2018, Cyber Command's primary operational force structure (the Cyber Mission Force) includes 133 teams of personnel that are tasked with specific mission pertaining to USCYBERCOM's guiding priorities. Each of these teams of cyber operators has between a dozen and about 80 personnel. While the Cyber Mission Force achieved operational capacity in 2018, many teams are, at time of writing, still being grown with the goal of a total personnel count around 6,000.

The structure of the Cyber Mission Force is an adaptation of previous inter-service cooperation and coordination under Joint Task Force Computer Network Defense (and, later, JTF-CNO and JTF-GNO). Each of the service branches is responsible for standing up a certain number of Cyber Command's operator teams. The Army, for instance, is responsible for 41 teams, the function of which is determined at the unified command level. Each service-specific element of the Cyber Mission force contains teams that are functionally the same. These include teams focus on the defense of DoD networks (Cyber Protection Teams), undertaking offensive operations or supporting conventional military operations (Cyber Combat Mission and Support Teams) and defending the homeland against major threats (National Mission Teams alongside National Support Teams).

USCYBERCOM has, since its founding in 2009, functioned as the directional heart of the U.S. government's efforts to address threats to government networks, mitigate "meaningful" threats to national security online and support conventional military operations in the cyber domain. Cyber Command is a combatant command. Prior to 2017, it was a joint operational unit that operated under the purview of United States Strategic Command. The founding of Cyber Command has been paralleled by other centralizing efforts beyond the DoD, such as the founding of discrete units for cybercriminal investigation under the Department of Justice and the creation of several coordinating offices – the DoD's Coordinator for Cyber Issues and a new Deputy Assistant Secretary of State, for instance – that both report on distributed efforts to departmental heads and liaise with the National Security Council at the White House. These developments have been at the heart of attempts to ensure interoperability of different military functions and government network defense efforts, and Obama administration actions in 2016 further put USCYBERCOM on a path to separate itself from other areas of NSA operation toward become a truly separate combatant command.

Enduring challenges

The result of American experiences with cyber conflict over the past three decades has been the development of a reasonably centralized system of institutions to underwrite government strategy on cybersecurity issues. But major developments, from Cyber Command to the creation of new departmental offices, only speak to one of the two primary government imperatives to secure the nation online. While DoD and close collaborators in the Department of Justice are arguably more effective than ever at meeting the requirements of executive strategy on cyber conflict functionality, the ability of government to provide for the defense of non-governmental elements of the country's national security apparatus is questionable. Certainly, programs like the Defense Industrial Base Cyber Pilot program have allowed for greater coordination between DoD and elements of the critical community of defense contractors and critical infrastructure operators. But mechanisms for effective involvement of the government in the defense of the private sector and of civilian organizations remain largely non-existent. Indeed, the formation of Cyber Command itself has driven concerns beyond the government that the DoD and the NSA might essentially discount civilian efforts to organize for cyber defense. This has, according to some commentators, had the clear effective of reducing interest in public-private initiative for protecting civilian networks.

Box 8.4: The defense of federal networks

Just as the Pentagon and the intelligence community historically ran into problems of authority and direction – i.e. figuring out who was in charge of what – when it came to the defense of DoD and linked networks, the IT efforts of the government's civilian agencies were, prior to about 2003, fragmented and lacking in cohesive policy or access to resources. In the late 1990s, SOLAR SUNRISE, Moonlight Maze and other events sparked a rapid move toward institutionalization of cyber capacity in a joint task force in the Department of Defense. About three years later, the events of September 11, 2001, prompted a series of similar reviews, critiques and remedies aimed at simplifying the cybersecurity enterprise across other government departments.

The Comprehensive National Cybersecurity Initiative (CNCI), under the leadership of Melissa Hathaway, proposed a series of measures to be taken to better affect the protection of federal networks. Two developments emerging from CNCI are particularly noteworthy. First, alongside a series of executive directives to standardize authentication systems and IT practices, the federal government developed and implemented an intrusion detection system called EINSTEIN.[19] The idea behind EINSTEIN, as is the case with all intrusion detection systems, is that software placed strategically at exchange points (i.e. the point of information transfer between federal networks and the outside Internet) can scan incoming data packets to look for abnormal, potentially malicious activity and then notify IT administrators where needed.

Second, the work done for CNCI motivated further optimization of federal network features with the Trusted Internet Connections Initiative in 2007. The idea behind TICI was simple. Federal cyberspace was incredibly cluttered and the task of protection mechanisms like EINSTEIN was made more difficult by the existence of network bloat. As such, under TICI the number of exchange points between federal networks and the outside Internet was massively reduced from ~4,300 to under 50.

At time of writing, all non-DoD United States' government agencies employ, by mandate, EINSTEIN 2.0 in their network defense efforts. EINSTEIN 2.0 is an updated version of the earlier software, which was not mandated for use across all departments, that performs a more sophisticated analysis of incoming data packets. Whereas the 1.0 version simply analyzed header data (as a pen register/trap and trace device might), 2.0 performs a limited scan of packet content to see if there is a match for known malware signatures. Because real-time analysis is not possible, data is copied temporarily for near-real-time assessment before being either deleted or (if malicious content is found) reviewed by an IT operator. A future version, EINSTEIN 3.0, is being developed to perform yet more sophisticated analysis of groups of data packet content as they come in.

Naturally, one might think that there are some potential legal issues with the federal government's employment of a system like EINSTEIN. Specifically, the

Department of Justice recognizes (and dismisses) two potentially concerns. First, an argument could be made that EINSTEIN violates the Fourth Amendment to the Constitution of the United States, which guarantees protections from unreasonable search and seizure of property, communications, etc. The DoJ admits that EINSTEIN probably violates the rights of federal employees because all communication is being monitored as it transits federal networks. Second, an argument could be made that EINSTEIN is statutorily unlawful. Specifically, temporarily copying data packets constitutes surveillance of American persons in a limited fashion, as outlawed by the Foreign Intelligence Surveillance Act of 1978, as well as a warrant-less review of private communications as outlawed by the WireTap Act.

The legality of systems like EINSTEIN rest on a few fronts. In general, much of what DoJ relies on in justifying such a monitoring system is the tendency of American courts to emphasize technical realities when assessing the applicability of law to new technology. In addition to the fact that federal employees agree to having their Internet communications monitored when they sign that banner agreement to the terms of federal service upon accepting employment, DoJ asserts that the technology doesn't really allow for a better kind of defensive monitoring and so should be permitted an alternative of absolutely no security. More importantly, getting warrants to search all traffic on an ongoing basis is entirely infeasible and, since FISA defines surveillance as being a targeted action and EINSTEIN reviews all traffic entering federal networks, FISA doesn't apply.

The cornerstone of existing efforts to address the challenge of civilian network defense pertaining to national security is a focus on the protection of critical infrastructure. The Obama administration – and presumably administrations to follow – in 2009 made the uncontroversial statement that CI was a "strategic national asset" that required deft government collaboration with private industry to ensure systems security, particularly when it comes to the national telecommunications infrastructure that underwrites IT functionality across every other sector. Specifically, the cornerstone of such efforts emerged from an Obama administration executive order in 2013 that established the NIST Framework for ensuring best practices by CI providers.[20] The NIST Framework essentially has three component parts. First, the Framework outlines different types different kinds of cybersecurity activities that relate to the function of private sector entities. The goal in doing so is largely (1) the setting of a standard "common language" surrounding practices and procedures in public-private dialogue and (2) the identification of a series of specific best practices for the kinds of private firms the Framework is aimed at. Second, the Framework includes a methodology for figuring out how compliant a specific actor is with the best practices outlined in the first part. Then, finally, the Framework provides a flexible roadmap for "underperforming" actors to reach higher levels of security capabilities as described in earlier sections of the Framework.

The NIST Framework *does* fill a gap in the U.S. government's focus on security of civilian elements (i.e. non-traditional elements) of the infrastructure of national security. And, in truth, the condition of the framework as one that outlines profiles for private sector actors that can be improved upon over time is broadly attractive. Though it faces the criticism of establishing American jurisdiction via government decisions and legal procedures, there is even reasonable support for the Framework as a possible template for standardizing international cybersecurity efforts. However, the Framework has clear drawbacks. A number of private sector leaders point out that the document is technically daunting and any firms that commit to functioning in line with the guidelines are likely to focus on actuarial threats at the expense of attempting to understand unique threat profiles of potential attackers. Moreover, there is a complete lack of incentives offered by the Framework for private actors to adopt its provisions. While it may serve as a good set of guidelines, the notion that it cannot possibly be adaptive enough to consider the morphing of cyber threats leaves private sector actors without any motivation to commit to the approach. When combined with the traditional disincentives industry actors have when it comes to information sharing (leaking of intellectual property or potential loss of reputation from admission of disruption, for instance), this dynamic places government efforts in a position of restrained progress – civilian engagement is better than ever but remarkably still constrained in the potential for meaningful threat mitigation *in the long term.*

Moreover, Russian cyber espionage and influence operations during the period of leading up to elections in the United States in 2016 continue to expand the notional footprint of critical national security assets that require the protection of the government.[21] Beyond the industrial and infrastructural targets that are the primary concern of the U.S. government under current procedures, data stolen from civil society organizations – in this case, the Democratic National Committee – was used in efforts to meddle with the political processes of the nation. Regardless of the degree to which such operations were successful (something as yet unknown), the contours of such an event suggest that societal functionality beyond CI require government oversight as part of a mission to ensure the integrity of national political processes. Particularly given the focus of the NIST Framework on actuarial categories for determining risk, it seems reasonable to say that the United States remains relatively underdeveloped in terms of its ability to safeguard civilian networks. Certainly, American governmental efforts have improved the prospects for CI protection in recent years and there is a clear, enhanced ability to achieve traditional military objectives in cyberspace in the function of USCYBERCOM and related actors. But a large number of questions, from how to important safeguard civil society networks and information to how to treat attribution in the context of international diplomacy, remain to demonstrate that existing approaches to cybersecurity are still a number of steps from addressing evolving threats to national security online.

Europe, NATO and cybersecurity

All national experiences with cyber conflict or upheaval are, regardless of similarities between incidents, inherently different from all others. This is quite simply because governments and, more broadly, national society copes with crises in different ways. States possess governments that take on a range of electoral formats and host unique arrays of institutions for ensuring national security. Perhaps more significantly, different countries approach questions of national security theorization, policy and practice in remarkably different ways.

Authoritarian states represent a stark departure from the United States in terms of how cybersecurity imperatives have been viewed and how cyber capabilities have been developed. Such states – in this case Russia, China, Iran and North Korea – are discussed in the following sections. By contrast with such a stark departure, however, have been the cyber conflict experiences of American partners in the West. Such states have not necessarily viewed cybersecurity at times acutely through the lens of global security commitments – the war against terrorism, actions in Iraq and Afghanistan, or standoff with North Korea, for instance – in the way that U.S. military and government stakeholders have. Likewise, many U.S. partners have approached cybersecurity issues with international cooperation more organically intended than has been the case with America's Department of Defense. Particularly in Europe, common defense is the cornerstone of national security policymaking. And American partners have rarely – at least in early days in the 1980s and 1990s – been faced with large-scale threats to the national security establishment. Instead, from France to the United Kingdom and Japan, threats have usually been highly specific in their potential impact.

This section describes the experiences of two Western security actors – the United Kingdom and the North Atlantic Treaty Organization (NATO). The point in doing so is to highlight the less tumultuous development of American partners and Western democracies in general. The United Kingdom's experiences with cyberspace roughly parallel those of other major American allies, like Japan, Germany, South Korea and France. At the same time, the United Kingdom's efforts reflect the transnational focus that is arguably more common among non-American Western states. Discussion of NATO, thus, is intended to illustrate how national efforts have culminated in unique and evolving mechanisms for cyberspace governance and conflict prevention by the international community.

The United Kingdom's experiences in cyberspace

The United Kingdom has a strong institutional heritage when it comes to technology and cryptanalysis. Whereas many other countries have incorporated cybersecurity procedures and abilities into national intelligence and counterintelligence missions, the United Kingdom has, since the mid-1900s, been able to rapidly and almost seamlessly meld intelligence service capabilities with advanced information systems operations by borrowing from legacy institutional mechanisms and practices. This ability stems from the strong traditions of ensuring inter-service operability going back to the efforts of researchers at Bletchley Park during World War II, who worked to crack the encryption of Nazi Germany' Enigma devices and give the Allies a critical edge in the war effort.[22] Moreover, this legacy has not only led to the maintenance of mechanisms for the rapid blending of capabilities between different arms of the security services; additionally, there has arguably been a more organic institutional and cultural awareness beyond even cybersecurity practitioners of the need to update regulations and operating dynamics over time within the United Kingdom.

As mentioned earlier, the United Kingdom has faced less in the way of highly visible or disruptive realization episodes that have shaped national policy and institutional development than has the United States. Specifically, the 1980s and 1990s saw remarkably few direct threats to the informational function of national security institutions and no threats to infrastructure. The first major engagement of the U.K. government with cyber conflict topics took place in 1998 in a Defense Review.[23] The first national Cyber Strategy was not published until 2009. Moreover, a more complete threat analysis of cyber threats was not incorporated into the National Security Strategy document until the following year, the first inclusion of the topic that led directly to initial funding for cybersecurity programming within the government. In short, the United Kingdom's experiences with cyber threats have been limited to such a degree that the government – as a whole, rather than the enduring efforts of intelligence outfits like the Government Communications

Headquarters (GCHQ) – has only turned to consider cybersecurity a national security imperative in the past decade.

Again, this does not mean that the government of the United Kingdom has not acknowledged the importance of the digital domain until recently. Rather, Britain has focused on cybersecurity much more in terms of low-level political and economic challenges than as a major conflict modifier. In many ways, the revolution in information technologies that initially drove economic growth through the 1980s actually links directly to Britain's heritage of technology-focused intelligence and production work during World War II. Not only did code breakers in the United Kingdom successfully break Germany's Enigma encryption to give the Allies an unprecedented advantage in the later years of the war, but also the elements of the scientific and military research communities funded or directed by the government during the war produced, among other things, new radar technology that revolutionized aerial combat. Following the war, personnel leaving government development programs continued to innovate and provide the country with unique advances in information technology. One officer's (Arthur Clarke) discovery of geosynchronous orbits enabled the development of satellites for commercial and military purposes, and others helped develop guided munitions technologies. The information-driven military ecosystem that emerged from such advances would help Britain quickly end the Falklands War and provided unprecedented command and control abilities to allied forces fighting in the First Gulf War in 1991. Following the Gulf War, information-driven military operations became so operationally synonymous with military functionality that the United Kingdom's abilities in this vein were officially labeled the "Network Enabled Capability." Though the end of the Cold War initially prompted military planners to think that the need for advanced warfighting capabilities was diminished, new challenges led to programming and procurement efforts to continue to upgrade the country's electronic and information warfare arsenal.

Most of these upgrades remained in the research and analysis phase throughout the 1990s. Arguably as a result of both the lessons of ELIGIBLE RECEIVER and American crises experiences with Moonlight Maze and SOLAR SUNRISE, policy documentation was not produced to address the need for information technology modernization of Britain's military until the Strategic Defense Review (SDR) of 1998. The SDR sought to centralize network defense and warfighting capabilities. In actions that mirrored those being taken across the Atlantic, institutional authority for cyber operations was centralized in the Permanent Joint Headquarters that controlled a joint budget for a number of sub-agencies. The Defence Communications Services Agency (DCSA) was set up in a fashion similar to DISA in the United States to coordinate operator activities spread across different services.[24] Likewise, Britain set up the Defence Computer Incident Response Team (DCIRT) to undertake a cyber defense mission. Specifically, DCIRT helped coordinate changes to the architecture of government information systems through the early 2000s to make accountability and threat assessment a simpler prospect. The task was not easy. In reality, a series of initiatives through the 2000s demonstrated that use of information systems and hardware across the government was incredibly fragmented. Particularly as the consumer market for ICT grew more sophisticated, different institutions pushed back against efforts to streamline architecture in a holistic sense for the government. This eventually prompted the government to appoint Chief Information Officers across agencies in an attempt to improve interoperability within the United Kingdom and in collaborations with the other four of the five "eyes" – intelligence counterparts in the United States, Canada, Australia and New Zealand.[25]

Britain's cybersecurity efforts in the era following this initial set of institutional developments have been shaped by cyber conflict incidents. However, again, they have largely been either low-intensity or those directly affecting other countries. Events like Russian attacks on Estonia and Georgia, Moonlight Maze and BUCKSHOT YANKEE in the United States, and the relative success of the Stuxnet worm have all provided narrative lessons to policy elites in the United Kingdom, namely that threats to national infrastructure and secrets are increasingly sophisticated. Direct institutional adaptation, however, has largely emerged from low-level challenges important enough to demand executive-level responses. Prior to the turn of the new millennium, for instance, concerns among military and industry practitioners that anomalies with computer clocks in control systems could lead to catastrophic failures at midnight on December 31, 1999 – nicknamed the "Y2k bug" – forced the government to improve the ability of CIOs to (1) recommend collaboration between areas of industry functioning on wildly different architectural foundations, (2) better identify what infrastructure might be considered critical to national function and (3) suggest best practices moving forward.[26] The infection of a large number of Ministry of Defense (MOD) computers in 2003 with the Lovgate virus – a basic virus that hid in email servers and impeded basic computer usage – was caught by the DCIRT, publicized and used as the basis for driving conversation in Parliament about needed upgrades to national technology standards frameworks.[27] The MOD's experience with a self-imposed denial of service attack – inaccurately dubbed the "Amarillo virus" that was actually a viral video email circulated throughout the MOD – led to new programs for training of users and coordination between network managers across service branches.[28] These lessons were reinforced again in 2009 when the MOD was hit with the "Conficker" virus, a minimally disruptive virus that compromised unclassified computers across the Royal Air Force and that might have been prevented with better software updating protocols.

Arguably, the United Kingdom is where the United States was circa the early 2000s. That is to say that the United Kingdom has a strong heritage and robust institutional foundation for developing cybersecurity standards and best practices. At present, national cyber capabilities focus largely on information technology support for traditional operations under DCSA, protection of government networks and national assessment of threats by the Cyber Security Operations Centre (CSOC). CSOC is the result of enduring calls to better define infrastructure critical to national security and to better provide for government protection of those resources. However, in reality, the building of comprehensive public-private initiatives in the United Kingdom is, as of the time of writing this chapter, still a work in progress.[29] The Office of Cyber Security and Information Assurance (OCSIA) was set up between 2011 and 2013 with a budget in excess of half a billion pounds to drive national cyber strategy on several fronts.[30] This includes focus on cybercrime and Britain has certainly had some success with the incorporation of organizations like the Police Electronic Crime Unit (PECU). Moreover, the United Kingdom has increasingly focused on international coordination. The United Kingdom was a signatory of the Council of Europe's Convention of Cybercrime that outlines cross-border standards for cooperation and legal expectations on a spectrum of ICT-aided criminal activities. And Britain's experience as the target of Russian influence operations during the 2015–2016 Brexit referendum campaign has led to increased inter-service cooperation with both NATO and European Union partners. But, at present, it might be fair to say that Britain's capacity to deal with incipient cyber threats out into the future is dependent on a growing range of variables and that the United Kingdom has yet to achieve a measure of cohesion in its approach to mitigating risks associated with the digital domain.

NATO and cyberspace

The North Atlantic Treaty Organization is a mutual defense organization with roots in the needs of Western democracies to counter the threat of the Soviet Union during the Cold War. NATO is constituted of 28 member states and has formal relations with more than 30 others through dialogue and extended partnership programs. The Alliance stipulates a number of requirements for its signatories that underwrite the collective defense mission of the community. Most significantly, the NATO charter stipulates that all member states spend at least 2% of their GDP on national defense and that – as outlined in Article 5 – aggressive action against one member necessitates a response (though not necessarily a military one) by all others. This second feature of the Alliance is particularly important, as it provides the mechanical trappings of collective defense beyond mere rhetoric that might otherwise characterize strong relationships between international partners in saying that an attack on one will, generally speaking, be interpreted as an attack on all. This element of the NATO charter has only been invoked a single time, following the September 11, 2001, attacks on the United States by members of al Qaeda, in the history of the Alliance.

NATO's policy and practitioner purview is broad. The Alliance supports national defense efforts, expeditionary capabilities when required and acts to streamline the ability of companies to provide for the common defense. This includes acting as an advisor and active voice for the purpose of improving regulatory conditions for defense industry actors and critical infrastructure stakeholders across the community. Thus, while the Alliance's focus on cybersecurity issues is certainly on military capabilities and mitigating threats to member governments, so too is it on a crosscutting set of missions ranging from the interoperability of military services to protection of national economic assets.

Much like individual countries have implemented policy and adapted institutions in the context of unique wake-up calls, so too has NATO seen the use of ICT in different conflicts and taken distinct lessons to heart. Several such conflicts and resultant inter-member responses, in particular, are worthy of note. Foremost among these are the conflicts in Chechnya in 1994–1995, which did not involve NATO directly, and operations in Kosovo by NATO members operating under the auspices of the Alliance. In the Chechen war, Chechen rebels fought a bloody independence war against Russian forces. The conflict lasted much longer than Russian military planners originally thought it might and ICT usage was a primary contributing factor to the end of the war. Specifically, Chechen forces uploaded gruesome pictures of the conflict to the Internet. In addition to creating international sympathy for their cause, the web campaign shifted Russian public opinion away from support for the conflict. From this, NATO, alongside member states and the Russian government, became aware of the potential for new information dynamics to alter the course of a conflict entirely through manipulation of social thought.

NATO's experiences during the Kosovo conflict, where the Alliance undertook airstrikes against Serbian targets in order to protect Kosovar civilians during the country's fight for independence, emphasized the need for improvements in interoperability in the use of ICT between member forces. In a famous incident, NATO aircraft accidentally bombed the Chinese embassy in Serbia as a result of information poorly compiled and presented to pilots via PowerPoint presentation. More significantly, retaliatory cyberattacks by both Chinese and Serbian hackers over the course of the conflict struck at NATO servers and prevented the effective use of information services to coordinate actions between member forces engaged in airstrikes. In short, NATO's first experience in a conflict that involved the use of ICT for combat augmentation resulted in a somewhat humiliating lesson in the limits of Alliance abilities to organize and act effectively in the digital domain.

For the Alliance, perhaps the two most significant watershed moments that led to distinct institutional reorganizations around cybersecurity challenges are also arguably the two most significant assaults on member states by outside forces – the 9/11 attacks on the United States and the 2007 Russia-Estonia conflict. The aftermath of 9/11 saw the mobilization of Alliance resources in support of a coalition action in Afghanistan against the Taliban. Given the nature of the strike against the United States, NATO response naturally included consideration of how terrorists and state sponsors of terror might use the web to cause widespread disruption targeting Alliance members. It was in this period that the phrase "cyber Pearl Harbor" began to circulate within Western defense communities to signify a possible large-scale attack on critical infrastructure and security forces of a given member country. In response to such concerns, NATO initiated the Cyber Defence Program and the Computer Incident Response Capability in order to coordinate specific defense missions across member organizations. Soon thereafter, the Communication and Information System Services Agency was founded in Belgium as a headquarters unit with a mission of coordinating the actions of all national-level headquarters for cyber operations.

Whereas 9/11 led to the rapid development of institutions for conducting cyber defense missions at the level of the Alliance, the narrative of the Estonian conflict in 2007 is one of failure to respond. As the result of a feud between pro-Russian protesters and the Estonian government over the movement of a Soviet-era statue from Tallinn, Russian-based hackers launched a massive set of denial of service attacks against ISPs and government systems in Estonia. In effect, Russian hackers managed to substantially block Internet access to citizens and private companies in Estonia – which is, in many ways, one of the most web-connected countries in Europe – for between three and four days. In reality, the attacks continued in various forms for around a month and NATO, despite arguably being required to aid in response efforts surrounding such an attack, failed to provide intelligence or resources meaningful to the termination of the attacks.

The result of the Estonian conflict was a summit in Bucharest in April 2008. Despite the inability of NATO or the broader international community to sufficiently aid Estonian defense efforts in 2007, the effect of Russian-based attacks was monumental. The Bucharest Summit essentially resulted in the incorporation of two organizations designed to prevent another Estonia from taking place, at least from the perspective of NATO's relative impotence during the crisis. The first of these was the Cyber Defence Management Authority (CDMA). The CDMA was set up and located in Brussels to be a coordinative center of all cyber defense efforts for the Alliance. The second organization to emerge from the summit was the Cooperative Cyber Defence Centre of Excellence (CCDCOE) located in Tallinn, Estonia. In reality, the purpose of these organizations was much more than simply the coordination of incident response efforts. Together, the CDMA and the CCDCOE are tasked with implementing Alliance cyber defense policy and standards, sponsoring research and undertaking training across military and civilian government services. The purpose of these organizations is, thus, to establish a measure of resilience in the ability of the Alliance and of individual members to prevent and respond to the kind of challenge observed in 2007. Major projects sponsors by the two have included the annual International Conference on Cyber Conflict and an inter-member joint exercise for cyber defense operators, the International Locked Shields Exercises. Additionally, these organizations have produced the landmark Tallinn Manual on the International Law Applicable to Cyber Warfare (and an updated version in 2017) that brings together more than 30 cyber conflict experts in outlining common principles and legal opinion of the relationship between potential cyber conflict activities and existing international law.[31]

Though the Alliance has not faced another large-scale cyber assault on information infrastructure since 2007, NATO has continued to proactively improve its cyber defense and operations coordination capabilities. In particular, the cyber conflict between Russia and non-member Georgia in 2008 saw the provision of NATO advisors and technical assistance to the former Soviet republic. Thought NATO could not directly aid Georgia, assistance has since led to several institutional reorganizations and a closer relationship between the Georgian and broader NATO defense community. And NATO members have variously taken it upon themselves to act off the back of NATO initiatives and dialogues for the further improvement of transnational cyber conflict coordination in recent years. Following the Lisbon Summit in 2010, NATO leaders agreed to the development of Rapid Action Teams to aid incident response teams located in individual member states. Following the Chicago Summit in 2012, five NATO members further established the Multinational Cyber Defence Capability Development Project to the same end.[32] And, significantly, the Alliance agreed to adopt the Enhanced Cyber Defence Policy after the Newport Summit in 2014 that seeks to improve commitment to cyber defense initiatives by members and to coordinate security efforts with European Union bodies where possible.[33]

In many ways, the question that must be asked with regard to the Alliance is one of relevance. How important are efforts undertaken by a supra-national organization like NATO when it comes to something like cybersecurity, particularly given the unique contours of threat and opportunity faced by different national governance systems? NATO seeks to broadly involve itself in four areas – (1) the defense of NATO networks, (2) the coordination of data collection and consultation procedures, (3) the construction of useful coordination forums for member practitioners on various levels and (4) the conceptualization of useful procedures for integration Alliance approaches with member processes. The development of the organizations noted earlier and the development of projects under their auspices have entirely been in aid of these objectives.

From one perspective, NATO has been immensely useful for its member community when it comes to the development of cybersecurity best practices and resources. Many member states would not have access to incident response capabilities as sophisticated as they do today without the efforts of the CDMA and the sponsorship of the CCDCOE. Moreover, NATO has effectively moved the ticker forward with regard to the conceptualization of rules of conflict and conflict remediation in cyberspace. The Tallinn Manual, in particular, is the gold standard for efforts to link aggression in the digital domain to the traditional legal tenets of international relations.[34] NATO, in short, has thus far done a magnificent job acting as a convener and intellectual organizer that underwrites the ability of Alliance members to innovate and to more effectively place cybersecurity at the forefront of national security efforts.

And yet, NATO faces challenges that bring into question the ability of the Alliance to be effective out into the future. Naturally, as is true of all intergovernmental organizations, NATO requires member states to green-light standard-setting efforts across the gamut of possible activities. Though military cooperation and crises response are desirable areas of coordination for members, increased focus on protecting critical infrastructure of member states is likely to run up against the problem of secular interests among non-state stakeholders across the Alliance. In particular, the recent manifestation of influence operation threats that target civil society organizations and political processes is likely to pose a significant problem for the Alliance insofar as involvement in coordinating the defense of non-government systems might run up against political opposition by anti-integrationist factions in different countries. And given that NATO's efforts to affect cybersecurity cooperation such that all members benefit, it is enduringly likely that some states might prefer to buck pass and refuse resource commitments on the grounds that common defense steps taken will aid all members regardless of cost allocation. In short, NATO is enduringly limited in its ability to achieve progress on cybersecurity because members and stakeholders within member states are variously incentivized to protect parochial interests or resist new costs across the gamut of possible future threats the Alliance might seek to address.

Polarizing perspectives: Russia's and China's cyber experiences

Policy on cybersecurity issues takes similar form in countries around the world. Likewise, it seems fair to say, as noted earlier, that mechanisms for addressing different cyber challenges emerges from the unique responses of pre-existing institutions to country-specific realization episodes. This is no less true for countries like Russia and China – other states that lead the world in investment and development in cyber capabilities across the gamut of sociopolitical functions – than it is for the United States and her partners. However, while the experiences of the United States have, at least until recently, encouraged the national security establishment to think of information security in terms of militarized threats to networks, systems and critical pieces of content, both China and Russia have for some years now viewed the implications of the information revolution for national security processes in a remarkably different way. Specifically, both countries' notion of information security has more clearly embraced the ideas outlined in a later chapter, namely that the information revolution has been *both* about the digitization of infrastructure and fundamental changes in the dynamics of the global informational environment. Thus, information security policy that aims to address issues of both national security and political stability must address ideas as much as it must consider technical security. This section outlines this perspective in greater detail below following historical descriptions of both China's and Russia's experiences in the cyber realm.[35]

China's experiences in cyberspace

The People's Republic of China's (PRC's) investment in the ability to regulate domestic web usage and to employ the various tools of statecraft via cyberspace is arguably only second to that of the United States. Though use of the web and the development of government cyber institutions in China spiked as it from nowhere in the late 1990s, the country today boasts the most Internet users in the world today. Moreover, the Chinese government's abilities to cultivate China-based elements of cyberspace and to conduct broad-scoped cyber operations in aid of national security objectives are considered by most Western analysts to be extremely advanced. The Great Firewall and related mechanisms of national digital censorship, in particular, are considered to be without parallel in terms of state capacity for shaping national social and political engagement.

But where has present policy and state cyber capacity come from? Just as with other countries, China's focus on the opportunities and risks of greater exposure to the digital world emerged from the distinct shape of perceived threats to the state. And, much as was the case for the United Kingdom and other European countries, focus on the web was limited even through the mid-1990s. Indeed, with China, public access to the Internet did not occur until 1995 and was in fact blocked by Western organizations as late as 1992.[36] Prior to 1995, information technology adoption was limited in China, despite government commitment (in 1987) to the goal of transforming the PRC's manufacturing economy into an information economy. The first email from China was only sent in 1987. The first domestic personal computer was only shipped in 1990. And the mobile phone market only really came into being in 1994 with the establishment of Unicom. Even post-1995, China experienced what we might think of as normal rates of ICT adoption. Through around 2000, Internet usage was limited to no more than a few million citizens. Chinese telecoms were allowed to partially privatize in 1997 and the government made initial investments in state-owned exploratory organizations, like the Great Wall Technology Group.

Not counting the actions of patriotic hackers attempting to disrupt and protest the actions of anti-Chinese mobs in Indonesia in 1994, arguably the main event that changed the trajectory of development in digital affairs in the PRC was – somewhat like America's experiences with the Morris Worm or ILOVEYOU – a threat (in this case, a domestic one) from a non-state actor that manifested online. During the 1990s, a spiritualistic organization practicing a particular form of *qigong* – a breathing and exercise activity popular across China – rose to prominence in the country. Name Falun Gong after the group's approach to meditation and exercise (Falun Dafa is often used synonymously in reports that name the group), membership in China blossomed from a few practitioners in 1992 to tens of millions in 1996–1997. Naturally, the Chinese Communist Party (CCP) was wary of such an organization with large-scale social appeal and influence. However, for many years the CCP had simply sought to establish oversight of the group and Falun Gong even benefited from direct state sponsorship (along with a number of other traditional *qigong* organizations). This all changed in 1998 following Falun Gong's rejection of new CCP regulations that would have established formal ties (and, thus, a degree of internal power to oversee) with the Party. Government response was swift. Prominent members were arrested on a range of charges, and Li Hongzhi, the group's founder, was quietly exiled abroad. In response, more than 10,000 Falun Gong protesters marched in Beijing. Protesters were peaceful and even helpful to police forces trying to minimize disruption. Protest organizers met the PRC Premier and disbanded after a seemingly successful attempt at retrenchment with the government. However, in the days and months that followed, direct orders from the Chairman of the Politburo translated into an official banning and the persecution of Falun Gong across China. That persecution continues to this day and reference to Falun Gong remains the most closely censored terminology in the PRC.[37]

What is remarkable about Falun Gong and China's experience with cyberspace is that the group's march on Beijing, organized almost entirely online, was a complete surprise for a government quite capable of monitoring its own population. Use of email, web forums and chat rooms allowed group organizers to mobilize thousands in a matter of days and force the government into a confrontation much covered by world media. Indeed, many scholars argue that it was the surprise nature of the confrontation that forced Chinese leadership to choose hardline repression over reconciliation. Following the main event and subsequent persecution orders, Falun Gong continued to use ICT to dramatic effect, managing to use the web to organize a clandestine press conference for foreign journalists that broadcast their plight to the world. And Falun Gong websites, prior to many takedowns in 1999 and 2000, were numerous and had proven to be a major driving factor in organizing recruitment efforts even after Beijing's initial move to persecute. In short, the PRC learned that the web was a serious source of potential threats to state security.[38] Different from the United States and other Western countries, of course, such threats were really about the political stability of the communist regime and Falun Gong's banning was an organized effort to eradicate a potential ideational threat to the PRC's limited culture of civic engagement.

The lessons of Falun Gong were twofold. First and most importantly, the episode demonstrated to the Chinese government that influence and capability in the digital world were a necessity from the point of view of the survival of the CCP. Second, the particular manifestation of the threat in the form of a social irritant implanted in the heads of strategic planners the notion that information security threats were ideational and content-based as much as they were logistical or technically disruptive. While the first lesson drove what we might think of as the digitization of infrastructure and institutions that has produced the modern Chinese cyber apparatus, the second has critically influenced the shape of doctrine and policy in a way that simply has not been the case – perhaps until recently where the effect of Russians influence operations has garnered broad public attention in Europe and North America – in the West.

China's steps to ramp up cybersecurity capabilities have come in a number of forms and Chinese strategy has a number of component parts. Cyber warfare capabilities are distributed across a range of different military and paramilitary organizations.[39] Much like in the United States, these can generally be broken up into categories based on different objective portfolios – military units to defend military networks and undertake offensive operations, civilian government units under the Ministry of State Security (MSS) for the same, and non-governmental proxies that are retained for operations that are specialized or fall outside of those bounds. Given this, it seems fair to say that the country's capabilities are reasonably centralized around the strategic perspectives and machinations of two entities – the People's Liberation Army (PLA) and the MSS.[40] Indeed, the centralization of cyber capabilities in these two actors has in recent years been, to some degree, a source of friction between the civilian government and the military that acts as a complicating factor when it comes to analyzing Chinese cyber activities. In short, this operational tension between civilian and military organizations stems from the fact that civilians have only limited opportunity to amend actions taken by military strategists that might disagree (sometimes by taking actions) with approaches to foreign policy issues. This dynamic emerges from the nature of interconnectivity between the military and China's civilian government wherein formal contact is limited to high-level executive contact between the Politburo and an exceedingly small number of military officers jointly operating as service chiefs.[41] Because this is the case, only limited oversight of all military-side cyber functions by the civilian government exists. Particularly given the nature of cyber operations wherein authority to act often comes from managerial operators by necessity (i.e. not from executives), this can be troubling for those trying to analyze intention and authorization surrounding Chinese cyber belligerence.[42]

Box 8.5: Net neutrality

Should a network operator have the right to determine what content is and is not allowed to flow across its networks? With some obvious criminal exceptions, the answer generally given in the Western world is no. Given the broad interpretation of the web as a public resource, the fact that it is based on privately owned infrastructure matters little in the fact of government-backed interest in protecting consumer rights and in sheltering freedom of speech. This principle is called **net neutrality**. At the time of writing, regardless of some recent volatility in debate within the United States, most Western countries actively protect consumers from biased network practices on the part of network operators or are, at the very least, actively pursuing legislation. Naturally, the principle is not operational in authoritarian states where there is strong government control of the information ecosystem, like China. Here, we have one main reason for continued international disagreement on the ideal shape of global Internet governance procedures as working better when determined from **multilateral** forums as opposed to **multi-stakeholder** ones.

That cyber capabilities are concentrated in the security services of the government is unsurprising, as the need to rapidly develop information security protocols and abilities in China evolved in an environment where the balance of potential to act was even more lopsided than it has been in the West. In essence, the PLA and the modernization programs undertaken since the 1990s constituted the only opportunity for government development of appropriate cyber defense and conflict instruments. Moreover, government motivation to invest in the digital domain occurred against a backdrop of extremely limited private enterprise operation in the mid- to late-1990s. The prevalence of state-owned and – directed enterprises in the 1990s provided the public–private bridge for network security coordination that is so difficult to build in capitalist states, and this continues even today where China's significantly more private economy is subject to extreme oversight.

Whereas the MSS largely focuses on civilian government networks, the other two elements of China's cyber force – i.e. the military and irregular "patriot hackers" or contracted mercenaries – are responsible for undertaking offensive operations both domestically and externally. Within the military, most relevant cyber actors fall under the PLA General Staff Department's Third Department, an organization that broadly parallels the functions of the NSA in the United States. Beyond the Third Department's traditional intelligence missions and functions, computer network defense, exploitation and offense capabilities are employed by subdivisions of the Beijing North Computer Center (PLA Unit 61539). Other important units include the Second Bureau of the Third Army, which functions as a center for operations coordination, and the PLA General Staff Department's Fourth Department in conjunction with broader electronic warfare efforts. More specifically, in terms of actions undertaken, units spread across the PLA and what we might think of as the government's direct sphere of directive influence beyond state organizations engage in (1) information domination activities, (2) espionage and (3) disruption and degradation operations.

Information domination activities are those cyber actions taken to disrupt or direct the narrative environment surrounding China's security interests. These actions are rarely sophisticated and often emerge from the PRC's irregular cyber forces – non-state patriotic hackers or, reportedly, mercenaries acting as proxies for state interests. Examples of such activities include vandalism of websites and denial of service attacks undertaken in conjunction with a particular security issue. For instance, Chinese hackers vandalized a range of websites in the United States and Europe following the accidental bombing of the Chinese embassy in Serbia during the Kosovo conflict.[43] Likewise, PRC-based hackers launched a range of DDoS attacks against Taiwanese government and civilian services during cross-Strait standoffs in the 1990s.[44] China's abilities and approach in this sphere emerge organically from domestic experiences, where low-intensity cyberattacks and operations have historically played a role in disrupting the coalescing of anti-state narratives around, among other episodes, Uighur separatism, Tibetan rebellion and alleged social dissidence by members of Falun Gong.

Espionage activities include efforts to exploit and access networks for the purposes of observing foreign operators and stealing information. Such activities are covered more fully in other chapters of this book, sufficed to say that Chinese espionage efforts are without peer in terms of the scope of infiltration of Western networks. Since at least 2004, China has been the hub of more espionage behavior (insofar as large numbers of operators appear to be based in China) than any other country in the world.[45] Moreover, from Titan Rain to Ghostnet, the People's Republic of China has reportedly been behind some of the most sophisticated, concentrated efforts to gather foreign intelligence on industry operations, military procedures, research and design and more. Such information is then used for a range of purposes, from informing PRC security efforts and policy to offering state-owned enterprises in China illegitimate competitive benefits via the provision of stolen information.

Finally, disruption and degradation operations differ from information domination operations in that a state seeking to impose real logistical costs on the ability of opponents to act in security affairs. Specifically, such operations can cover the spectrum of techniques from low-intensity vandalism and denial actions to sophisticated malware seeding and network attack. The aims of such operations vary, but can include denying foreign adversaries the ability to control information systems and direct their own operations. In rare instances, aims might include physical damage, such as occurred in the much-described case of Stuxnet in 2010–2011.

Broadly, the PRC's cyber strategy revolves around uniquely Chinese understandings of the relationship between information systems and national welfare on three fronts. First, in terms of the potential for using cyber weaponry in conflict for a range of purposes, Chinese approach to strategy and development of resources has emerged from (1) the country's unique doctrinal history with asymmetric warfare following the communist revolution in 1949 and (2) PLA observations of how information technologies have enhanced America's competitive military edge in conflict since the Vietnam War. The result, even beyond consideration of the digital domain, has been the consistent development of doctrine that emphasizes joint operation and coordination between service branches. Particularly when it comes to scenarios related to engaging either the United States or regional neighbors (most often in relation to the Taiwan Strait or disputes in China's near seas to the south and east), PLA doctrine has pivoted on the development of operating principles that situationally meld the best options available on land, at sea and in the air. This has extended to the cyber domain in that Chinese leadership has consistently codified approaches to cyber operations in strategic documents as being entirely about supporting strategic objectives. Cyber operations are arguably not intrinsically valuable in terms of geopolitical gain and China appears to have embraced this notion. Instead, the development and employment of cyber resources is exclusively employed to better national position. This might explain why China has largely refrained from interfering in the politics of foreign nations, as Russia has done over the past decade, but is the most prolific state sponsor of espionage and low-level disruption attacks during diplomatic crisis in the Asia-Pacific.

Box 8.6: China's "informationization" of conflict in the digital age

There is, at least until the mid-2010s, a significant difference in the way that non-Western countries view information technologies' impact on warfighting. Whereas the United States' defense community and those of close partners often view cyber conflict as a series of interactions that occur on a new plane of operation and can affect the function of security assets on other planes (i.e. the fifth domain), China, Russia and others have consistently articulated an approach to doctrinal and asset development that views the advent of new information technologies as something that fundamentally changes the character – if not the actual underlying nature – of war.

The People's Republic of China, in particular, has since 2004 emphasized the need to "informationize" military forces and to approach questions of strategy, doctrine and tactics from more holistic perspectives.[46] This means, in essence, that China's military leaders and planners view evolving information technologies as systems and platforms that don't just enable new modes of action and reaction, but rather as forces that can change the nature of the warfighting environment. ICTs do not just augment and enhance traditional military power in their implementation; they also force service personnel to alter their approaches to problem solving, national populations to react to aggression differently and more.

In reality, China's "informationized" warfare concept is simply a poor translation for something more akin to "industrialization" than anything else. The People's Liberation Army simply recognizes explicitly that information technologies in the Internet age constitute a system shock to the global system that has both obvious impacts and recursive, amplifiable consequences for national societies. As such, military doctrine increasingly recognizes not only the need to counter conventional forces or the imperative to protect their cyber operations but also the value of disrupting information systems and deceiving foreign militaries for strategic gain.

Second, China sees the web and network technologies as a multiplier for the country's modernization efforts. Economic growth, resilience to global financial shocks and potential for innovation are all intrinsically tied to the effective integration and utilization of the web by citizens, industry and government. Thus, all manifestations of cyber strategy in China reflect an effort to accelerate retention of advantages from network integration. In non-security terms, this means effective enterprise management support and oversight, effective regulation of users, protection of end users and more. From a security perspective, this means the protection of intellectual property and the compromise of foreign advantages in both economics *and* security matters. In essence, cyber abilities present as a useful tool for leveling the playing field through the theft of IP from foreign companies and the mitigation of growth costs for Chinese companies. Though such actions do not easily allow Chinese industry to absorb information and bypass the process of innovation, cyber espionage *does* help to offset development costs, unfairly inform Chinese actors about their operating environment and discourage foreign competition where espionage means increased costs to investment.

Finally, and related to the focus of Chinese strategy on improving national welfare, China's approach to cybersecurity cooperation in an international sense reflects the belief that governments should remain the gatekeepers of operation in cyberspace. In contrast to the "multi-stakeholder" approach advocated by Western countries where non-state actors are considered equal partners for purposes of cyberspace governance, China is one of the leading voices in the international community advocating for a multilateral approach to governing the web. For China, the conclusion that states should be the final arbiters of policy and practice emerges again from the unique manner in which the communist government articulates governance goals and has experienced conflict online. Since social dissidence are seen as direct threats to state security and political stability in a way they are simply not in democratic states, cybersecurity policy emerging from core national security imperatives inevitably touches on the issue of sovereignty in a different format than it might in the West. Assurance of sovereignty in the United States and European countries, for instance, tends to manifest in government approach as the desire to protect domestic consumers, companies and assets in accordance with established legal precedent. For China, sovereignty more specifically means a flexible right to regulate and oversee societal functions. Doing so means denying the equal say of non-governmental stakeholders. Thus, China has enduringly supported international compacts and norms that de-emphasize the right of non-state actors to govern elements of the web free of direct government oversight.

Russia's experiences in cyberspace

Russia's development of cyber capabilities and the incorporation of cyber warfare into Russian foreign policy might seem, to the casual observer, to parallel China's experiences. And yet, in many ways, to think of Russia's and China's presence in the digital domain as similar is to think of their approaches in terms of their similarities as authoritarian states facing a range of potential security competitors in their near abroad. Both countries have invested heavily in the digital methods of sociopolitical control. Likewise, both countries view cyber capabilities as a means to obtaining geopolitical objectives and have adopted cyber techniques as low-intensity tools of preference in manipulating foreign affairs. But there are distinct differences in the sources of cyber strategy and practice between the two countries.

Specifically, to understand the way that Russia has approached the digital domain is to understand the country's unique relationship with organized crime and the geopolitical nature of the country's resurgence following the end of the Cold War. These dissimilarities to China's experiences explain various differences in approach in cyber operations between the two countries, including emphasis on economic versus political information in information theft operations between China and Russia – respectively – and the relative lack of interest in Beijing in the influence operations that have so clearly characterized Russia's cyber conflict approaches in recent years. Indeed, when it comes down to it, the two countries' only real agreement on approach emerges from the mutual desire to mitigate sources of social and political upheaval in censorship efforts.

Rather than attempt to understand Russian institutional and doctrinal commitment to cybersecurity efforts through the lens of distinct realization episodes, many of which are unclear or under-researched in Western analysis, it is instead best to think of Russia's approach as stemming from three dynamics: (1) pre-existing doctrine and strategic approaches to foreign security interactions, (2) experiences with social unrest and (3) the country's extensive criminal ecosystem. On the first count, Russia's approach to statecraft and politicking in Eurasia has – much as is the case with China – enduringly emphasized the important role of asymmetric operations in helping to cultivate favorable conditions for Russian government, culture and economics. Going back as far as the 1880s, the Russian empire and its communist and semi-democratic descendants have seen the cultivation of instability and sociopolitical subversion among neighbors as the key to unimpeded development of Russian interests. Thus, cyber techniques have been embraced by Russian security forces and by the Russian military as a toolkit for causing instability.

Perhaps more importantly, the importance of such operations has been reified by the geostrategic perspective – an enduring driver of security doctrine – of the Putin administration, which sees the instability of the current international community as a necessary condition for the resurgence of Russia as a major stakeholder in world affairs and for offsetting the natural vulnerabilities of the resource industry-dependent national economy. Specifically, Russian state strategy emerges not only from traditions driven by geographic location but also from the unique conditions the country faces post-Cold War. Again, the country has a contracting population, a lagging economy, economic vulnerability based on dependence on commodity prices, a contracted sphere of political influence since the fall of the Soviet Union and, until recently at least, a military in dire need of modernization. Given these circumstances, the clear path to the high reputation, expansive influence and cultural supremacy that Putin's government desires lies with the destabilization of contemporary geopolitical alignments. The retrenchment of countries of the Former Soviet Union (FSU) within the Russian sphere of influence requires the disorganization of foreign resistance to Moscow's overtures. And the reassertion of Russian influence in global terms requires the undermining of international institutions and norms. In all of these endeavors, cyber techniques are seen as powerful tools for augmenting Russia's traditional tools of statecraft.

At the same time, it is perhaps unsurprising – particularly given the authoritarian turn politics in the Russian Federation has taken since the late 1990s – that the government's presence online has been influenced by experiences with social unrest. In particular, the Pussy Riot collective – a large, decentralized network of anti-Putin, pro-democracy and anarchist groups – has been a constant thorn in the side of the Russian government since 2012, when the band Pussy Riot was put on trial for a range of crimes related to violation of public decency laws. Groups and individuals linked with the movement have used social media to organize mass protests and have been known to undertake low-intensity cyberattacks to vandalize the websites of political elites. Moreover, the movement has more recently been connected with the efforts of legitimate oppositions to the Putin government. The result of such experiences with opposition to the rule of the Putin administration and the current structure of Russian political processes has been the application of cyber force by the Russian government in a range of mitigation operations. Specifically, Russian security forces and non-state proxies have attacked opposition websites on hundreds of occasions, have planted incriminating information on the computers of opposition leadership and have actively censored social media communications between Russian citizens seeking to organize protest.

Finally, Russia's cyber conflict approach has significantly been influenced by the existence of a robust criminal ecosystem and set of traditions within the country. In particular, the government's development of cyber institutions and the specific modulation of capabilities between different units are linked to the enduring relationship between national oligarchs – most of whom benefit directly from government patronage or specific positions within government – and organized criminal enterprise.

Though Russian strategic imperatives are relatively simple to understand, the factors that go into planning of specific cyber operations are complex. In general, much as has been the case for China, a primary motivation of Russia finds itself requiring consistent and extensive access to foreign information. Intelligence on foreign companies, governments and civil society organizations are useful for a range of purposes, but specifically for (1) subversion of foreign political processes and (2) as fuel for necessary economic diversification in Russia. Likewise, much as is the case with China, Russian cyber capabilities are concentrated in three arms of government influence – a civil-military arm called the Federal Protective Service (FPO) dedicated mainly to government network protection, military-intelligence services and a large sphere of non-state proxies. Military and intelligence cyber responsibilities are split between those of the Federal Security Service (FSB) and the Military Intelligence apparatus (GRU). Those organizations collectively undertake a range of espionage, disruption and information operations directed through cyberspace.

The third arm of Russian cyber capabilities is a broad sphere of non-state actors whose services are employed in times of particular crisis. In many ways, Russian use of non-state proxies is unique insofar as the ecosystem of both criminals and patriotic hackers that respond to state direction is mostly independent. It is also highly organized around criminal syndicates that control botnets and underwrite the marketplace of malicious code globally. The logic behind massive reliance on non-state proxies for foreign cyber operations is twofold. First, as described in Chapter 11, non-state actors are cost-effective in that the state involved need not commit institutional resources to equipment or training. Moreover, they need not be maintained at all times and actually make money for themselves through criminal enterprise, mitigating the need for the government to funnel in resources to ensure quality of abilities. Second, and relatedly, non-state proxies are themselves highly qualified in the context of Russia's organized criminal ecosystem. Involvement in global malware markets and the cyber mercenary industry ensure consistent updating of skills and equipment. Thus, when needed, non-state proxies can be employed effectively and rapidly.

In efforts to destabilize foreign powers and to undermine the institutions of prevailing world order (as led by Western democracies), Russia has broadly been tied to three formats of large-scale cyberattack aimed beyond its borders in recent years. First of all, Russia engaged in 2007 and 2008 in what we might call cyber blockade behavior in disputes with Estonia and Georgia. In both cases, Russia became tangled in disputes with former Soviet countries over issues pertaining to the wishes of ethnic Russians in either country – regarding the removal of a Soviet-era statue in Estonia and over the wishes of two provinces to secede in Georgia. In both instances, Russian hackers, many of whom were clearly non-state patriotic hackers and criminals, undertook days-long assaults against both countries' information infrastructures. For the most part, these attacks took the form of widespread denial of service attacks against government services and vandalism of government websites. But the massive and persistent nature of the attacks partially succeeded in both cases in hampering the ability of citizens and governments to access the Internet. This had the effect of hampering the function of national economies and caused significant problems for state efforts to deal with non-cyber elements of each crisis.

Russia has also been behind more highly targeted cyberattacks aimed at disrupting specific services or infrastructure in FSU states. Specifically, Russia is known among military and intelligence analysts for the development of complex malware that, often when enabled by low-level techniques like spearphishing emailing or direction from Twitter accounts, allow for espionage against hardened targets and direct disruption of infrastructure. An example of one such employment is Black Energy, where a piece of malicious code – a Trojan – was employed in 2015 and 2016 against Ukrainian electricity infrastructure. This action, though notable as the first cyberattack to actually take out part of a national energy grid, is one of dozens of cyberattacks undertaken by Russian security services and affiliated hackers in support of Russian-aligned forces fighting in the ongoing Ukrainian civil conflict. These attacks are less concerted in terms of strategic objectives than were Russian actions in 2007 and 2008. Nevertheless, they are a common feature of Russian relations with those FSU and nearby countries not aligned with Moscow's interests.

Finally, Russia has broadly engaged in information warfare operations augmented by various kinds of cyberattacks in recent years. While these operations have certainly taken place in FSU countries, primary targets of interest for Western analysts have included the United States, France and the United Kingdom in 2015–2016, 2015/2017 and 2016, respectively. Strategically, the purpose of attacks in these countries has been to destabilize the tenets of political operation among democratic leaders of the international community. Tactically, such operations aim target the operation of legitimate political actors and manipulate events so as to produce outcomes favorable for Russian foreign policy. Broadly, information operations undertaken by Russia in these cases have aimed (1) to establish sources of disinformation, (2) to use social media for targeted subversion and (3) to create crises of credibility around traditional countervailing institutions of democratic operations (i.e. traditional media outlets, political parties and expert voices). Primary methods for doing this have included cyberattacks to steal sensitive information, which is then leaked, monitoring of political stakeholders and the creation of false information outlets. Though the effects of such meddling is unclear and under-researched still at the time of writing, such operations against election processes in the United States and France – and against the Brexit referendum process in the United Kingdom – Russia has shown itself adept at engendering broad-scoped instability through immensely simple uses of ICT. While most attacks in these efforts consist of simple phishing followed by the delivery of Remote Access Trojan (RATs) that allow hackers to evade detection and steal information from email accounts/databases, the results have variously been paradigm-shifting national debates about both political events and – even where subversion directly fails – the viability of institutions previously considered inviolable. In short, even where Russia has failed to tamper with elections, there is clear evidence of crises of constitutionality and fundamental political function emerging from information operations since 2014. Naturally, particularly wherein the reputation of Russia as villainous is already partially cemented in Western debate on foreign affairs (and therefore not at stake), these outcomes are as desirable to the Putin administration as direct election of pro-Russian elements in the West might be.

On the periphery: the experiences of Iran and North Korea

Beyond states and regions we might think of as traditional hubs of power and influence in world affairs, governments across the full gamut of interests and alignments are moving into the cyber sphere and developing conflict capabilities. In particular, countries that we often consider to be rogue or at least ardently opposed to the present state of international order have increasingly utilized the digital domain to combat forces, from regional opponents to India, the United States and China, they perceive as arrayed against them. Two of the more notable states in this category are the Islamic Republic of Iran and the Democratic People's Republic of Korea (DPRK or North Korea). These states occupy what is arguably a unique position of opposition to mainstream world politics. This section describes both countries' experiences with developing cyber capabilities, though it does not describe in-depth the geopolitical positioning or historical motivations of these countries in regional affairs.

It should be noted up front that knowledge about the experiences of these countries and their approach to cybersecurity operations and governance is incomplete. To a degree, this is unsurprising for two reasons. First, these countries, like China and Russia to lesser degrees, operate from belligerent foreign policy positions and are constituted of authoritarian institutions far less open than counterparts in Western democracies. Second, as is broadly the case in countries around the world, cybersecurity responsibilities are difficult to gauge. In part, this is because information technology issues are crosscutting and states incorporate the development of cyber approaches into existing institutional infrastructure. This produces a fog of operational approach that is hard to pierce for analysts abroad, particularly where there are no constitutional stipulations regarding transparency.

Iran's experiences in cyberspace

Iran is a country of major concern to the United States, European partners and regional opponents like Israel and Saudi Arabia. Though adoption of and investment in information technology for military purposes was limited through the 1980s, 1990s and even the 2000s, Iran seems notably capable of using cyber instruments to further its interests in the Middle East.

Three distinct episodes constitute the Islamic Republic's use of cyber weapons to aid geopolitical objectives. First among these is the takedown of a U.S. drone in 2011. In December of that year, an unmanned RQ-170 Sentinel aircraft was brought to the ground near Kashmar in northern Iran "with minimal damage" sustained. The government of Iran announced that the takedown was the work of its "cyber warfare unit" and that the drone had violated sovereign airspace. Regardless of the veracity of that claim (most likely true), the incident demonstrated that Iran had a clear interest in communicating cyber warfare capabilities. In truth, it is unclear as to what caused the drone to go down in Iran. The possibility that a cyber electronic warfare attack took the plane down is distinct, but there are a number of problems with the most common explanations in this vein. For one thing, the drone seems to have landed itself (and was subsequently damaged from having to land on irregular terrain), suggesting the GPS system on board could have been spoofed by cyberattack. And yet, the RQ-170 is not entirely reliant on GPS and would likely not needed to make such a landing. Another possibility is that the drone's command and control systems simply malfunctioned. Regardless, the incident signaled Iran's clear interest in making their abilities in this vein known.

More specifically relevant to the development of state cyber capabilities, U.S. reports point to the Iranian government as having been behind a series of denial of service attacks between 2011 and 2013 against American banks. The attacks consumed significant resources and inflicted unknown financial costs on the banks involved. Responsibility for the attacks was claimed by the Izz ad-Din al-Qassam Cyber Fighters, but U.S. government reports quickly pointed the finger at Iran claiming that the attacks were in retaliation for Stuxnet.[47] Regardless of motivation, the attacks demonstrated Iran's nascent abilities to organize a series of cyberattacks – though, admittedly, unsophisticated ones – against foreign targets.

Finally, and most notably, Iran allegedly launched a virus called Shamoon against the computers of Saudi Aramco, an oil conglomerate based in Saudi Arabia (Iran's most significant regional competitor).[48] The Shamoon virus was nowhere near as sophisticated as Flame or the sections of that program that were tailored to produce Stuxnet. However, the virus struck tens of thousands of computers, wiping data from nearly 30,000, destroying system files and preventing the reboot of a great number of machines. The infrastructural costs were significant, as were losses from data deletion. Perhaps more importantly, Shamoon was indigenously developed in Iran. Not only did unknown government operators manage to successfully launch a viral attack of this nature, but also the 2012 Saudi Aramco incident demonstrated again Iran's rapidly maturing domestic capability to develop relatively sophisticated cyber capabilities.

Strategically, Shamoon also demonstrated Iran's willingness to use cyber weapons more freely in aid of securing geopolitical objectives than they might employ other mechanisms of state power. Indeed, each of these three major cyber conflict episodes involving Iran fit with the strategic perspective on cyber operations that the Islamic Republic – following China, in particular – seems to have adopted. Namely, this includes a focus on cyber weapons as aids to asymmetric warfare against entrenched opponents that allows for belligerent behavior without risking violent retaliation. Cyber capabilities allow for unique disruption akin to that practiced by Iranian state proxies spread throughout the Middle East – destruction and disruption of infrastructure with only plausibly deniable linkages to the Iranian state. Likewise, cyber capabilities are particularly use for information domination and cultivation purposes, and allow Iran to signal and drive favorable normative conditions through the tailored employment of cyberattacks.

The parallel between Iran and China is actually particularly noteworthy insofar as it helps explain similar approaches to the incorporation of cyber capabilities into foreign and military policy. As is broadly true of all states, focus on the potential of information technology for political purposes in Iran was absent until the government was faced with unique crises in which the dynamics of the digital domain became part and parcel of state abilities to ensure national stability. Much like China, the first real episode of this kind had to do with the potential for social and political unrest. In Iran, this took the form the actions of the Iranian Green Movement, a civil movement that mobilized around calls to remove the President of Iran from office in 2009–2010. Much as happened in the case of Falun Gong in China, the Green Movement relied heavily on social networks supported by information infrastructure. Indeed, prior to the late 2000s, the country's main claim to cyber fame actually lay in the existence of Iranian hacker collectives and the massive adoption of programs like Dynaweb (ironically developed by Falun Gong exiles in the West) that allowed individuals to bypass what limited state restrictions on Internet access existed in the 2000s.[49] These elements came to bear on the legitimacy of the Iranian state following Mahmoud Ahmedinejad's election in that protesters were able to mobilize and march in Tehran with relative ease against the efforts of state security services. Moreover, more so than had been the case with Falun Gong, the Green Movement was able to broadcast its message and gained support from Iranians and others both around the country and around the world. Naturally, this episode was a wake-up call to the country's rulers, who saw such civil actions as a potential threat to national political instability.

Unlike China, Iran quickly experienced realization episodes linked with the integrity of and dangers to national security infrastructure in addition to the threat to political stability. In particular, the cyber campaign of the United States and Western allies (plus Israel) against Iran's nuclear weapons development infrastructure – known as Olympic Games – came to a head in 2010 when a malicious computer worm named Stuxnet physically damaged a number of centrifuges at the Natanz uranium enrichment facility.

Stuxnet, discovered in Europe by a forensic computer scientist called Sergey Ulasen, was spread via Microsoft Windows using known vulnerabilities to exploitation and targeted the Siemens industrial control systems in use at Natanz. Over time, Stuxnet was able to silently proliferate such that it was introduced (likely via USB) to the computers at Natanz. Introduction in this way defeated the facilities air gap defenses, which quite simply defend a network through a lack of connection to the broader Internet.[50] Once inside, Stuxnet altered the operational parameters of centrifuges and, by most estimates, caused damaged to about 10% of the facility's machines.[51]

Iran's response to these wake-up calls has been diverse. Much like China, Iran's cyber capabilities are split between three arms of state operators – civilian government actors, military actors and non-state proxy agents. Likewise, the latter two actors are those most worthy of interest for students of cyber conflict, largely because civilian capabilities are limited to intra-department functions (i.e. even government network defense falls to the military). Since the early 2010s, Iran has developed reasonably sophisticated cyber conflict capabilities. Saudi Aramco, which is likely the incident most widely known to those interested in Iran's cyber conflict portfolio, was actually a reasonably unsophisticated operation to disrupt the capabilities of a regional opponent. Since 2012, Iranian hackers have struck at targets in Israel repeatedly, defacing websites and intruding to disrupt service and set up backdoors for future operations. In some cases, culprits have appeared to be Iranian government operators – so assumed given forensic evidence linking intrusions to IP addresses near Iranian government facilities – and non-state proxies.

Iran's technical capabilities are hard to gauge beyond a general trend toward greater sophistication. The development of institutional infrastructure in Iran focused on cyber warfare capabilities is somewhat easier to deduce. At the highest level, cyber policy in Iran is driven by the dictates of the Supreme Council of Cyberspace, set up by Ayatollah Khamenei for the express purpose of coordinating ICT governance approaches across a range of areas. The Council's membership broadly includes the President of Iran alongside representatives of the judiciary and legislature and the leaders of Iran's police forces, intelligence community and departments of telecommunications, science and culture. In the military, Iran's primary cyber force is organized under the Cyber Defense Command, which was also established by Ayatollah Khamenei in 2010. The CDC in Iran specifically began life with the goal of responding to Stuxnet.

Iran's future ability to employ cyber techniques for political and geopolitical gain is unclear. However, responses to Stuxnet and the social unrest around the 2009–2010 election season have clearly included some development of capacity to act in the digital domain. In addition to cyber strikes against regional opponents, Iran's censorship regime has borrowed from the lessons learned by Gulf states during the Arab Spring. The country's approach increasingly incorporates a mixed approach to social expression that seeks to provide a pressure valve for dissident sentiment alongside strict censorship of speech that advocates assembly and protest. Combined with clear focus on using cyber weapons for disruption and destruction detached from the traditional risks of using state assets to attack regional opponents, it is clear that Iran has at least notionally committed to becoming a first order cyber power in world affairs.

North Korea's experiences in cyberspace

Whereas the realization episodes experienced by most countries are reasonably clear to the casual analyst, the drivers of North Korea's programmatic development of cyber conflict capabilities are not clear. More than is the case even with other authoritarian regimes, information about the institutions and internal functionality of the Hermit Kingdom is extremely hard to come by. Though some information exists about how cyber institutions are organized, what is known about North Korea's cyber arsenal and strategic approach ICT usage for conflict purposes emerges entirely from the experiences of those attacked by the DPRK online over the past decade.

Unlike other countries, where there is a distinct division of focus between military matters and the regulation/repression of civil society, North Korea's cyber institutions almost entirely focus on military and intelligence matters.[52] In large part, this emerges from the poor ICT penetration in North Korea. Simply put, few members of the population have access to the Internet and those non-state proxies employed by the government are either directly employed at military facilities or abroad in China. According to a number of Western intelligence and academic analyses, cyber operations undertaken by the DPRK is generally overseen by two entities that employ around 5,900 "cyber warriors" – the Reconnaissance General Bureau (RGB) and Korean People's Army General Staff Department (GSD).[53] RGB was formed in 2009 as an amalgamation of a range of intelligence units and special operations organizations. The RGB is the primary organization tasked with foreign operations, specifically influence operations and the provocation of foreign governments. RGB directly answers to the North Korean leader. By contrast, the GSD is primary tasked with network defense operations and is somewhat less clearly organized. By comparison with the relatively centralized RGB, GSD is somewhat like the Department of Defense in the United States circa 1995 – operationally able to bring network defense resources to bear on incipient challenges, but relatively decentralized. The primary way that GSD and RGB work together is in the provision of intelligence information from GSD to RGB for the purposes of planning cyber operations abroad.[54]

North Korea is a notable cyber power largely because of the degree to which the country has effectively incorporated operations in the digital domain into its existing doctrine. Nestled in Northeastern Asia between U.S. allies and a friendly China that is nevertheless a major stakeholder in the international community, the Hermit Kingdom has for many decades resisted any form of liberalization and domestically encourages a cultural narrative of national victimization. North Korea, in the words of the country's leaders, is a bastion of cultural purity and security amidst foreign powers that seek to destroy Korean culture. In reality, North Korea's isolation is the result of a standoff between moral and security considerations on the part of Western and Western-backed opponents. The country's brutal repression of its own people is unacceptable. At the same time, the country's militaristic stance – particularly its development of nuclear weapons – means that few world leaders have advocated forced regime change.

Given this dynamic, North Korea's foreign policy doctrine has always been one of foreign destabilization, low-risk provocation and deniability. The degree to which the country provokes versus joining other states at the bargaining table traditionally oscillates in direct proportion to the country's economic woes. Where the DPRK's struggling economy portends a real threat to the regime's rule, the Kim dynasty uses the promise of receding provocation as a means for obtaining international aid and release of long-term sanctions. Upon relief, provocation – at least insofar as the country can avoid major conflict – continues.

Cyber techniques have enabled North Korea increasingly to strike out against foreign states in low-risk/high-gain ways and have even been the object of new efforts to drive different informational narratives related to the DPRK in international affairs. Of note, North Korean hackers have prosecuted a range of attacks against South Korean and Japanese entities over the past half-decade. In 2013, North Korean hackers hit South Korean banks with denial of service attacks and vandalized websites. In the same episode, DPRK hackers were behind attacks on South Korean media stations and malware named "DarkSeoul" was discovered across a range of financial services firms. In 2014, North Korean hackers were, according to the President of the United States and U.S. intelligence, behind an attempt to coerce Sony Pictures into not releasing *The Interview*, a comedy about an attempt to assassinate the leader of North Korea. In that episode, the DPRK threatened cyberattacks against Sony if demands were not met, an effort that was ultimately unsuccessful and resulted in a major ten hour long cyber assault against North Korean ISPs by U.S. government operators.[55]

Ultimately, whereas Iran might considered to be a rising cyber power, the increasing sophistication of North Korea's cyber conflict capabilities must be considered in the context of the DPRK's geopolitical position. Cyber capabilities reflect a unique addition to the North Korean arsenal of provocation and destabilization. However, disruption abilities lend little additional to the DPRK in terms of the geopolitical dynamics of East Asia. The North Korea remains a pariah state and even minimal evidence of DPRK belligerence is met with universal condemnation. Thus, the potential for anything more than annoying actions is nearly zero. Rather, it is likely that the main threat potential accrued by North Korea's continuing development of online capabilities comes from the rising potential for success in espionage operations. North Korean operators have shown a limited ability to exhilarate information during cyberattacks, as well as a proclivity for coercive blackmail as a viable tool of statecraft. Greater sophistication in cyber techniques could, out into the future, provide the DPRK with access to technologies or the resources of other global belligerents to enhance state power in meaningful ways.

Notes

1 A more extensive treatment of cyber policy categories and issues is available in Bayuk, Jennifer L., Jason Healey, Paul Rohmeyer, Marcus H. Sachs, Jeffrey Schmidt, and Joseph Weiss. *Cyber security Policy Guidebook*. John Wiley & Sons, 2012.
2 See Healey, Jason, and Karl Grindal, eds. *A Fierce Domain: Conflict in Cyberspace, 1986 to 2012*. Washington, DC: Cyber Conflict Studies Association, 2013.
3 Ibid.
4 Ibid.
5 Ibid.
6 Young, Mark D. "National Cyber Doctrine: The Missing Link in the Application of American Cyber Power." *Journal of National Security Law & Policy* 4 (2010): 173.
7 See *inter alia* Lynn, William J. "Defending a new domain: The Pentagon's cyberstrategy." *Foreign Affairs* 89, no. 5 (2010): 97–108; and National Security Strategy, White House 2018.
8 See, *inter alia*, Barrett, Matthew P. *Framework for Improving Critical Infrastructure Cybersecurity Version 1.1*. No. NIST Cybersecurity Framework, 2018; Shen, Lei. "NIST Cybersecurity framework: Overview and potential impacts." The *SciTech Lawyer* 10 (2013): 16; and Shackelford, Scott J., Andrew A. Proia, Brenton Martell, and Amanda N. Craig. "Toward a global cybersecurity standard of care: Exploring the implications of the 2014 NIST Cybersecurity Framework on shaping reasonable national and international cybersecurity practices." *Texas International Law Journal* 50 (2015): 305.
9 Stoll, Clifford. "The cuckoo's egg: Tracking a spy through the maze of computer espionage." (1989).
10 Ibid.
11 Orman, Hilarie. "The Morris worm: A fifteen-year perspective." *IEEE Security & Privacy* 99, no. 5 (2003): 35–43.
12 DeNardis, Laura. "A history of internet security." In *The History of Information Security*, pp. 681–704. 2007.

13 Healey and Grindal (2013)
14 For perhaps the best discussions on these developments, see Verton, Dan, and Jane Brownlow. *Black Ice: The Invisible Threat of Cyber-Terrorism*. Osborne, 2003; Carter, Ashton B. "The architecture of government in the face of terrorism." *International Security* 26, no. 3 (2002): 5–23; and Kaplan, Fred. *Dark Territory: The Secret History of Cyber War*. Simon and Schuster, 2016.
15 Adams, James. "Virtual defense." *Foreign Affairs* (2001): 98–112.
16 See Moteff, John, and Paul Parfomak. "Critical infrastructure and key assets: Definition and identification." Library of Congress, Washington, DC, Congressional Research Services, 2004; and House, White. "The national strategy to secure cyberspace." Washington, DC: White House, 2003, respectively.
17 Kaplan, Caren. "Air power's visual legacy: Operation Orchard and aerial reconnaissance imagery as ruses de guerre." *Critical Military Studies* 1, no. 1 (2015): 61–78.
18 See Lewis, James A. "Computer espionage, Titan Rain and China." *Center for Strategic and International Studies-Technology and Public Policy Program* (2005): 1; and Thornburgh, Nathan. "Inside the Chinese hack attack." *Times, August* 25 (2005).
19 EINSTEIN has been assessed extensively from both usage and legal perspectives. See *inter alia* Bradbury, Steven G. (2009) Legal Issues Relating to the Testing, Use, and Deployment of an Intrusion-Detection System (Einstein 2.0) to Protect Unclassified Computer Networks in the Executive Branch. DoJ, Opinions of the Office of Legal Counsel, in Vol. 33 (January 9, 2009); Bradbury, Steven G. "The developing legal framework for defensive and offensive cyber operations." *Harvard National Security Journal* 2 (2011): 629; Coldebella, Gus P., and Brian M. White. "Foundational questions regarding the federal role in cybersecurity." *Journal of National Security Law & Policy* 4 (2010): 233; and Oree, William L. *Analysis of the United States Computer Emergency Readiness Team's (US CERT) Einstein III intrusion detection system, and its impact on privacy*. Monterey CA: Naval Postgraduate School, Dept of Information Sciences, 2013.
20 Sedgewick, Adam. *Framework for Improving Critical Infrastructure Cybersecurity, version 1.0*. No. NIST-Cybersecurity Framework. 2014.
21 See U.S. Computer Emergency Readiness Team, *GRIZZLY STEPPE – Russian Malicious Cyber Activity*, February 10, 2017.
22 Discussed in Chapter 4. Additionally, for a good resource that aims to catalogue the national legacy of signals intelligence efforts in the United Kingdom and the United States in the context of later cyber developments, see Corera, Gordon. *Cyberspies: The Secret History of Surveillance, Hacking, and Digital Espionage*. Pegasus Books, 2016.
23 Strategic Defense Review, 1998.
24 Healey and Grindal (2013).
25 Ibid.
26 Ibid.
27 Ibid. Also see Symantec, Report on Lovgate, 2003.
28 "'Amarillo' Video Crashes MoD PCs." BBC.co.uk. May 17, 2005.
29 Cornish, Paul, David Livingstone, Claire Yorke, and Dave Clemente. *Cyber Security and Critical National Infrastructure: Chatham House Report*. Royal Institute of International Affairs, 2012.
30 Ibid.
31 For more on NATO's re-orientation toward cyber defense efforts following these experiences, see *inter alia* Hughes, Rex. "NATO and cyber defence." *Atlantisch Perspectief* 33 (2009); Herzog, Stephen. "Revisiting the Estonian cyber attacks: Digital threats and multinational responses." (2011); Czosseck, Christian, Rain Ottis, and Anna-Maria Talihärm. "Estonia after the 2007 cyber attacks: Legal, strategic and organisational changes in cyber security." *International Journal of Cyber Warfare and Terrorism (IJCWT)* 1, no. 1 (2011): 24–34; Healey, Jason, and Leendert Van Bochoven. *Nato's Cyber Capabilities: Yesterday, Today, and Tomorrow*. Atlantic Council of the United States, 2012; and Fidler, David P., Richard Pregent, and Alex Vandurme. "NATO, cyber defense, and international law." *John's Journal of International & Comparative Law*. 4 (2013): 1.
32 Joubert, Vincent. *Five Years After Estonia's Cyber Attacks: Lessons Learned for NATO?*. NATO Defense College, Research Division, 2012; and Hunker, Jeffrey. "NATO and cyber security." In *Understanding NATO in the 21st Century*, pp. 178–199. Routledge, 2013.
33 Deni, John R. "NATO's new trajectories after the Wales Summit." *Parameters* 44, no. 3 (2014): 57.
34 Schmitt, Michael N., ed. *Tallinn manual on the international law applicable to cyber warfare*. Cambridge University Press, 2013; and Schmitt, Michael N., ed. *Tallinn Manual 2.0 on the international law applicable to cyber operations*. Cambridge University Press, 2017. See also Schmitt, Michael. "International

Law in Cyberspace: The Koh Speech and the Tallinn Manual Justaposed." (2012); Fleck, Dieter. "Searching for international rules applicable to cyber warfare – A critical first assessment of the new Tallinn manual." *Journal of Conflict and Security Law* 18, no. 2 (2013): 331–351; and Kessler, Oliver, and Wouter Werner. "Expertise, uncertainty, and international law: A study of the Tallinn Manual on cyberwarfare." *Leiden Journal of International Law* 26, no. 4 (2013): 793–810.

35 For other works that seek to do the same, see Thomas, Timothy L. "Nation-state cyber strategies: Examples from China and Russia." *Cyberpower and National Security* (2009): 475–76.

36 See, among other sources, Liang, Bin, and Hong Lu. "Internet development, censorship, and cyber crimes in China." *Journal of Contemporary Criminal Justice* 26, no. 1 (2010): 103–120; Lum, Thomas. *Internet Development and Information Control in the People's Republic of China*. Washington, DC: Library of Congress, Congressional Research Service, 2006; and Lu, Wei, Jia Du, Jin Zhang, Feicheng Ma, and Taowen Le. "Internet development in China." *Journal of Information Science* 28, no. 3 (2002): 207–223.

37 For further description of this period, see Bell, Mark R., and Taylor C. Boas. "Falun Gong and the Internet: Evangelism, community, and struggle for survival." *Nova Religio: The Journal of Alternative and Emergent Religions* 6, no. 2 (2003): 277–293; Hachigian, Nina. "China's cyber-strategy." *Foreign Affairs* 80 (2001): 118; Deibert, Ronald J. "Dark guests and great firewalls: The Internet and Chinese security policy." *Journal of Social Issues* 58, no. 1 (2002): 143–159; Chase, Michael S., and James C. Mulvenon. *You've got Dissent! Chinese dissident use of the Internet and Beijing's Counter-Strategies*. Rand Corporation, 2002; and Yang, Guobin. "The co-evolution of the Internet and civil society in China." *Asian Survey* 43, no. 3 (2003): 405–422.

38 Tong, James W. *Revenge of the Forbidden City: The Suppression of the Falungong in China, 1999–2005*. Oxford University Press, 2009.

39 Lindsay, Jon R., Tai Ming Cheung, and Derek S. Reveron, eds. *China and Cybersecurity: Espionage, Strategy, and Politics in the Digital Domain*. Oxford University Press, USA, 2015.

40 Ibid.

41 Ibid.

42 Indeed, Lindsay argues that such factors can inhibit accurate assessments of China's actions online. See Lindsay, Jon R. "The impact of China on cybersecurity: Fiction and friction." *International Security* 39, no. 3 (2015): 7–47.

43 Porterfield, Jason. *Careers as a Cyberterrorism Expert*. The Rosen Publishing Group, 2011.

44 Wu, Xu. *Chinese Cyber Nationalism: Evolution, Characteristics, and Implications*. Lexington Books, 2007.

45 See Thornburgh, Nathan, Matthew Forney, Brian Bennett, Timothy J. Burger, and Elaine Shannon. " 'The Invasion of the Chinese Cyberspies (And the Man Who Tried to Stop Them)." *Time Magazine* 29 (2005): 1098961–1; Norton-Taylor, Richard. "Titan Rain: How Chinese hackers targeted whitehall." *The Guardian* 4 (2007); and Thornburgh, Nathan. "Inside the Chinese hack attack." *Times* August 25, 2005.

46 For more on this, see Fritz, Jason. "How China will use cyber warfare to leapfrog in military competitiveness." *Culture Mandala: The Bulletin of the Centre for East-West Cultural and Economic Studies* 8, no. 1 (2008): 2; Ventre, Daniel, ed. *Chinese Cybersecurity and Defense*. John Wiley & Sons, 2014; and Zhi-jia, Y.A.N. "Exploration on the cyber corps training in small-medium sized enterprises in the construction of informationization." *Modern Computer* (2011): Z1.

47 See Horgan, John. "From profiles to pathways and roots to routes: Perspectives from psychology on radicalization into terrorism." *The ANNALS of the American Academy of Political and Social Science* 618, no. 1 (2008): 80–94; and Iasiello, Emilio. "Cyber attack: A dull tool to shape foreign policy." In *Cyber Conflict (CyCon), 2013 5th International Conference on*, pp. 1–18. IEEE, 2013.

48 See Bronk, Christopher, and Eneken Tikk-Ringas. "The cyber attack on Saudi Aramco." *Survival* 55, no. 2 (2013): 81–96; Bronk, Christopher, and Eneken Tikk-Ringas. "Hack or attack? Shamoon and the Evolution of Cyber Conflict." (2013); and Dehlawi, Zakariya, and Norah Abokhodair. "Saudi Arabia's response to cyber conflict: A case study of the Shamoon malware incident." In *Intelligence and Security Informatics (ISI), 2013 IEEE International Conference on*, pp. 73–75. IEEE, 2013.

49 Gutmann, Ethan. "Hacker nation: China's cyber assault." *World Affairs* (2010): 70–79.

50 See Lindsay, Jon R. "Stuxnet and the limits of cyber warfare." *Security Studies* 22no.3 (2013): 365–404. Also see Aleksandr Matrosov, Eugene Rodionov, David Harley, and Juraj Malcho, "Stuxnet under the Microscope." eset, *White Paper*, January 20, 2011; and Zetter, Kim. "How digital detectives

deciphered stuxnet, the most menacing malware in history." *Wired Threat Level Blog* July 11, 2011, www.wired.com/threatlevel/2011/07/how-digital-detectives-deciphered-stuxnet.

51 Albright, David, Paul Brannan, and Christina Walrond, "Did stuxnet take out 1,000 centrifuges at the natanz enrichment plant?." *Institute for Science and International Security*. December 22, 2010, 3–4. See also Langner, Ralph. "To kill a centrifuge: A Technical analysis of what stuxnet's creators tried to achieve." *The Langner Group* November 2013.

52 Jun, Jenny, Scott LaFoy, and Ethan Sohn. *North Korea's Cyber Operations: Strategy and Responses*. Rowman & Littlefield, 2016.

53 Ibid.

54 Ibid.

55 We discuss this episode in previous chapters. Broadly, however, for further detail, see Haggard, Stephan, and Jon R. Lindsay. "North Korea and the Sony Hack: Exporting instability through cyberspace." (2015); and Whyte, Christopher. "Ending cyber coercion: Computer network attack, exploitation and the case of North Korea." *Comparative Strategy* (2016).

Further readings

Corera, Gordon. *Cyberspies: The Secret History of Surveillance, Hacking, and Digital Espionage*. Pegasus Books, 2016.

Healey, Jason, and Karl Grindal, eds. *A Fierce Domain: Conflict in Cyberspace, 1986 to 2012*. Washington, DC: Cyber Conflict Studies Association, 2013.

Jasper, Scott. *Russian Cyber Operations: Coding the Boundaries of Conflict*. Georgetown University Press, 2020.

Kania, Elsa B. "The ideological battlefield: China's approach to political warfare and propaganda in an age of cyber conflict." In *Information Warfare in the Age of Cyber Conflict*, pp. 42–53. Routledge, 2020.

Kaplan, Fred. *Dark Territory: The Secret History of Cyber War*. Simon and Schuster, 2016.

Russell, Alison Lawlor. *Cyber Blockades*. Georgetown University Press, 2014.

Shane, Peter M., and Jeffrey Allen Hunker, eds. *Cybersecurity: Shared Risks, Shared Responsibilities*. Carolina Academic Press, 2013.

Valeriano, Brandon, Benjamin Jensen, and Ryan C. Maness. *Cyber Strategy: The Evolving Character of Power and Coercion*. Oxford University Press, 2018.

Council on Foreign Relations. In*dependent Task Force Report No. 80: Confronting Reality in Cyberspace: Foreign Policy for a Fragmented Internet*. July 2022. www.cfr.org/report/confronting-reality-in-cyberspace

Domingo, Francis C. *Making Sense of Cyber Capabilities for Small States Case Studies From the Asia-Pacific*. Routledge, 2022.

9 States at cyberwar

The dynamics and national strategies of cyber conflict

In this chapter, we build on Chapter 7's discussion of the empirical trends and dynamics of cyber conflict and discuss cyber warfare as it pertains to state security interests. Specifically, we undertake to describe the various elements of the 21st century national security apparatus that are vulnerable to different forms of cyber-enabled warfare, before addressing core questions about how useful cyber conflict is to state actors.

What is vulnerable in the age of cyber conflict?

If there is one thing that the brief history of interstate cyber conflict outlined in the past chapter indicates, it is that the apparatus of national security is multifaceted in the digital age, far more so than has been true – regardless of whether you delineate an era in terms of wars fought, organizational changes achieved or technological advances made – of eras past. Simply put, national security in the digital age revolves around more than just military capabilities. To some degree, of course, this has always been the case. A state's power is typified by its warfighting capabilities, its institutional pull and its ability to affect norms of behavior in international affairs. National security planners must consider the protection of those processes that undergird all of the above. These include (1) actual security and military procedures, (2) those systems that govern national economic and social function (i.e. the various realms of critical infrastructure), (3) the political-institutional and real informational foundation of the innovation economy, (4) the regulations of coordination between government and private industry and (5) the ability of the domestic political process to function according to design and prevailing expectations. This final one is more pressing for democratic nations, of course, than it is for semi-democratic or authoritarian ones. Of these different imperatives, the nature and function of a nation's regulatory environment – both in how it operates in the domestic and international arenas – is unpacked in greater detail in other chapters. The remaining four are discussed in some detail now.

Military systems

In many ways, military systems constitute the simplest element of the national security apparatus insofar as actual cybersecurity procedures include the employment of computer network attack, exploitation and defense techniques to safeguard critical platforms/information and augment traditional military capacity. A range of operations are possible for military forces, including the use of attack techniques to disrupt enemy systems, the use thereof to enhance the function of kinetic forces or the use of exploitation techniques to pave the way for greater effectiveness in future operations.

DOI: 10.4324/9781003246398-9

In reality, it is arguably most accurate to say new challenges in the digital age for military forces are not necessarily bound up in new technical possibilities for attack and defense – except possibly to admit that military forces are, quite obviously, vulnerable to cyber disruption – so much as they are about the organization of military assets and procedures such that multifaceted service branches are able to consistently take advantage of new methods for defense and offensive operational success. Military cyber conflict considerations are inevitably not only about defending from foreign cyberattacks. Neither are they entirely concerned with how best to incorporate new disruptive techniques into military procedures. Considerations also include using digital techniques for reconnaissance purposes, coordinating with non-military agencies that aid national security operations (including the intelligence community, the technology industry, defense contractors and ISPs) and ensuring that continued technological innovation is both possible and free of malicious meddling from foreign actors. And on top of this, of course, military forces have had to adapt to the special characteristics of conflict in the digital age by developing new rules of engagement.

At this point in time, the norm for development of centralized authority on network warfare as a coordination structure for military forces across countries appears to be the incorporation of a specific joint forces structure that has jurisdiction over the cyber domain. In the United States, this was the Joint Functional Component Command for Network Warfare, which since 2009 is now extant in the form of U.S. Cyber Command. In this, the United States – and this format is mirrored across a range of partner nations in Europe, East Asia and Latin America – has essentially identified **cyberspace** as a domain that is unique and discrete from other traditional operating domains (air, land, space and sea). Such joint force structures are given broad purview and control of coordinating procedures for any functions of military forces that include network warfare. For the United States, this means that Cyber Command broadly has responsibility for supporting and undergirding the network functions of other Combatant Commands in operations.

Military forces around the world face a unique set of issues when it comes to developing appropriate rules of engagement for interactions with the security elements of other states. In particular, military organizations must contend with the issue of aligning the dynamics of conflictual interactions in the cyber domain with pre-existing norms and rules of engagement outlined in adherence to the various laws of armed conflict, which will be discussed in more detail in Chapter 12. Cyber attack is not inherently violent, but it is aggressive and can disrupt the lawful operation of foreign governments, industry and society. Military forces must determine (1) the conditions under which the undertaking of cyberattacks is permissible, (2) the specific profile of targets that can be attacked, (3) the duration of an assault, (4) necessary communications surrounding an attack and (5) lines of communication within the government necessary to authorize different forms of action.

In reality, this final challenge of lawfare is particularly worrisome for analysts considering the prospect of cyber conflict thresholds – i.e. the barriers that may or may not exist to prevent the outbreak of digital or digitally augmented conflict. Authorization for cyberattacks may not necessarily come from the executive level. Indeed, in many cases there is simply no link between high-level officials and those personnel responsible for authorizing either initial or reactive attacks. In a situation where a system has been compromised, for example, the responsibility for retaliating in order to delete stolen information may in some instances fall to the officer on duty or his direct commander. Thus, the threshold for military-to-military or military-to-nonstate actor interactions in cyberspace is normally variable based on the condition of military procedures in place across countries. Understanding decision-making as emerging more from this set of dynamics than from centralized strategic planning, democratic debate or some other catalyzing process is called the **cybernetic model** of decision-making by scholars of strategic studies, wherein the clash of procedures principally determines the shape of conflict incidents.[1]

For the U.S. military, such procedures – in the context of defensive operations – are called **response actions** (RAs).[2] RAs set specific guidelines about what kind or intensity of attack warrants a response. Likewise, it outlines what legitimate targets look like, such as an originating machine for an attack but not a "zombie" computer that likely does not belong to the attacker. RAs are perhaps fuzziest with regard to the geography thresholds that typically bear on decisions to respond to an attack. In conventional military operations, kinetic assets are often limited in their ability to retaliate over large distances. More importantly, doing so may compromise broader strategic security conditions, and so response procedures set strict limits on response boundaries. No such considerations exist with cyberspace, at least in terms of purely geographic considerations. Instead, RAs must be based on a shifting understanding of the nature of foreign threats as they relate to international and national laws, the jurisdiction of foreign governments and the jurisdiction of other agencies at the national level, such as the **Federal Bureau of Investigation**.[3] The purpose in structuring guidelines in this way is twofold. First, such procedures prevent the employment of military assets in such a way that there would be broader, unintended ramifications to diplomatic relations between states. Second, military forces are able to rely on other assets before committing to reactionary defensive operations that may be wasteful, may be deemed heavy-handed or may reveal counter-force abilities prematurely.[4]

Two further digital age challenges for military establishments bear mention – that of talent acquisition and that of weapons development. In both cases, a unique condition of conflict in the information age acts as the factor driving a need for adaption. That condition has to do with the source of the technological innovation that defines and continually transforms the characteristics of the cyber domain in which militaries must operate. Simply put, much of the inherent ability of a state's military to remain on the cutting edge of electronic and information warfare extends from the health of the innovation. Unlike with other military assets, such as fighter jets or submarines, innovation relevant to the armed forces is not – indeed, *cannot* be – isolated in a neat complex of defense contractors. A nation's broader technology industry and the development of global technology infrastructure matter and must be factored into planning processes.

One challenge particular to military forces is that of relevant talent in an establishment's workforce.[5] Personnel with unique abilities to aid in systems administration, asset development and more must be retained over long periods of time. To do this, the armed forces must compete with the trappings of a burgeoning global marketplace for such specialized talent. In other words, militaries have to entice employees that might otherwise go into the private sector.[6] This invariably means not only improving economic incentives for would-be government workers but also liberalizing cultures within the armed forces and civilian government ranks. A related challenge is the acquisition of actual systems useful for defense, detection and offense in the cyber domain. In many ways, of course, effective realization of such a development program extends organically from solving labor force issues and from effective coordination with the private sector (to be discussed further in the next few chapters). But militaries do face training and planning issues in a macro sense that are additionally worthy of note here. Such issues are no more complicated than the fact that large numbers of service personnel must constantly be trained to incorporate new requirements and new abilities of operation into planning processes. At the highest levels, where relatively higher numbers of older personnel are concentrated, this means the effective reimagining of ongoing education initiatives and necessary emphasis on effective articulation of new operation imperatives to civilian government.[7] Otherwise, militaries can face additional issues of asset acquisition that emerge from an ineffectual articulation of development priorities in the statements of military and civilian leaders, government budgets and executive-level strategies.

Critical infrastructure

When it comes to national security and cyber issues, it is difficult to avoid the topic of critical infrastructure. "Critical infrastructure" is a term of art – first promulgated in a series of commission reports on infrastructural vulnerabilities in the United States following the Oklahoma City bombings in 1995 – that describes different sectors of national systems so important that the destruction or debilitation of one would have a major impact on national security and welfare.[8] Critical infrastructure in the United States specifically describes the 16 infrastructural sectors that, for lack of a better phrase, allow the country to run. These sectors are diverse and include a range of societal disparate functions, from water systems and agriculture to transportation, dams and nuclear energy.

The notion that critical infrastructure might qualify as a component part of a conceptually broadened national security apparatus is relatively uncontroversial. Indeed, public infrastructure has been targeted and protected by states in wartime and peace for thousands of years. The health of a nation's roads, railroads, energy infrastructure and productive abilities are, much as is discussed later when we turn to the issue of intellectual property, clearly correlated with national capacity to act in the national interest. With regard to the information revolution, however, the threat to critical infrastructure has become significantly more pronounced in any imagining of conflict with aggressive foreign actors.

Cyberspace and information technologies are crosscutting. Every sector of critical infrastructure relies on information systems to function effectively. Some, of course, rely on ICTs far more than others. Jeffrey Hunker describes this variation in terms of a concept from the engineering field – tightly or loosely coupled.[9] Tightly coupled systems are those where change or disruption in one sector means a direct and rapid effect elsewhere. Loosely coupled systems describe a slower moving set of effects. With regard to critical infrastructure, we might consider sectors like agriculture to be loosely coupled. Disruption has a clear effect, but it is not necessarily drastic or immediate, particularly if rapid reversal is possible. Banking, energy infrastructure and transportation, by contrast, are tightly coupled. Naturally, attacks against tightly coupled critical infrastructure are considered to be more significant in the context of national security processes than might be intrusions targeting loosely coupled sectors.[10]

Though cyber technologies are crosscutting, there are effective hubs of national abilities to ensure digital functionality for the country writ large. Specifically, the telecommunications sector of critical infrastructure governs the ability of other sectors to function in full. Therefore, telecommunications infrastructure is a particularly serious consideration for states when preparing for a range of cyber conflict scenarios. Relationships with critical infrastructure sectors will be discussed further in the chapters focusing on national experiences, including with public-private dialogues across countries.[11]

For now, though, it is perhaps worthwhile to note that one of the major issues that states face with regard to infrastructure protection comes from the fact that governments need to adapt policy to suit massive sections of national economies. For non-authoritarian states, this is not easy. Different economic sectors have remarkably different and remarkably entrenched perspectives on information technology issues. Not only that, but effective state command of the national security apparatus inevitably means the ability to affect control of different processes in crisis scenarios. This will be discussed further in the next chapters, but in short there is a clear issue in terms of the abilities that governments want with regard to infrastructural function. Perhaps the simplest example is that of the kill switch option. A kill switch would be a mechanism of governance set up to allow a state executive (or perhaps a small legislatively delegated body) to effectively shut down the national Internet through control of ISPs. Naturally, this option is unacceptable to a range of national industrial and infrastructural parties, regardless of their commitment to national security integrity, because it directly clashes with a range of competing behavioral motivations, most notably the need to function as private businesses and responsibilities to the public. In the context of national and international security imperatives, governments must develop policy that brings as much capacity for effective control as possible within reach while still responsibly meeting the demands of society and industry. This imperative is further exacerbated by nuanced formats of relationships between government and industrial sectors, wherein some particularly tightly coupled ones – like the banking sector – make natural partners for government that do not act impartially in national-level conversations regarding appropriate buy-in to cybersecurity initiatives.

Intellectual property

As mentioned in the previous section, the nature of the innovation economy directly determines the operational capacity of major elements of a state's national security infrastructure, including the functionality of militaries in warfighting. Though this is perhaps more true within the context of cyber conflict than it is in the aggregate, the idea that economic and innovative potential links to national power is not new. To the contrary, scholars of IR of all stripes have regularly argued that the dynamic operation of the innovation economy is part of the basis of long-term power potential.[12] This is not the entire picture, of course, but it is certainly the case that industrial efficiency is linked to economic growth and development. Digital intrusion on a massive scale to steal data from governments and private companies, whether coordinated or not, portends a significant redistribution of innovative processes that might feasibly lead to unwanted imbalances of power. Even though there are challenges in absorbing massive amounts of stolen information, the theft and reapplication of intellectual property is one commonly discussed way that countries can change the dynamic of their own industrial potential in the short term. Thus, much as the British and French Empires concentrated considerable effort on the task of trade and economic administration, states in the digital age focus attention on the issue of low-level incursions with the aim of minimizing future security costs.

Naturally, the exact costs of such redistribution are difficult to pinpoint. In the aggregate, reporting has held that costs to global industry from the theft of intellectual property are annually perhaps as low as $2 billion and as high as $400 billion. Given that global GDP of $94 trillion as of 2021, these figures do not necessarily suggest major disruption. But it is worth considering the long-term challenges to national security from the direct and indirect effects of such information theft, both criminal and political.

Perhaps the most commonly cited deleterious outcome of intellectual property theft is the direct transfer of sensitive technological or intelligence secrets from the infiltration of government, intelligence community or defense contractor systems. In 2015, a series of leaked reports claimed that Chinese authorities had stolen more than 50 TB of information from the U.S. government that included blueprint information for weapons platforms like the F-35 Lightning II fighter craft.[13] Certainly, cyberattacks undertaken during Moonlight Maze, Titan Rain/Byzantine Hades, Shady Rat and other episodes gathered large amounts of both unclassified and classified information useful to foreign governments.[14] Chinese authorities have levelled reciprocal accusations against the United States, accusing it of building an "axis of cyber theft" in June 2022 based on a recent report claiming the U.S. government stole 97 billion pieces of Internet data in 30 days alone.[15] Naturally, such one-way exchanges of information risk the national security of the victim state by enhancing foreign potential to compromise the victim's capabilities.

With regard to the innovation economy itself, broad-scoped problems of intellectual property loss can cause massive problems with investor confidence on a number of fronts. Start-up businesses, which are the heart of many technology-focused sectors that ultimately undergird large-scale industrial programs that support government security establishments, can experience shrinkage as the costs of initial investment – or even just perceptions of costs involved – rise directly as a result of information redistribution. Likewise, national industry can suffer the loss of first-mover advantage after the costs of research, development, and operation are weighed against the costs of foreign companies that benefit from illicit information transfer. Moreover, such intrusions can *discourage* strong industry-government cooperation on meaningful cybersecurity practices in that companies may feel unwilling to share information about incidents that would result in lost consumer and shareholder confidence.

Information warfare, influence activities and democratic integrity

We cannot characterize security issues in the wake of the information revolution as entirely relating to the digitization of infrastructure. Much as Gutenberg's printing press, the telegraph, and the television did in eras past, information technologies have had gargantuan effects on the information environment in which global politics takes place.[16] Individuals, institutions and countries access and create information in new and unique ways. New paradigms of access to information about the world around us have had apparently lasting psychological effects regarding the way that we, as humans, problem solve (i.e. we more naturally consider network-shaped solutions to complex issues and perceive issues of global and national security in different ways than we have in the past). Likewise, information itself is more open to manipulation and multifaceted framing by a range of actors.

Why discuss such extended changes to the environment of international politics here? The fact of the matter is that, much in the same that economic considerations like the innovation economy link directly to national potential to act in security affairs, the marketplace in which ideational contestation and debate occurs can be highly impactful when it comes to state approaches to foreign policymaking. This is, of course, particularly true of democratic countries where governments are designed so that political processes interact with social discourse in as healthy a fashion as possible to then direct national trajectory.

In democracies, the legitimacy of governments extends from popular mandate. This does not only mean that a government elected through a robust and free set of elections has the ability and responsibility to act on behalf of the people. It also means that governments must – broadly speaking, regardless of whether officials are perceived as pure representatives or delegates – respond to national discourse on particular policy issues and perspectives. In political theory, a range of theories speak to the manner in which popular discourse coalesces to become policy positions that are then either accommodated or responded to by the state. Marketplace of Ideas theory, for instance, is a set of arguments going back to John Stuart Mill that says democracies tend toward centrist, prudent policy positions as a result of the ideational tendencies of the marketplace.[17] Much as happens in economic markets, extreme positions are countered by less extreme ones as individuals seek new information and change preferences to find a position that seems most beneficial. This does not necessarily mean that democratic bodies always adopt "true" or "right" policy stances, but they do tend to be reasonably prudent. The success of this marketplace depends on the power of countervailing institutions of democracy. When an actor like the U.S. President – who has an unusual advantage in speaking about security issues because of his direct control of intelligence resources – states a position on a subject, democratic process tends to see his rhetoric countered by the information-seeking behaviors of a free media and the self-serving rhetoric of other elected officials who are interested in being seen to represent constituent interests.

Regardless of whether or not one buys this particular imagining of democratic process (other scholars, for instance, have variously argued that people seek value resonance or are risk-averse when it comes to specific costs to be incurred in conflict), there is a reasonable consensus that the dynamics of interaction between different elements of democratic societies drives foreign policy and defines policy trajectories in unique ways. Naturally, then, the integrity of this process is a critical part of any state's security apparatus, and a healthy domestic marketplace of ideational discourse is an asset to be protected. We return to this point in detail in Chapter 10 in discussing cyber-enabled forms of unconventional warfare.

Cyberwar: how likely to take place?

How likely is the occurrence of "cyberwar," by itself, between two countries? In other words, how likely is it that a major interstate conflict might occur entirely in the digital domain? From the perspective of military strategists and security scholars, cyber conflict operations do relatively little in isolation to improve national warfighting capabilities. This is not the same as saying that cyber weaponry adds little to the warfighting toolkit of states. In reality, there are a great many things that Internet technologies and the computer systems they connect do to add to the security portfolio of states; they augment military potential, they offer opportunities for interfering in foreign countries' internal affairs and they offer new ways by which a government might favorably affect the global information environment. And yet claims about the possibility of "victory" in cyberwar or other low-level cyber conflicts imprecisely describe the benefits to be had from the construction and employment of a cyber arsenal.[18] In fact, despite over 30 years of cyber conflict, the strategic utility of cyber operations remains unclear.

Box 9.1: Cyberwar scenarios

Across the scholarly and practitioner literature on cyber conflict, three distinct scenarios have variously been linked with the notion of "cyberwar," a conflict of exceptional proportions fought between states entirely – or almost entirely – via digital means.

The first among these is the least disruptive to states, as it focuses on military systems alone. Simply put, scholars have occasionally envisioned cyberwar as a large-scale debilitating attack on complex military systems. The only extant example of such an attack taking place, many note, is the Stuxnet attack on the Natanz uranium enrichment facility in Iran in 2010. There, as is described in the previous set of sections, a worm of remarkable sophistication was employed to sabotage the legitimate functions of the Natanz facility, namely of the centrifuges used to produce enriched isotopes of weapons-grade uranium. In other scenarios, scholars have envisioned a broad set of cyberattacks aimed at multiple military bases and at the nuclear command/control infrastructure of the United States. Though (as this section argues) these acts would make little strategic sense if employed without other force additionally being brought to bear, such complex disruption of military functions would undoubtedly – if only achieved via cyber means without subsequent action in other domains – constitute a cyberwar.

The second and third cyberwar scenarios constitute the more realistic set of possibilities, at least according to a wide range of security practitioners. On the one hand, cyberwar may take the form of effective **cyber blockade**. The blockade, which is discussed elsewhere in this book, takes the form of a massive denial of service attack against the Internet infrastructure of an entire country. Russian-based hackers achieved such an effect in Estonia in 2007 and, partially, in Georgia in 2008. The U.S. hackers blockaded North Korea in 2014 in response to the attacks and threat of further attacks leveled against Sony Pictures, Inc. During a blockade, Internet access from within the target country is ideally entirely cut off. At the very least, a partial blockade would allow a country to hamper commerce, key government functions and the full range of societal activities.

Finally, cyberwar may take the form of broad-scoped attacks against national critical infrastructure (CI). This scenario, popularized in movies like *Die Hard 4*, is the one typically held up as most concerning by cybersecurity practitioners. A successful disruption of one or several CI sectors could cause immense damage to the national economy, and the testimony of various stakeholders since 2001 suggests that a variety of foreign irritants, from al Qaeda to the Russian Federation, have taken active steps to map U.S. CI vulnerabilities. It is important to note that this scenario is highly variable and *does* include a cyber blockade of Internet infrastructure. After all, as later chapters describe, Internet and telecommunications critical infrastructure act as a plane upon which all other CI sectors function. If those networks are systematically compromised, then other CI sectors, from financial services to transportation, will feel the immediate effects.

Here, we might consider what traditional theories of warfighting say about the translation of political interests to favorable outcomes through aggression and violence. In short, such theories argue that aggression and violence can be useful to states for two reasons. First, states can use the threat of violence, implicit or otherwise, to achieve security goals.[19] This is the domain of coercion, wherein states can either compel or deter action by opponents. Compellence describes successful attempts to force a foreign power to change their behavior and take an action via threat of force that they would not have otherwise taken. Deterrence, by contrast, describes successful attempts to force such an adversary to avoid taking action that they would otherwise take. Second, states can employ capabilities for violence to directly take control of an opponent's territory and political systems. Here, military and other forces of a state are employed to physically occupy enemy positions and force – through conquest – changes in foreign behavior.

Cyber compellence and deterrence are complex subjects which are covered in more depth in the next two subsections. Nevertheless, there is an overall principle of cyber conflict that is relevant when we consider the simplest scenario in which a state threatens cyberattacks in an attempt to change the behavior of a foreign actor – cyber incursions are, with few exceptions, aggressive but *not violent*.[20] Physical damage from cyberattack is rare and requires a confluence of circumstances, such as those specific to the case of Stuxnet, that is uncommon in world affairs. Thus, in the bulk of cases, disruption and damage brought about by a cyber offensive will be temporary. Computer systems can be repaired and secured, in some cases in minutes. Disruption to sensor systems or even critical infrastructure systems can be modified to compensate and to close known vulnerabilities in short order. On top of this, much of the value of cyber weapons is in the relative secrecy of their development. Cyber exploitations are often "one shot," meaning that the use of a particular technique often notifies the victim of the mode of assault and provides a basis for better defense in the future.

The upshot of this dynamic is quite simple. Without the use of additional tools of statecraft and violence, cyber capabilities do not promise to states the ability to achieve lasting victories, to occupy territory or to force changes in an opponent's behavior. In this way, we might think of state cyber conflict capabilities as an adjunct modifier of conflict that relies on other elements of a state's security infrastructure – such as traditional military forces, intelligence assets or diplomatic channels – to achieve meaningful foreign policy outcomes.[21] Of course, this argument only applies to state adversaries acting within the confines of commonly held assumptions about the balance of power in international affairs. As the next two chapters note, state actors sponsoring conflict below the threshold of war and terroristic non-state actors may have significant incentive to engage in cyberwar insofar as widespread disruption might aid constitutive objectives.

Does cyberspace make states more likely to attack one another?

Does the existence of a new domain of possible human interactions in cyberspace mean that states are more likely to attack one another than in eras past (i.e. than in eras before the Internet existed)? In this section, we consider this question in two ways. First, given the opportunities for mayhem and given the special characteristics of human operation online, is there unprecedented motivation for states to attack one another *entirely through digital means*? Second, does the existence of cyberspace and of extensive cyber arsenals employed by state security services mean that there is a new risk involved in states who face off in the real world? In other words, does cyberspace introduce new risk to conventional interstate security relationships?

Through cyber alone: how inclined are states to employ cyber weapons?

In the literature on IR and strategic studies, the Security Dilemma (SD) is a commonly cited concept that describes how two actors – usually countries – can move toward the brink of conflict entirely without the intention of doing so.[22] Also called the "spiral model" or the Thucydides Trap, the SD essentially describes a situation in which efforts to enhance the ability of one state lead one or more foreign states to attempt to mobilize their own security forces in response. For instance, if State A decides that it must develop and produce new advanced tanks to replace obsolete models, State B might interpret this as being a direct threat to national security. Even if there is no knowledge of duplicity or ill-intent, the fact that (1) there is no way to actually gauge foreign intentions and (2) no global police force to call on if you are attacked without warning, forces State B to take actions to compensate (for instance, with the production of its own new tanks or the mobilization of border fortifications). This balancing effort, mirrored in turn by the first mover who undertakes the same calculated assessment of State B's mobilization, leads to a spiral of hostilities in which rising tensions emerge from no original intention for conflict.

Box 9.2: The attribution problem revisited

Perhaps the simplest unique characteristic of interactions in the cyber domain, the attribution problem, is the core focus of much literature on conflict in the digital age. Simply put, it is difficult for defenders to identify attackers during cyber incidents.[23] This is true on several fronts.

First, not all cyberattacks are detectable. Effective cyber warfare certainly includes a set of actions wherein the attacker is able to elude the detection apparatus of the defender entirely. Likewise, disruption is often an effect to be avoided in cyber conflict, particularly where the goal is information monitoring or exfiltration. It is also worth noting in this vein that some cyberattacks do not achieve the intended result. Where knowledge of the victim's systems is imperfect or the design of cyber weapons is shoddy, there is great possibility for modes of intrusion that sidestep the original objective and are potentially more difficult to detect, because detection procedures are designed around catching specific types of intrusions.

Second, even where attacks are detected, there is a long and often complex pathway toward full attribution (i.e. a set of technical information that allows a defender full information on the nature of the attack and the exact actions, down to location, of the attacker). Well-targeted intrusions can occur rapidly. As happened variously during Moonlight Maze, well-executed attacks that occurred over the course of mere minutes limited the opportunities available to investigators to observe attack behavior and draw inferences. It should be noted that attribution is not impossible at all in the cyber domain. Quite the reverse. Most incidents leave tell-tale markers of one kind or another. However, unraveling the full story of the details of a particular attack requires specialized equipment and human resources, as well as – often – the cooperation of a number of entities indirectly involved in a given episode (including ISPs, other government agencies, private companies and foreign organizations). Attribution investigations are sometimes hampered further by the need to obtain information that exists in formats protected by law as private, such as personal computers.

Finally, even where technical attribution is possible, there is a distinct difference between identifying the origin point/details of an attack and being able to effectively assess responsibility. Intent and direction are political phenomena, and it is inappropriate – though certainly indicative – to argue that basic descriptive information proxies for an understanding of responsibility. During Moonlight Maze, for instance, the successful entrapment of the intruder through the download of an altered Adobe Acrobat software package did not provide investigators with strong enough evidence to move forward with diplomatic efforts to end the conflict episode. The identification of an IP address in Moscow may have been yet another "zombie" computer being used as a beachhead or could have been the home computer of a non-state hacker without links to the Russian government.

Naturally, attribution uncertainties prompt all manner of challenge for state security actors, from technical analysts through military managers to the policy practitioners and diplomats that must act on threat intelligence in times of crisis. Where the stakes might be relatively high in potentially accusing a foreign power of aggression, what level of certitude enables practitioners to be comfortable pointing the finger. Typically, attribution is broken down into three categories based on the robustness of the case that evidence can make – essentially being able to prove "who dunnit" (1) to oneself, (2) to the attacker and (3) to the broader global community. Being able to attribute cyber sophisticated attacks directly to foreign state actors in the eyes of the global community is difficult, not least because the connection between government actions and the actions of hacking units is often nigh impossible to demonstrate. Being able to demonstrate knowledge of an attack to one's attacker may be of some use behind closed doors as a threat or a bargaining tool, as arguably was the case with Moonlight Maze. But insufficient evidence to move the conversation into the public sphere makes such attribution cheap – attackers are not motivated to stop or admit responsibility because the victim has no power (other than counterattack) to produce punitive consequences.

The existence of an SD between states is particularly sensitive to two factors: (1) the nature of military technology as offense or defense dominant and (2) *perceptions* held by either party about the utility of that same technology. In other words, whether or not a country possesses technology that is most useful in a first strike scenario (vs. a defensive one) and whether or not other countries can tell the difference determines whether or not the SD is acute or muted (i.e. whether the risk of conflict is high or low).

So is there unprecedented motivation for states to attack one another *entirely through digital means* alone? Is the security dilemma acute online? The question with cyber is threefold. First, is cyberspace offense dominant or not? Second, can we tell the difference between offensive and defensive abilities given the constant development of systems and techniques? And, third, is it possible for us to quantify foreign perceptions of cyber conflict dynamics? In other words, is it possible to tell what other countries think about cyber conflict and adapt our doctrine accordingly?

Offense/defense differentiation is hard. Because of the attribution problem, we might consider strategic calculations on the part of intruders to be low-risk, high-gain at all times. In reality, of course, this is variably dependent on the abilities of the defender. It is a common meme in the literature on cybersecurity that the domain is "offense dominant," meaning that the current state of technology produces a singular value in attacking. Would-be intruders have all of the cards and there are few risks involved in engaging in aggressive activities for a range of purposes.

Specifically, cyberspace is an offense dominant domain because of several specific reasons, so the argument goes. First, the effectiveness of different techniques or combinations thereof in intruder approaches almost always relies on secrecy prior to the attack. Success is, therefore, a function of *not* communicating with a defender in any way (either directly or through probing attacks) prior to an assault. Cyber "weapons" also have volatile half-lives in that their effectiveness might be lost entirely as a result of, for instance, a basic systems update or patch on the part of the defender. Moreover, cyber weapons, as will be discussed later, are not best suited for coercion where there is a need to send messages of intent and threat to a defender. The result of all this is that cyber "weaponry" is use-or-lose in that its primary characteristic is effectiveness from obscurity. This incentivizes first strike doctrine, which is reified in the lack of executive control often found with such operations. In short, authorization to hack rarely comes from the top and often emerges from mid-level managerial entities (or individuals) in charge of incident planning and response. Again, defenders and attackers both know this, which encourages the mutual development of first strike practices.

As anyone taking a second glance at the issue might conclude, this reading of the situation *somewhat* misstates the issue.[24] The attribution problem also provides a range of unique abilities to defenders in the cyber domain. In particular, the ability to *deceive* attackers and to encourage unique behaviors that compromise the ability of intruders either to effectively achieve objectives or to remain hidden is a powerful one. One need only look at early cyber conflict incidents wherein a honeypot was employed to entrap intruders – Cuckoo's Egg and Moonlight Maze, in particular – to see the tools available to defenders trying to even the score and examine assault conditions. This is not to say that defense is easy. But there are inherent advantages held by the defender, including the ability to set up traps based on expected intruder behaviors that nullify expected gains and the fact that defense analysis is not a digital set of processes.

Two other points in this vein bear mention. First, it is quite arguable that attackers have significant incentive to restrain their aggressive actions.[25] In fact, the more likely an attacker is to undertake a major assault, the less likely that same attacker should be to engage in regular, broad-scoped disruptive activity. The logic here is that, while probing attacks and reconnaissance might be necessary to design and implement a major attack, regular intrusion is likely to give defenders significant warning and opportunities to disrupt the exact actions necessary to achieve more sophisticated intrusions. Second, and relatedly, it should be noted that cyberattack is a forced interaction.[26] With planning, intrusion can be desirable from the perspective of the defender, as it gifts new knowledge about network vulnerabilities and offers avenues for hack-back in which defenders can infect or disrupt specific would-be attackers.

There is certainly a good argument that the domain is offense dominant. However, additional thought suggests that it might perhaps be more accurate to say that cyberspace is an *offense enabling* domain. The technology *does* allow attackers unique abilities such that tactical calculations take on a low-risk, high-gain flavor. Moreover, these calculations are not affected across the board by the actions of defenders, particularly because of the high costs involved in constructing effective cyber defense and forensics capabilities. Fluctuations in decision-making on the part of attackers are entirely dependent on case-by-case circumstances. At the same time, however, defenders are not *defenseless*. Given the right infusion of capital and appropriate design, systems can be made secure such that intruders are effectively deterred from attack. For one, militaries and intelligence units regularly and necessarily engage foreign foes non-combatively to signal ability. Such intrusions cannot be differentiated from pre-strike reconnaissance, leaving defenders with an unfortunately broad scope of possibilities to consider in their analysis. Moreover, good cyber defense is not just perimeter defense, but rather includes active measures that – though a standard definition of "active defense" continues to elude practitioners and industry alike – necessarily include preemptive intrusion to gauge adversary intentions. The result of this is, quite simply, that differentiating between incoming attack and standard defensive operations is extremely difficult in practice.

Given all of this, can we tell the difference between offensive and defensive abilities given the constant development of systems and techniques? And is it possible for us to quantify foreign perceptions of cyber conflict dynamics? The answer to these questions is simpler, though not particularly mollifying. As discussed previously in several places, differentiating between exploitative intrusions, aggressive attacks and cyber actions designed to actively aid defensive efforts is immensely difficult. Likewise, a quantification of how others think about cyber conflict is prone to imprecision because of the degree to which there is likely going to be variation at the level of different operational units. Among other ways, quantifying assessments are done through in-depth studies of foreign doctrinal developments, Professional Military Education programs and joint exercise experiences undertaken by members of the international community. However, as is enduringly true of intelligence analysis and military assessments of human factors in conflict, assessing perception and intention is always difficult due to the task of having to predict psychological factors. This, again, is even harder with cybersecurity, as decision-making rarely occurs at only the executive level and as operational behaviors are likely to vary across units of the security services.

The suggestion that emerges from such a read of the dynamics of cyber conflict is that, regardless of the technical realities of cyber capabilities, the SD is acute in interstate relations, at least insofar as cyber operations alone are concerned. Perception is uniquely difficult to quantify, and the nexus of actions required to maintain effective defensive procedures and to gather information relevant to national security policy planning means that government security establishments are likely to face the challenge of intention analysis linked to different kinds of cyber intrusions in perpetuity.

Do cyber weapons affect the likelihood of conventional warfare?

Beyond cyberspace, a core concern among strategic planners and scholars of IR is that cyber warfare lends itself organically to greater instability between states interacting in the real world, particularly in crisis situations. Crisis stability refers to the potential for sudden conflict during a particularly heated standoff between two states over a specific issue. Examples of a crisis might historically include the assassination crisis that sparked World War I, the Sudetenland crisis that led to the appeasement of (rather than conflict with) Nazi Germany at Munich and the Cuban Missile Crisis, wherein the United States launched a naval blockade of Cuba to prevent further construction of nuclear missile facilities there by Soviet forces intent on responding to the United States' previous placement of missiles in Turkey. In contemporary security discourse, the most commonly cited potential crises that would involve great powers include a Sino-American standoff over the issue of Taiwan's independence, a re-launched conflict on the Korean Peninsula, an escalating conflict between China and Japan over maritime territories in the East China Sea, and a range of scenarios involving Israel and belligerent neighboring states.[27]

It should be noted up front that cyber warfare likely *cannot* cause crisis instability, where two states find themselves on the brink of direct military engagement, by itself.

The reason for this is simply that cyber weapons do not contain a unique feature of other weapons that might be employed in crisis scenarios – they are not "use it or lose it" (although the fear or losing an ability to utilize zero-day exploits could have a somewhat destabilizing effect). Some crisis situations are defined by the relative military weakness of one side to the other due to the situational deployment of a particular military force or the geographic conditions of the standoff. In such situations, the value of the military asset that portends this first mover advantage is greatly increased in the present because future action might allow an enemy to engage on more even terms. Therefore, the advantaged side is incentivized to strike first. Cyber weapons simply do not function this way. Though additional time *does* often allow a defender more opportunity to potentially mitigate the effects of an intrusion, the fact is that cyber weapons are not inherently able in most situations to degrade the kinetic ability of enemy militaries to strike back.

Nevertheless, cyber weapons as an additional feature of *existing* crisis scenarios do potentially make for greater instability. The first major reason as to why this may be the case is the same reason that cyber weapons cannot themselves be the cause of a crisis.[28] Because a cyberattack invariably awards defenders the ability to analyze an intrusion and adapt defenses in a short period of time, employment of a cyber weapon incentivizes quick decisions to act in other ways while there is an advantage – perceived or actual – from the initial effects.

Such a set of incentives is bolstered further by the fact that specific cyber techniques are difficult to duplicate in terms of their intended effect. There are two main reasons for this. First, defenders are able to more effectively shore up vulnerabilities once an initial attack vector and target have been identified.[29] Second, intrusions in interstate conflict situations often rely on complex vulnerabilities in sophisticated systems.[30] Finding more than one vulnerability that allows for the achievement of a similar effect is no easy task. Thus, early use and early commitment to other actions to escalate a crisis based on an initial attack make significant sense.

Another reason why cyber warfare is potentially dangerous in crisis situations has to do with intention and strategic assumptions made about the other side. Cyber weapons do not always produce clear or easily verifiable results.[31] Attackers know what was meant to happen but don't necessarily have a good line on where the effect was achieved. Likewise, defenders are able to see what happened in general but have no understanding of the parameters of the operation as set by the intruder. Particularly where some attacks will have unintended or secondary consequences – such as a situation in which information theft efforts unexpectedly deny network access briefly to a military unit in the field – opponents will have to make assumptions about the intentions of the other side. Given the tension likely to characterize a crisis scenario, it seems logical that intentions will be considered as belligerent by default. Moreover, such uncertainty simultaneously acts to incentivize cyber probing of an opponent, as knowledge that many attacks will go unnoticed in some way reinforce the low-risk, high-gain conditional reality faced by belligerents.

Finally, there is the possibility for increased crisis instability from employment of cyber weapons inherent in the organizational setup of military network warfare procedures.[32] Where information warfare operations are broad in both scope and severity, authorization for a range of actions lies not at the level of national executives or even with high-level military commanders. In many situations, this spells out a troubling situation in which unit commanders must respond to a particular intrusion effort or must coordinate with intelligence operations across institutional lines to achieve surveillance and reconnaissance mission objectives. And, of course, this decentralization of authority, which exists even in situations where national network warfare operations run through a specific command entity, also worsens the calculation made to assess foreign intent in that it invokes a traditional psychological challenge of interstate relations – decision-making is not centralized; however, one side simultaneously (1) expects that the other knows this and (2) still tends to think of an opponent as a monolithic, centrally controlled enemy organization.

Can states use cyber weapons for coercion?

In general, cyber capabilities offer little opportunity for either state or non-state actors to coerce foreign governments. In particular, the knowledge that any "victory" of disruption will be temporary nullifies much of the potency of any threat made with coercion in mind. Of course, it is possible that the threat of a broad-scoped cyber intrusion might convince some actors to change their behavior. As is covered in this section, however, even in such instances it is clear that the possibility of disruption from the employment of digital weapons alone does not dictate the potential for success in coercion.

First, it is important to consider the dynamics of coercive efforts targeted at the state in international affairs. At the most basic level, an actor's ability to coerce might stem from the application or threat of physical violence, diplomatic pressure or the use of indirect mechanisms to apply economic or social pressure. Often, though a state is the recipient of coercive pressure, coercive actions target individuals or organizations associated with or operating under the jurisdiction of a particular regime. Coercion most commonly takes the form of strategic attempts to either deny or punish a competitor. Strategies of denial essentially promise to significantly deny access to a market, a piece of territory, an allied partner and their forces or another such geopolitical resource. By contrast, strategies of punishment threaten direct repercussions for the competitor if behavior is not altered. Such strategies can usually be broken up further into policies that threaten the punishment of military or related governmental assets (including strategic territorial possessions) and policies that threaten retaliation against the civilian population or civilian infrastructure. In all cases, coercion is more based on an appraisal of the *power* an actor has to hurt,[33] rather than the application of force itself; coercion is achieved through only the threat or *limited* application of military or other forces.[34] After all, the point is that the target has a choice available to them – stay the course or, if the conditions are compelling, change course.[35]

Failure of coercion occurs in one of several ways. First, failure occurs when the coercive policy fails to produce any change in the behavior of the targeted actor.[36] This failure mode is somewhat hard to observe, as time is a significant variable. Coercion might yet occur if the applied forces remain viable and the external situation has not changed the range of choices available to the target.[37] Nevertheless, long-term non-effect is a sure sign of the failure of coercion. Second, failure occurs when the "sender" of the coercive signal abandons that policy without any perceptible result.[38] This occurs before any deadlines that have been set in the conveyance of the coercive signal. By contrast, and finally, failure can occur following the period of time signaled to the target in which change was meant to take place.[39] In this situation, failure occurs because the signaler shrinks away or finds itself unable to carry through on the threatened action. Regardless of whether this stems from political unwillingness or technical capacity, the result is the same. It is important to note that failure to coerce is not synonymous with the necessity of the use of force.[40]

For coercion to succeed, it is useful to think of coercive signals as requiring three basic components. First, as Rich summarizes,[41] any attempt at coercion demands clarity in the expected result.[42] Desired changes in the behavior of the targeted competitor must be clearly interpretable in the manner in which the coercive action is made. Vagueness undermines the attempt being made and is itself a common cause of the failure to coerce. As one might expect, this required component presents varying degrees of challenge to states that favor the use of one type of coercive instrument – sanctions or diplomatic statements, for example – over another, like the mobilization of military units in a particular area.

Second, any demands made of a target must be accompanied by some signal of urgency that hints at swift repercussions in the absence of imminent changes in behavior.[43] Without such a signal, coercive demands – even implicit or veiled ones – might lack the necessary time-based component that allows the "sender" to approximate a schedule for its next move.[44] Moreover, the target might ignore general threats as not credible.[45] After all, the power of a particular signal lies principally with its credibility. While this depends to some degree on the capabilities of the signaling actor, coercive power most critically depends on the ability of that actor to make short-term intentions clear.[46]

Finally, the targeted actor must believe itself in real danger of repercussions given inaction.[47] In particular, the target must be made to believe that it will be worse off if the threat of action is implemented than if it turns away from the current course. Here, it is easy to think that the most potent relevant element is the capabilities of the signaling actor.[48] However, it is important to consider the geopolitical context within which coercion is undertaken. The position of the signaler in the international system plays a significant role in determining the potency of a threat and the degree to which the target considers itself in danger.[49] Is the signaler socially bound by the expectations of either order or partnerships to reaffirm a particular precedent? Are there constraining domestic variables at work or has the signaler historically followed through with threats? And are there secondary circumstances to consider in international affairs – such as a related conflict – that might place constraints on future interactions?[50]

The bottom line is that success in coercive efforts depends on the ability of that competitor to process information and to understand the intended consequences of the promised course of action. As may be clear by now, this is problematic when it comes to the employment of cyber arsenals – that is, without the parallel use of some other tool of statecraft – to such an intended end.

With regard to the question of coercive strategies, particularly strategies of denial and punishment, cyberattack strategies come in a series of flavors. Though the prospective scope of resources required in such an undertaking is enormous, a cyber blockade is the tactic that proxies as perhaps the most direct corollary to more conventional tools used in strategic efforts to punish. Cyber blockades describe a large-scale set of efforts made to essentially cut off a country's access to the Internet, whether for minutes, days or longer, with far-reaching effects not only for industry, government and citizenry but also for foreign social and economic target dependents.[51] When attempted in a limited fashion, such blockade-style tactics present in the form of directed attacks that might seek to either take control of a central system – such as government servers or a critical infrastructure control system – or disrupt information processes in an ultimately harmful way.

Beyond direct damage that might be made to digital systems, prosecutors of a cyberattack that aims to coerce might also focus on the content of information and the socioeconomic or political value it holds. More in line with the mafia-flavored metaphors that strategists often use to describe basic compellent concepts, computer network infiltration techniques allow for access to private, propriety information that is often valuable *only* due to its protected nature. Control of such information or the ability to damage it provides cyber coercers with the means to manipulate a target on the level of socioeconomic operation – i.e. to bribe a target actor.

It is important to note that the range of threats possible through digital means is in constant flux. Though the strategic parameters of computer network assault are unlikely to change in the abstract sense, it is undoubtedly the case that constant evolutions – and even revolutions – across various aspects of the incredibly varied cyber ecosystem in the world mean additional vulnerabilities for states over time.

So how viable is coercion enabled by cyberattack? One consideration is that the high degree of innovative ICT evolution present in industrial sectors in developed economies plays an important role in determining the efficacy of attempts to coerce online. A high rate of adaptability on the part of prospective targets variably affects the strategic parameters of any possible or threatened assault. How can an attacker be sure that a victim is incapable of quickly adapting to effectively defend against disruption? Here, of course, we also have to consider the fact that any intrusion is likely to produce temporary gains for an attacker. Again, unless the purpose of the assault is to enable a non-digital form of disruption such as the release of sensitive data, cyberattack promises only short-term problems for a victim. Likewise, the threat of attack itself potentially diminishes an actor's ability to make demands. After all, past the initial stages it is increasingly likely that a system flaw exploited in an attack will be unusable or of significantly reduced value in the future. In short, effectiveness in cyber assaults clearly relies heavily on secrecy and a lack of technical foreknowledge. Loss of this advantage affects the ability of a coercer to credibly communicate intentions and to convince the target that danger is both real and hard to defend against.

Of course, no discussion of the use of cyber weapons for compellent means would be complete without recognition of the attribution problems that naturally accompany digital, as opposed to more conventional, forms of attack.[52] Due to the nature of computer systems as effectively detached from physical actor identifiers, attribution of a particular attack is difficult. More specifically, attributing agency to a particular state or major actor is often difficult in a geopolitical sense. On the one hand, it is not easy to trace an attack or, if an attack is traced, to link it to a particular actor with only geographic IP address information to go on. On the other hand, even if the target produces a comprehensive portfolio of evidence that includes geographic information and correlated likelihoods of agency based on prior observation of exploitative activities by a specific actor, it is remarkably difficult to know whether the actions of an individual or small organization is representative of a broader policy. Indeed, public discussion thereof is often deemed harmful for a series of political and economic reasons.[53] While this might sound favorable for an attacker, however, the coercer necessarily requires a degree of communication and understanding as part of the compellence process. Thus, coercers face the unenviable task of attempting to either create conditions under which attribution is not problematic or produce a strategy that retains non-attribution benefits *and* effectively communicates both demands and intentions.

Finally, there inevitably exists the possibility that cyber coercion – even that which aims to use obtained information in efforts to change an opponent's behavior – will suffer because it is difficult to bound the intended audience when making threats. The 2014 hack and threatened disruption of Sony, Inc. systems by North Korean hackers in response to the release of *The Interview* provides a good example of how imprecise and shoddy threat behavior can backfire on a would-be coercer. Here, the perpetrators of the coercion campaign chose the substantive target of their operation poorly. While Sony may have fit the requirements for an appropriate target organization, the "Guardians of Peace" placed too much focus on a single piece of content – the movie, *The Interview*. Doing so moved the scope of the compellent effort beyond Sony, though not in the details, and expanded the parameters of the operation to societal circumstance not flexibly changeable with the tools that were being brought to bear. When the specific reference to the film was made, the campaign's perpetrators essentially changed how Sony's inevitable non-release would be cast – from an issue of organizational decision-making to one of free speech and civil liberties.

The "Guardians of Peace" incident also invited the jurisdictional interests of law enforcement agencies by expanding the scope of their own efforts to include threats of personal injury, both on specific persons and on theater-going citizens. Such a move further pushed the parameters of what is realistic for such an effort and weakened the operation on a number of fronts. First, threatening a broader range of actions, particularly ones that are generally open-ended on a national scale, reduces the threat credibility of the compellent effort enormously. Second, such threats muddy the timeframe as seen by the target and the calculations the target might make. Additional threats imply not only that the coercer is desperate to affect change in the near term but also that alternative outcomes might suffice. In this situation, the target may be increasingly unsure as to whether its own actions will suffice or whether the broader mechanisms invoked when the coercer issued new threats might either save it the cost of altering course or cushion the costs of standing firm. Finally, and perhaps most importantly in the long run, widening the scope of the operation invites the application of security measures beyond those originally under consideration. This can, and did in the case of North Korea's campaign, have the effect of making attribution significantly more likely, in both technical and social terms. Social and political context clearly matters a great deal when it comes to cybersecurity issues. The case of North Korea and Sony, in particular, highlights the way in which social capital and democratic norms can interact with what might otherwise have been a case of simple actor-target criminal coercion.

A final observation about coercion by cyber means seems prudent. Simply put, the coercion described earlier is a form of complex political maneuvering. In spite of the fact that coercion via cyber means appears to be of minimal use for state actors writ large, it would likely be disingenuous to say that cyber tools are not useful for the simple act of signaling opponents. A common tactic employed by intelligence units is that of "burning a vulnerability" wherein a cyber intrusion is affected – often where the technique is likely to be of limited use for much longer – to generally demonstrate the ability to do so. In short, cyber techniques can be used to send simple messages, from communicating the broad extent of an actor's capabilities to demonstrating knowledge of an ongoing hack to an opponent. In this way, at the operational level, cyber weaponry can be remarkably useful as coercive tools.

Can states deter cyber aggression?

Though these discussions of coercion and SD address the idea that states will be more or less motivated to launch cyberattacks, we do not directly engage with the notion that states might be able to actively deter their adversaries from cyber action. We do so here with a discussion of deterrence theory as it pertains to cyberspace.

Deterrence theory is nothing new but deterring offensive cyber operations is. The challenges to applying deterrence theory to cyber warfare relate to marked uncertainty

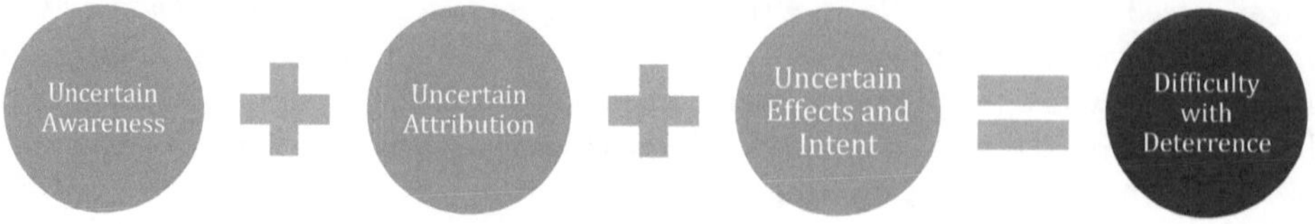

Figure 9.1 Challenges in applying deterrence theory to cyber warfare

with respect to, first, awareness and attribution of an attack, and, second, the uncertain effects of any attack (as depicted in Figure 9.1).

The difficulties surrounding attribution and control of its effects make deterrence of offensive cyber operations (OCO) uniquely difficult. In some cases, lack of control makes the application of the weapon both enticing for the attacker but also risky due to blowback onto his or her own interests, society and economy, and those of his or her allies, and the risk of escalation by the defender, if, indeed, he or she is able to determine the attacker. Peter Singer of the Brookings Institution and others have identified this lack of attribution as the key factor that prohibits the direct and immediate application of deterrence theory to the cyber realm.[54] If an attack is attributable, then traditional deterrence applies, including the possibility of a kinetic response. If an attack is not attributable, or the attacker believes it will be falsely attributed, it may be so enticing a weapon as to be irresistible.

This is an old problem – if you could do something bad and get away with it, would you? This issue has been considered in various guises by philosophers and political leaders throughout history. In *Republic*, Plato provides the example of *Gyges' Ring*, which made its wearer invisible.[55] Would a man wearing *Gyges' Ring* be righteous; alas, no, he concluded. The temptation of being able to get away with something malicious without attribution would be too great, and even a moral man would be corrupted by such power. Cyber weapons give a nation state a *Gyges' Ring*, and, increasingly, we witness the consequences. The implications of this uncertainty illustrate the need to develop an approach to improve the ability to apply deterrence to cyber conflict.

Overview of deterrence theory

Modern deterrence theory is largely associated with nuclear policy. During the Cold War, the United States and the Soviet Union adopted a survivable nuclear force to present a 'credible' deterrent that maintained the 'uncertainty' inherent in a strategic balance as understood through the accepted theories of major theorists like Bernard Brodie, Herman Kahn and Thomas Schelling.[56] Theories of deterrence were largely developed early in the Cold War by academics coming to grips with the intellectual conundrum and novelty of the political and military impact of nuclear weapons, and arguably prevented a world war by allowing policymakers to understand how nuclear weapons affected traditional tools of statecraft – deterrence and coercion – and the risks associated with nuclear war.

Ultimately, deterrence is the manipulation of the cost/benefit calculation an adversary undertakes related to a given action. A nation can convince its adversary to avoid taking a specific action by reducing the prospective benefits and/or increasing the prospective costs. Cyber deterrence is therefore simply the manipulation of an adversary's cost/benefit analysis of a given cyber activity.[57] Keeping someone from doing something you do not want him or her to do may be brought about by threatening unacceptable punishment if the action is taken. This is called deterrence by punishment or reprisal (the power to hurt). Convincing the opponent that his or her objective will be denied to him or her if he or she attacks, is known as deterrence by denial (the power to deny military victory).[58] In the nuclear context, complete defense was impossible so deterrence by punishment was the primary approach.[59],[60] Further, nuclear deterrence sought to deter any nuclear attack (along with other major aggressive behaviors, such as a Soviet invasion of Western Europe with conventional forces), but in the context of cyber, a threshold or subset of cyber activity is the target of deterrence, such as OCO against critical infrastructure or economic targets. Both deterrence by denial and deterrence by punishment may apply in the case of a cyberattack; however, two major problems exist.

Awareness of cyberattack and attribution

The first major problem posed by most cyber weapons is the challenge of becoming aware of the attack and properly attributing the attack once it has occurred. These problems are extremely difficult to resolve as a result of the tremendous difficulty in conclusively determining the origin, identity, and intent of an actor/attacker operating in this domain, compounded by the fact that defenders generally lack the tools needed to reliability trace an attack back to the actual attacker. As Rid argues, all cyberattacks to date have been examples of sophisticated forms of sabotage, espionage, and subversion and are reliant on this attribution difficulty.[61] Cyberspace is truly global, and nearly all action passes through networks and ISPs in multiple countries. Additionally, the hardware used to conduct cyber warfare can be owned by innocent noncombatants, illicitly harnessed for malicious use through the use of computer viruses. The 2007 Estonian, 2008 Georgian and 2013 Ukrainian experiences highlight the challenges associated with uncertainty and attribution in cyberspace. For all three of these attacks, while the role of Moscow was suspected, evidence of direct involvement was lacking and therefore plausible deniability was possible.[62] Millions of devices continue to be compromised and used illicitly as part of various networks – 'botnets' – utilized to conduct cyber-attacks.[63] This also provides plausible deniability to state-sponsored activity.[64] However, the lack of expected Russian cyberattacks in the first few months of the invasion of Ukraine in 2022 may illustrate that this dynamic is shifting in the other direction with more tools and ability to attribute malicious cyber behavior.

Finally, if quality evidence tracing an attack back to its origin is obtained, it still may not lead to attribution of the attack. Knowing the originating IP address of an attack vector will not necessarily indicate who the attacker was or if they were acting with state support or direction. Sometimes an analysis of the malware itself can provide clues, but these could just as easily be deliberate decoys intended to lead investigators astray and are unlikely to result in firm attribution of a cyberattack. The challenges of attribution in cyberspace make it very difficult (although not impossible) to attribute hostile action in cyberspace to a particular individual, organization or state and so make cyber warfare particularly appealing for an adversary that wants to execute an attack anonymously or at least with reasonable deniability. This poses significant challenges for achieving offensive deterrence against cyberattack, as an adversary can have some reasonable expectation that it may be impossible to fully attribute the attack and impose reliable costs for the action.

Uncertainty regarding cyber weapon effects

The second major characteristic of cyber weapons that significantly impacts the logic of deterrence is the uncertainty regarding their effects. Due to the potential for IT network evolution as well as IT interdependencies, it is difficult to predict the precise effects of an attack. In cyberspace, the targeted actor is capable of literally flipping a switch and instantly changing the network, or even unplugging it altogether. This factor is a destabilizing force as it rewards immediate hostile action to prevent network modification if cyber reconnaissance-targeting intrusions are later detected. In essence, it is the opposite of stable deterrence and akin to nuclear crisis instability where nuclear deterrence may fail because it incentivizes a first strike. Defenders may also have unknown automated countermeasures that negate the desired effects of cyberattacks, such as instantaneous network reconfiguration or firewalls. For example, the Stuxnet attack is likely no longer able to continue to attack Iranian nuclear facilities as the zero-day exploits it utilized have been plugged by Iranian officials. In addition to network/target evolution, cyber weapons themselves can also be unpredictable and can evolve. A cyber weapon can adapt – as was seen with the Conficker virus. Conficker included a mechanism that employed a randomizing function to generate a new list of 250 domain names, which were used as command and control rendezvous points, on a daily basis. Thus the virus remained adaptable and stayed ahead of those seeking to shut down or hijack the illicit Conficker-enabled network.[65]

Network interdependencies are another dynamic contributing to the potential for collateral damage that is characteristic of cyber weapons. Because the Internet is made up of hundreds of millions of computers connected through an elaborate and organic interwoven network, and it is the backbone of much of the global economy, there is the potential for significant unintended and collateral impacts from cyber action. This interconnected nature of IT systems has led to real-world collateral damage. For example, the 2007 Israeli cyberattack on Syrian air defense systems as part of Operation Orchard, was believed to have also damaged domestic Israeli cyber networks.[66] Fear of this kind of cyber collateral damage has had a profound effect on military planning. As another example, in 2003, the United States was planning a massive cyberattack on Iraq in advance of any physical invasion – freezing bank accounts and crippling government systems. Despite possessing the ability to carry out such attacks, the Bush administration canceled the plan out of a concern that the effects would not be contained to Iraq but instead would also have a negative effect on the networks of friends and allies across the region and in Europe.[67] The adverse consequences of such unintended results were powerful deterrents for the United States. Of course, this is not to say that other states would be similarly deterred from such actions, especially states that do not have the alliance obligations and responsibilities of the United States. For example, the 2017 NotPetya attack, which is of suspected Russian origin, was aimed at Ukraine but spread to many international companies, causing over $10 billion in damage.

The uncertain effects cyber weapons coupled with the availability of defenses and the need for secrecy and surprise, reduce their ability to serve as a strategic deterrent in their own right. Available defenses and the potential for network evolution to mitigate the effects of an attack given early warning requires cyber attackers to rely on surprise for much of their effectiveness. To achieve surprise, secrecy is required, reducing the ability of a state to make credible threats without compromising their cyber warfare capabilities. Credible threats regarding specific means of attack or targets invite the threatened state to take protective actions which could blunt the deterrent value of a threat. Essentially, although cyber weapons have the potential to inflict unacceptable damage against an adversary, they are likely unable to offer states a credible, consistent and 'assured' capability for doing so. This deficiency significantly undermines their suitability as a deterrent tool, and instead they are more likely to support an intelligence, surveillance and reconnaissance mission, or to be used as a first strike weapon, preemptively, or as force multipliers.

Beyond the two key challenges identified, other factors complicate the deterrence of OCO. For example, these include the blurred spectrum and confusion over key concepts such as Computer Network Exploitation (CNE), Computer Network Attack (CNA), Information Operations (IO) and other concepts, "inherent instability" associated with the perception of offensive dominance in cyber conflict due to "Zero-Day" exploits, the relative perishability and fragility of cyber weapons, and uncertainty over the relative balance of power. Cyber scholars have long focused on these challenges of applying deterrence theory to cyber conflict.[68] Few scholars have focused on the tactical and operational issues associated with applying deterrence theory in practice – namely the nuclear war planning experience – to the cyber problem set. The exception is Austin Long's article titled "A Cyber SIOP? Operational Considerations for Strategic Offensive Cyber Planning."[69] Long's article highlights the value of moving beyond solely theoretical analysis to focus on operational issues, such as planning, targeting and command and control. Thus even if a theoretical construct for cyber deterrence does not yet fully exist, there may be additional lessons at the operational and tactical level from nuclear deterrence. Finally, the most recent Commander's Vision for U.S. Cyber Command (2018) signaled that the United States was moving beyond the attempt to deter cyberattacks and instead focused on persistent engagement and efforts to actively thwart and defend in cyberspace. Persistent engagement is the idea of constant contact with adversaries so as to be in a proactive posture to defend and reflects an evolution away from traditional models of deterrence and restraint.[70] Namely, the concept of persistent engagement describes the fundamental difference between international engagement online and in the real world where digital interactions are constant – in an extreme sense – and numerous in contrast with kinetic ones. Therefore, American strategy today focuses on building deterrence via the idea of "defending forward," or actively intruding on non-American networks so as to find and blunt threats to American interests before they activate. Alongside conventional attempts to build broad constraining norms on cyber aggression – discussed in Chapter 12 – this conflict posture is intended to produce deterrent effects that simply have not been seen yet with cyber conflict.

Cyber power and power in the digital age

What makes a state "cyber powerful"? What makes a state powerful in the digital age? Are these two things different from one another? When it comes to cyberspace, much talk of power revolves around the ability of actors to use new digital tools to change opponents' behavior. Many works that have forwarded working definitions of cyber power commonly focus on "hard" abilities – i.e. the technical capabilities (or *means*) of both state and non-state actors in international affairs. Spade, for instance, simply describes cyber power as "the employment of computer network attack and computer network exploitation," before going on to argue that the term means advantage or influence gained through the use of their abilities.[71] Betz and Stevens likewise, though they define power as depending on social and political conditions external to cyberspace, argue that cyber power is a manifestation of power in cyberspace itself. They tie their cyber power framework to capacity in cyberspace in saying that "any actor with access to cyberspace and the requisite skills and knowledge can, hypothetically, enact compulsory cyber-power against another."[72] Betz, in particular, likens cyber power to airpower insofar as the potential for single powerful strikes is common, but strategic power to employ cyber resources in-depth demands infrastructure at the level of the state (implying, of course, that rich states will inevitably be more powerful).[73] And Pope, in his brief discussion of the meaning of cyber power, falls back on the notion of coercive capacity and the idea that the ability to coerce largely constitutes actor power.[74] Thus, cyber power itself is, according to many, constituted by the means – i.e. the technical capacity – to coerce and ultimately change the behavior of one's opponents over time.

Box 9.3: Where does state power come from?

In a seminal work on the nature of power in IR, Barnett and Duval forward a definition of power as being "the production, in and through social relations, of effects on actors that shape their capacity to control their [circumstances and] fate." In phrasing a broad definition of power in this way, the authors attempt to make clear that we are dealing with a hierarchical logic of being. Power is not merely the capacity to affect favorable outcomes. Rather it is those things involved in producing that capacity and is centrally defined by the nature of those social relationships that characterize the international system.

Barnett and Duvall argue[75] that power is constituted across four different categories (Table 9.1). These categories are determined by variation across two analytic dimensions. The first of these dimensions imagines that power emerges from different features of the international system. Power can work in the context of active relationships between agents and objects, such as the interaction of states in bargaining, diplomacy and conflict or the interaction of countries in the context of international institutions (formal or otherwise). At the same time, however, power emerges from the consequences of sociopolitical dynamics that precede active relationships between actors in world affairs, what the authors refer to as the "constitution" of relationships. In other words, the context of an actor's geography, sociocultural positioning and more affect the circumstances under which states will more directly engage others. Though such a differentiation might be understood as social constructivist in nature, the reality is that this basic supposition is broadly accepted in IR literature in the distinction between extant and latent sources of power. Geographic location, access to natural and human resources and more define the nature of "social" relations – meaning simply the substantive shape of actors' positioning relative to one another – and affect the degree to which states understand one another as more or less powerful. Understanding this dimension of power is perhaps of most relevance in this article's attempt to flesh out conceptual understanding of cyber power because most preceding efforts to define or engage with cyber power as a distinct concept fall squarely under the heading of "power emerging from active relationships."

Table 9.1 Barnett and Duvall's taxonomy of power

		Relational Specificity	
Power Emerges From		*Direct*	*Indirect (or Diffuse)*
	Specific Interactions of Actors	Compulsory	Institutional
	Constitutive Relationships Between Actors	Structural	Productive

The other dimension of power is concerned with the directness of the relational link between actors (and their interests) and their peers. Power can work through clear connections between states and objects (such as international institutions or specific conflicts). In this vein, power is defined by actors' positioning in the context of an active relationship. On the other hand, power can work through indirect conditions that define the relationship of one actor to both other actors and objects. In layman's terms, there exists a power relationship between, say, the United States and Burkino Faso *even where and when there exists no active relationship*. Here, power is defined by those factors that constitute the networks in which states are embedded. Again, though there is clear suggestion of social constructivism in this assertion, this notion is nevertheless broadly accepted and employed in IR theories beyond that perspective, including tangentially in the cyber conflict field in Nazli Choucri's employment of lateral pressure theory to the development of cyberspace.

Taken together, Barnett and Duvall argue that these analytic dimensions of power suggest a fourfold taxonomy of power that is far more inclusive than the threefold levels of analysis understanding of power often employed among security scholars (i.e. power via force, over rules and over preferences). **Compulsory power** describes the ability of agents to achieve direct control over one another via a range of mechanisms. **Institutional power** describes the ability of actors to control distant peers via manipulation of or control over rules of the game. **Structural power** describes direct constitutive control over others, such as Actor A's ability to determine Actor B's power through the medium of economic dependence on Actor A. And **productive power** describes indirect constitutive control over others in the context and construction of those broader socio-technical relationships within which all states are embedded.

Most scholars agree that this focus on technical capability does make some sense if the question is about power that emerges from cyber capabilities – i.e. about being "cyber powerful." But any conversation about the nature of power in world politics must recognize a basic argument made by political scientists – that power has much to do with the institutional and normative context within which actions are taken. Joseph Nye's early effort to describe the nature of cyber power points out the flaw in presuming that the basis of power in any category is the technical capability of the instrument(s) involved.[76]

Much as might be considered the case with, for example, nuclear weapons, power from possession of such weapons derives not only from the ability to use or threaten use. Possession, which with nuclear weapons is a relatively clear-cut status beyond a certain point, can be framed and maneuvered so as to manipulate the preferences of domestic and foreign audiences, can be used as a buy-in chip to position in different formal and informal rule-making positions and more.[77] In this way, power is conceptualized not only as a capacity to force behavioral changes but also as an ability both to influence the "rules of the game" – thus tacitly constraining others – and to alter preferences.[78] Capabilities-focused definitions thus fail to account for a great deal of variation in outcomes emerging from the nexus of abilities surrounding a particular set of instruments. This is particularly the case with cyberspace, where possession and the potential for use of a diverse range of abilities is relatively difficult to assess and characterize in the aggregate.

In many ways, it is useful to think of power in the context of how the information revolution has systematically changed the dynamics of human interaction in world affairs. In general, we might think of three principal types of changes that have occurred as a result of the global integration of information technologies across all levels of societal functioning. These are described later. Taken together, these different kinds of changes effectively describe the changing context within which state behavior emerges, not only in terms of the digitization of global infrastructure but also in terms of the institutional and social trajectory of international order. In short, the information revolution of the past several decades has done more than just advance technical frontiers; it has fundamentally transformed communications infrastructure, how people (and states) communicate and how ideas spread.

How global infrastructure functions

Perhaps more obviously than other changes, the ubiquity and sophistication of information technologies has augured profound changes to the infrastructure of the everyday functioning of global system processes. These changes to how industries, governments and militaries function is the focus of much scholarship on cyberspace and IR.

For international security studies, in particular, the digitization of global infrastructure challenges both state security and a wider range of normative, legal and economic interests. The strategic rationale behind the design and deployment of advanced "cyber weapons," for instance, can broadly be found in the twofold digitization of information and control dynamics around the world – that is, the digitization of security systems and the digital inter-connection of previously discrete functions. Stuxnet, for example, was designed to circumnavigate "air-gap" defenses that would otherwise have rendered attempts at network intrusion impotent.[79] Likewise, Stuxnet – alongside other programs like Flame and, responsively, efforts such as Byzantine Foothold – was designed to take advantage of the widespread inter-connectedness of computer systems in recent years, transmitting component parts of itself via otherwise innocuous network or media transfers over time.[80] As some work has argued, the massive integration of ICTs across societal and industrial sectors also constructs unprecedented security obstacles for national security in economic[81] and legal[82] terms, with broad-scoped inter-connection of systems allowing for varying degrees of access and control of information.[83]

Beyond the traditional purview of the security studies field, the digitization of global infrastructure has had a more direct impact on the shape of the global economy and on processes of global governance. Across industry, government and public organizations at almost every level, means of financial transaction, recordkeeping and utilities procurement look markedly different than they might have in the 1980s, even if the specific aims and parameters of a given set of processes remain the same.[84]

How people (and societies) interact with each other

By contrast, the change to the global system more focused on by the more loosely defined body of work on cyberspace and political organization has less to do with technical conditions than it does to do with agent behavior. Related to the broad digitization of global infrastructure, recent decades have seen unprecedented changes in the ways that global constituents (individuals, communities, organizations, etc.) communicate and consume information.[85] This certainly might be thought of as a consequence of global infrastructure digitization. However, changes to the nature of inter-constituent communication are both unique and fundamentally related to the dynamics of the global system in which specific actors are embedded. In short, new communication and information consumption modalities affect preference sets in a very basic manner.[86] Individual and organizational approaches to problem solving, self-representation and other fundamental political activities continue to adapt to match the network realities of an increasingly transnationally oriented – rather than nationally oriented – international system. This is a natural outgrowth of the proliferation of information linkages across most societal functions.

How people (and societies) see each other

Finally, the global adoption and integration of ICTs across the full range of societal functions has augured significant changes in the way that ideas are communicated, disseminated and presented. These changes are not limited to inter-personal communications.[87] Just as fundamental changes in the nature of communications' possibilities for people and institutions around the world have affected how preferences are constructed, dynamics of ICT utilization and development across both public and private sectors have had a unique impact on patterns of ideational inter-connections across numerous types of boundaries in the digital age.[88] Though drastically understudied in comparison as a type of systematic change fueled by global ICT integration, a diverse body of scholarly work in the social sciences has for some years consistently demonstrated that the market-specific nature of ICT development has had noticeable effects on patterns of political organization and expression in world affairs. In one vein, for instance, patterns of public opinion and information consumption on different topics has been linked to the specific dynamics of Twitter usage.[89] The rise of virtual, polycentric communities centered around the use of specific social media platforms and activated by community attention to a topic – rather than more traditional governmental or old media focus on an issue – has, at least to some degree, altered the dynamics of gatekeeping and agenda setting in IR. One immediate effect of this fundamental change to the global system is that political actions in the information age reach, affect and interest a much broader subset of global constituents than they may have in decades past, while the dynamics of specific areas of state interest – such as aspects of national security – have expanded to interact with societal functions at many levels.

These revolutions in the topology of infrastructure, information and preferences in the global system affect the dynamics of international interactions and human organization in a number of meaningful ways. Specifically, these changes to the global system have been and increasingly are the cause of new types of relationships and institutions in international affairs, including increasingly complex links across borders that do not rely on state accommodation. Shifts in national security, economic security and civil liberties' preferences are increasingly forcing existing institutions to strike new institutional equilibriums, to interpret old rules in new ways and to reinterpret patterns of interaction. Moreover, new institutions are needed for a host of issues related to ICT-derived preference sets, and both new and existing institutions are having to adapt to the conduct of IR constituted of new forms of social organization, including virtual communities and new global media patterns. In other words, the trajectory of global ICT adoption and development has affected distinct and meaningful changes to not only the technical context of international affairs but also the social and institutional networks in which human interaction occurs.

The meaning of global changes for cyber power

So, given these changes to the international environment in which states operate, what does power in the digital age – a state's true "cyber power" – look like? Among scholars, there are two emerging points of view. The first argument is essentially that cyber power is multifaceted and highly variable. It manifests in an actor's abilities to control the continued development of technical, regulatory and normative conditions surrounding ICT usage in favorable ways. Cyber power and cyber power strategies are not just about the use or threat of use of cyber weapons.[90] Cyber power certainly might be constituted of different abilities to intrude, disrupt, exploit and safeguard information systems attached to a range of societal functions. But it can equally derive from an ability to influence formal and informal rules regarding cyberspace or the capacity to induce changes – whether through soft or hard power means – in the preferences of peer constituents in the global system. While cyber power strategies might include the development of disruptive capacity across a range of information systems, it might also include actions concerned with, for instance, the condition of norms of network usage or case-specific efforts to influence foreign political conditions to catalyze favorable policy outcomes (i.e. states might undertake to promote "Internet access as a human right" organizations because the action is normatively beneficial).

The second argument is essentially that cyber power will enduringly be tricky to pin down for those interested in understanding some kind of balance of global cyber power. This perspective is not necessarily mutually exclusive from the argument that cyber power is simply constituted of an incredibly diverse set of determining factors. For many, being cyber "powerful" is about broadening the spectrum of possible ICT-aided actions that can meaningfully secure national interests through (1) the development of technical capacity and infrastructural redundancy, (2) the construction of rules of the road favorable to an actor's portfolio of capabilities and (3) the development of norms of behavior to underwrite the strategic perspective of the actor involved. But viewing power in the digital age as relating to strategic goals achieved in the context of how ICTs have changed the international system also moves the focus from inherent state attributes to the shifting context of that continuing transformation. State cyber power would essentially be contingent on the context of an actor's strategic interests and the nature of broad systemic developments – such as the expansion of Internet infrastructure or linguistic shifts in the content of the web – that favor that actor. Given that assessments of power will then radically vary depending on an understanding of diverse factors and that divergent perceptions of the abilities of others will invariably make for diverse assessments of just how "cyber powerful" others are, how useful a concept is "cyber power" really? To understand the abilities of states and potential for conflict, it is often better simply to focus on the derivative elements of power – i.e. on how cyber capabilities might be used at any given time to win wars or to coerce or deter others.

Implications for policy and statecraft

There are clear implications of what has been outlined in this chapter for policy and statecraft. Some of this is covered in other chapters. It is worth mentioning here, however, that foremost among these is a great need for national-level dialogue about the scope of what constitutes cyber conflict and the degree to which the Internet has altered the character of warfare. The need for such conversations – which will emerge naturally over a span of many years as governments and their populations are exposed with ever-increasing frequency to the externalities, if not the direct effects, of cyber conflict – is multi-form. In particular, where popular responsiveness will dictate the extent of a state's reaction to cyber conflict (i.e. in democratic societies), greater information and knowledge will help produce prudent policy outcomes. Specifically, greater conversations about cybersecurity issues will also act as a stabilizing agent for certain elements of the interstate cyber conflict set of calculations. Clear delineation of and dialogue on the subject of what constitutes critical assets and red lines for cyber aggression can help clarify the parameters of deterrent strategies and discourage foreign entities in efforts to coerce.

At the same time, it seems only logical that states should publicize how they plan cyber conflict issues and specific crisis situations, at least when it comes to major disruption scenarios. Such a move would help mitigate fears from escalating from the various uncertainties involved in cyber conflict and would blunt the effectiveness of cyber conflict oddities of operations, including regarding the independence of different network warfare arms of the government and lines of authorization during conflict. Such an effort would also help with definitional issues that emerge in translating operational principles to either the public dialogue or foreign analytic setting.

Policymakers would likewise do well to recognize that there are great efficiencies in cybersecurity policy endorsed by the broadest number of actors possible in a given country. Naturally, this gets into the territory of public-private dialogues and the relationship between government branches. In part, this is the topic of the previous chapter on national experiences with cybersecurity governance. Nevertheless, it is worth further noting that public conversation to produce more effective and representative policy positions should produce egalitarian ends.

Finally, states should apply appropriate lessons learned and processes utilized in other arenas of international security affairs but should be careful to avoid applying existing procedures where utilization might produce inefficiencies. Specifically, states would naturally do well to encourage national conversations in the context of the existing consensus that is the law of armed conflicts. For Western countries in particular, aligning national conversations with the ethical and moral imperatives of prevailing international legal conditions is critical if the trappings of cyber conflict governance are to organically fit with the contours of the present liberal international community. Likewise, states should recognize that existing mechanisms for the normalization of the outputs of such conversations could be effective when it comes to cybersecurity. Wargames and joint professional military (and civilian) training, in particular, can be effective for standardizing operational expectations among state security organizations. Of course, there is a challenge involved in setting up effective collaborations in this manner, as security actors may be incentivized to hide capabilities. Nevertheless, such mechanics of normalization of interstate affairs have significant value when it comes to building engagement and process expectations between those communities that will inevitably clash in the future in cyber conflict episodes.

Notes

1 See Mintz, Alex, and Karl DeRouen Jr. *Understanding Foreign Policy Decision Making*. Cambridge, UK: Cambridge University Press, 2010.
2 National Research Council. *Technology, Policy, Law, and Ethics Regarding US Acquisition and Use of Cyberattack Capabilities*. Washington, DC: National Academies Press, 2009.
3 Ibid.
4 Ibid.
5 Libicki, Martin C., David Senty, and Julia Pollak. *Hackers Wanted: An Examination of the Cybersecurity Labor Market*. Santa Monica, CA: RAND Corporation, 2014.
6 Ibid.
7 Ibid.
8 Marsh, Robert T. *Critical Foundations: Protecting America's Infrastructures*. Washington, DC: The President's Commission on Critical Infrastructure Protection, 1997.
9 Shane, Peter M., and Jeffrey Allen Hunker, eds. *Cybersecurity: Shared Risks, Shared Responsibilities*. Durham, NC: Carolina Academic Press, 2013.
10 Ibid.
11 Ibid.
12 See, for instance, Whyte, Christopher. "Developed States' Vulnerability to Economic Disruption Online." *Orbis*, Vol. 60, No. 3 (2016), 417–432; and Whyte, Christopher. "Power and Predation in Cyberspace." *Strategic Studies Quarterly*, Vol. 9, No. 1 (2015), 100–118.
13 Gorman, Siobhan, August Cole, and Yochi Dreazen. "Computer Spies Breach Fighter-Jet Project." *Wall Street Journal* Vol. 21 (April 21, 2009). www.wsj.com/articles/SB124027491029837401.
14 See Lewis, James A. "Computer Espionage, Titan Rain and China." *Center for Strategic and International Studies-Technology and Public Policy Program* (2005), 1.
15 See Siwei, Zhao. "Exclusive: Report Reveals how US Spy Agencies stole 97b Global Internet Data, 124b Phone Records in Just 30 days." *Global Times* June 13, 2022.
16 The printing press is the key focus of Anderson's seminal work on information technology as a force shaping sociopolitical identity and resultant institutions. See Anderson, Benedict. *Imagined Communities: Reflections on the Origin and Spread of Nationalism*. New York: Verso Books, 2006.
17 See *inter alia* Kaufmann, Chaim. "Threat Inflation and the Failure of the Marketplace of Ideas: The Selling of the Iraq War." *International Security*, Vol. 29, No. 1 (2004), 5–48; Pentland, A., and Juan Carlos Barahona. "Marketplace of Ideas." In *ICT4D Conference*. 2003; and Thrall, A. Trevor. "A Bear in the Woods? Threat Framing and the Marketplace of Values." *Security Studies*, Vol. 16, No. 3 (2007), 452–488.
18 Jensen, Benjamin, Ryan C. Maness, and Brandon Valeriano. "Cyber Victory: The Efficacy of Cyber Coercion." In *Annual Meeting of the International Studies Association*, 2016.
19 For an in-depth overview of the topic, beyond that which is provided later in this chapter, see Valeriano, Brandon, Benjamin Jensen, and Ryan C. Maness. *Cyber Strategy: The Evolving Character of Power and Coercion*. Oxford, UK: Oxford University Press, 2018.
20 This is the primary argument Gartzke offers in deconstructing what he calls "the myth of cyberwar." Gartzke follows on here from Rid's seminal article and book that argue that cyber conflict is something "not war" – it is sabotage, subversion, espionage and activism, but not actually warfighting. See Gartzke, Erik. "The Myth of Cyberwar: Bringing War in Cyberspace Back Down to Earth." *International Security*, Vol. 38, No. 2 (2013), 41–73; Rid, Thomas. "Cyber War Will Not Take Place." *Journal of Strategic Studies*, Vol. 35, No. 1 (2012), 5–32; and Rid, Thomas. *Cyber War Will Not Take Place*. New York: Oxford University Press, 2013.
21 Ibid.
22 For seminal work on the topic, see *inter alia* Jervis, Robert. "Cooperation under the Security Dilemma." *World Politics*, Vol. 30, No. 2 (1978), 167–214; Axelrod, Robert, and William Donald Hamilton. "The Evolution of Cooperation." *Science*, Vol. 211, No. 4489 (1981), 1390–1396; Keohane, Robert

O. *After Hegemony: Cooperation and Discord in the World Political Economy*. Princeton, NJ: Princeton University Press, 2005; and Wendt, Alexander. "Anarchy Is What States Make of It: The Social Construction of Power Politics." *International Organization*, Vol. 46, No. 2 (1992), 391–425.
23 For significant literature on the problem of attribution of cyberattacks, see Tsagourias, Nicholas. "Cyber Attacks, Self-Defence and the Problem of Attribution." *Journal of Conflict and Security Law*, Vol. 17, No. 2 (2012), 229–244; Rid, Thomas, and Ben Buchanan. "Attributing Cyber Attacks." *Journal of Strategic Studies*, Vol. 38, Nos. 1–2 (2015), 4–37; and Lindsay, Jon R. "Tipping the Scales: The Attribution Problem and the Feasibility of Deterrence against Cyberattack." *Journal of Cybersecurity*, Vol. 1, No. 1 (2015), 53–67.
24 Gartzke, Erik, and Jon R. Lindsay. "Weaving Tangled Webs: Offense, Defense, and Deception in Cyberspace." *Security Studies*, Vol. 24, No. 2 (2015), 316–348.
25 Ibid.
26 Ibid.
27 See Gompert, David C., and Martin Libicki. "Cyber warfare and Sino-American Crisis Instability." *Survival*, Vol. 56, No. 4 (2014), 7–22.
28 Ibid.
29 Ibid.
30 Ibid.
31 Ibid.
32 Ibid.
33 Ibid.
34 See George, Alexander. *Forceful Persuasion: Coercive Diplomacy as an Alternative to War*. Washington, DC: United States Institute of Peace Press, 1991; Jakobsen, P. *Western Use of Coercive Diplomacy after the Cold War: A Challenge for Theory and Practice*. New York: St Martin's Press, 1998; Mearsheimer, John. *The Tragedy of Great Power Politics*. New York: W.W. Norton, 2001; Pape, Robert. *Bombing to Win: Air Power and Coercion in War*. Ithaca, NY: Cornell University Press, 1996.
35 See Byman, Daniel, and M. Waxman, *The Dynamics of Coercion: American Foreign Policy and the Limits of Military Might*. Cambridge, UK: Cambridge University Press, 2002
36 Schelling, 1966.
37 Ibid.
38 Ibid.
39 Ibid.
40 Ibid. Also see Tarzi, S. "Hypotheses on the Use and Limitations of Coercive Diplomacy," *International Studies*, Vol. 36, No. 1 (1999), 63–75.
41 See Rich, Jason. "Adding Some Context: The Systemic Constraints on Coercion." *Politics*, Vol. 33, No. 1 (2013), 37–46.
42 See Blechman, Barry, and Tamara Wittes. "Defining Moment: The Threat and Use of Force in American Foreign Policy." *Political Science Quarterly*, Vol. 114, No. 1 (1999), 1–24.
43 See George, Alexander. *Bridging the Gap: Theory and Practice in Foreign Policy*. Washington, DC: United States Institute of Peace Press, 1993; and George, Alexander, David Hall and William Simons. *The Limits of Coercive Diplomacy*. Boston, MA: Little, Brown & Co, 1971.
44 Ibid.
45 Regarding credibility and the opportunity costs of poor signaling during coercive efforts, see Freedman, Lawrence. "Strategic Coercion" In Lawrence Freedman (ed.), *Strategic Coercion: Concepts and Cases*. New York: Oxford University Press, 1998; and Gregory Treverton, *Framing Compellent Strategies*, MR-1240-OSD, Santa Monica, CA: RAND Corporation, 2000.
46 Ibid.
47 See Byman and Waxman, 2002, George, Hall and Simons, 1971.
48 See Byman and Waxman, 2002.
49 This is an extremely common theme in the literature on coercion and interstate behavior more broadly. See Sterling-Folker, Jennifer. "Realist Environment, Liberal Process, and Domestic-Level Variables." *International Studies Quarterly*, Vol. 41, No. 1 (1997), 1–25.
50 For perhaps the broadest ranging set of descriptions and analyses on the context of coercive diplomacy, see Robert J. Art and Patrick Cronin (eds.), *The United States and Coercive Diplomacy*. Washington, DC: United States Institute of Peace Press, 2003.
51 See Russell (2014) for the definitive description of cyber blockades and of the challenges involved in their implementation.

52 For the most complete description of the attribution problem and extended work on the history and contemporary puzzles involved, see Rid, Thomas, and Ben Buchanan. "Attributing Cyber Attacks." *Journal of Strategic Studies*, Vol. 38, Nos. 1–2 (2015), 4–37. For earlier work, see Clayton, Richard. *Anonymity and Traceability in Cyberspace*, Vol. 653, Technical Report, Cambridge: University of Cambridge Computer Laboratory, 2005; Brenner, Susan. "At Light Speed: Attribution and Response to Cybercrime/Terrorism/Warfare." *The Journal of Criminal Law & Criminology*, Vol. 97, no. 2 (2007), 379–475; and Wheeler, David A., and Gregory N. Larsen, *Techniques for Cyber Attack Attribution*. Alexandria, VA: Institute for Defense Analysis, 2003.

53 See Betz, David, and Tim Stevens, *Cyberspace and the State*, Adelphi Series, London: IISS/Routledge, 2011.

54 Singer, Peter W., and Allan Friedman, *Cybersecurity and Cyberwar: What Everyone Needs to Know*. Oxford, UK: Oxford University Press, 2013, 144–148.

55 Plato, *Republic*, trans. by Allen Bloom. New York: Basic Books, 1968, 37–38.

56 For reasons examined in this section, it may be necessary to update Herman Kahn's six desirable characteristics of deterrence: (1) frightening; (2) inexorable; (3) persuasive; (4) cheap; (5) non-accident prone; (6) controllable, to include a seventh for cyber deterrence: Recognized.

57 Samantha Ravich, and Annie Fixler. *Framework and Terminology for Understanding Cyber-Enabled Economic Warfare*. Foundation for Defense of Democracies. February 23, 2017. www.defenddemocracy.org/media-hit/annie-fixler-framework-and-terminology-for-understanding-cyber-enabled-economic-warfare/.

58 In the context of deterring cyberattacks, deterrence by punishment can be through retaliatory cyberattacks (deterrence-in-kind) or other means, such as a kinetic or diplomatic response (cross-domain-deterrence).

59 Andrew May. *The RAND Corporation and the Dynamics of American Strategic Thought*, 1946–1962. Doctoral Dissertation, August 6, 1998, 225.

60 This idea of the impossibility of defense began to change with President Reagan's pursuit of the Strategic Defense Initiative and improved missile defense technologies.

61 Rid, "Cyber War Will Not Take Place." 5–32.

62 Markoff, John. "Before the Gunfire, Cyberattacks." *The New York Times* August 13, 2008. www.nytimes.com/2008/08/13/technology/13cyber.html?_r=0.

63 Greenemeier, Larry. "Estonian 'Cyber Riot' Was Planned, But Mastermind Still A Mystery," *Information Week* August 3, 2007. www.informationweek.com/estonian-cyber-riot-was-planned-but-mast/201202784.

64 Svensson, Peter. "What Makes Cyber Attacks So Hard to Trace? *Brisbane Times* March 22, 2013. www.brisbanetimes.com.au/it-pro/security-it/what-makes-cyber-attacks-so-hard-to-trace-20130322–2gkfv.html.

65 Burton, Kelly. "The Conficker Worm," *SANS* October 23, 2008. www.sans.org/security-resources/malwarefaq/conficker-worm.php.

66 Lewis, James. "The Korean Cyber Attacks and Their Implications for Cyber Conflict." *Center for Strategic and International Studies* October 23, 2009. http://csis.org/publication/korean-cyber-attacks-and-their-implications-cyber-conflict.

67 Markoff, John, and Thom Shanker. "Halted '03 Plan Illustrates U.S. Fear of Cyber Risk." *The New York Times* August 1, 2009. www.nytimes.com/2009/08/02/us/politics/02cyber.html.

68 For example, see Libicki, Martin. *Cyber Deterrence and Cyber War* (2009); Nye, Joesph. "Nuclear Lessons for Cyber Security," in *Strategic Studies Quarterly* (2011); Buchannan, Ben. *The Cybersecurity Dilemma* (2014); Goldman, Emily. *Cyber Analogies* (2014); Defense Science Board. *Report of the Task Force on Cyber Deterrence* (2017); and Nye, Joesph. *Deterrence and Dissuasion in Cyberspace in International Security* (2017).

69 Long, Austin. "A Cyber SIOP? Operational Considerations for Strategic Offensive Cyber Planning." *Journal of Cybersecurity* December 20, 2016.

70 For a more detailed explanation and assessment of persistent engagement, see Schneider, Jacquelyn. "Persistent Engagement: Foundation, Evolution and Evaluation of a Strategy." *Lawfare* May 10, 2019.

71 Spade, Jayson M. *Information as Power: China's Cyber Power and America's National Security*. Carlisle, PA: United States Army War College, 2011.

72 Betz, David J., and Tim Stevens. *Cyberspace and the State: Toward a Strategy for Cyber-Power*. The International Institute for Strategic Studies, December 1, 2011.

73 Betz, David J. "Cyberpower and International Security." *Foreign Policy Research Institute*, e-Notes June 2012.

74 Pope, Billy. "Cyber Power: A Personal Theory of Power." Center for International Maritime Security May 2014. http://cimsec.org/cyber-power-personal-theory-power/11436.

75 Barnett, Michael, and Raymond Duvall, eds. *Power in Global Governance*, Vol. 98. Cambridge, UK: Cambridge University Press, 2004.

76 Nye, "Cyber Power" (2010).

77 For the seminal description of the different political uses of and incentives for the acquisition of nuclear weapons, see Sagan, Scott D. "Why Do States Build Nuclear Weapons? Three Models in Search of a Bomb." *International Security*, Vol. 21, No. 3 (1996/97), 63–65.

78 A significant literature in political science, spread across various subfields, holds that power means conceptually different things depending on how one assesses the format of favorable outcomes. Starting with Dahl, various authors have pointed to power as being tri-dimensional. Yes, (1) actor A can use force or coercion to force actor B to change their behavior in a way they might not otherwise have done. But it is also possible for actor A to (2) utilize its position to influence rules and norms in such a way that actor B finds itself constrained and impelled to behave in a way other than it might have under anarchy. Likewise, it is (3) possible for actor A, through soft power means or through precedence in certain relationships, to fundamentally change the preferences of actor B such that certain behaviors become a default choice. See Dahl, Robert A. *Who Governs: Democracy and Power in an American City*. New Haven, CT: Yale University Press, 1961; Bachrach, Peter, and Morton Baratz, "Decisions and Nondecisions: An Analytical Framework." *American Political Science Review* (1963), 632–642; Lukes, Steven. *Power: A Radical View*. London: Palgrave, 2nd ed., 1974; Gaventa, John. *Power and Powerlessness: Quiescence & Rebellion in an Appalachian Valley*. Chicago, IL: University of Illinois Press, May 1, 1982; and Nye, John S. *Soft Power: The Means to Success in World Politics*. New York: Public Affairs Press, 2004.

79 For detailed accounts of the Stuxnet episode from technical, policy and political perspectives, see Matrosov, Aleksandr, Eugene Rodionov, David Harley, and Juraj Malcho. "Stuxnet under the Microscope." *ESET, White Paper* January 20, 2011; Lindsay, Jon R. "Stuxnet and the Limits of Cyber Warfare." *Security Studies*, Vol. 22, No. 3 (2013), 365–404; Albright, David, Paul Brannan, and Christina Walrond. "Did Stuxnet Take Out 1,000 Centrifuges at the Natanz Enrichment Plant?" *Institute for Science and International Security*, December 22, 2010, 3–4; Langner, Ralph. "To Kill a Centrifuge: A Technical Analysis of What Stuxnet's Creators Tried to Achieve." Dover, DE: The Langner Group, November 2013; and Zetter, Kim. "How Digital Detectives Deciphered Stuxnet, the Most Menacing Malware in History." *Wired Threat Level Blog* July 11, 2011. www.wired.com/threatlevel/2011/07/how-digital-detectives-deciphered-stuxnet.

80 This has elsewhere been described as a weapon of mass effect, differentiated from digital weapons that cause massive disruption of systems by the widespread deployment but low-intensity nature of the eventual effect. See Whyte, Christopher. "Power and Predation in Cyberspace." *Strategic Studies Quarterly*, Vol. 9, No. 1 (2015), 100–118.

81 Among others, see Cornish, Paul, David Livingstone, Dave Clemente and Claire York. *On Cyber Warfare*. London: Chatham House, 2010; Vaishnav, Chintan, Nazli Choucri and David D. Clark. *Cyber International Relations as an Integrated System*. MIT Political Science Department Research Paper No. 2012–16 (June 14, 2012); Valeriano, Brandon, and Ryan Maness. "A Theory of Cyber Espionage for the Intelligence Community." *EMC Conference Paper* (2013); and Lewis, James, and Stewart Baker. *The Economic Impact of Cybercrime and Cyber Espionage*. Washington, DC: Center for Strategic and International Studies, July 22, 2013.

82 For debating various aspects of international-oriented legal challenges, see Owens, William A, Kenneth W. Dam, and Herbert S. Lin, eds., *Technology, Policy, Law, and Ethics Regarding U.S. Acquisition and Use of Cyberattack Capabilities*. Washington, DC: National Academies Press, 2009; Clark, David D., and Susan Landau. "Untangling Attribution," In *Proceedings of a Workshop on Deterring Cyberattacks: Informing Strategies and Developing Options for U.S. Policy*. Washington, DC: National Academies Press, 2010; Toren, Peter. "A Report on Prosecutions under the Economic Espionage Act." *Paper Presented at the American Intellectual Property Law Association annual meeting*, Trade Secret Law Summit, Washington, DC, October 23, 2012; Germano, Judith. *Cyber Security Partnerships: A New Era of Public-Private Collaboration*. New York University School of Law: The Center on Law and Security, 2014; Germano, Judith, and Zachary Goldman. *After the Breach: Cyber Security Liability Risk*. New York University School of Law: The Center on Law and Security, 2014.

83 For a fuller discussion of issues of access and control, see Lindsay, Jon R., Tai Ming Cheung and Derek Reveron. *China and Cyber Security: Espionage, Strategy, and Politics in the Digital Domain*. Oxford, UK: Oxford University Press, 2015.

84 Choucri, *Cyberpolitics in International Relations*, 2012.

85 For prominent work in this vein, see Bimber, Bruce. *Information and American Democracy*. Cambridge, UK: Cambridge University Press, 2003; Bennett, Lance, and Shanto Iyengar. "A New Era of Minimal Effects? The Changing Foundations of Political Communication." *Journal of Communication*, Vol. 58, No. 4 (2008), 707–731; and Earl, Jennifer, and Katrina Kimport. *Digitally Enabled Social Change*. Cambridge, MA: MIT Press, 2011.

86 Though earlier, Chafee and Metzger present one of the most notable descriptions of one major change in global constituent preference sets—that people have increasingly come to expect to receive the types of information they want to receive, rather than the information that the media might offer without popular input. See Chaffee, S., and M. Metzger. "The End of Mass Communication." *Mass Communications and Society*, No. 4 (2001), 365–79.

87 For seminal works in this tradition, see Galtung, J. and M. Ruge. "The Structure of Foreign News: The Presentation of the Congo, Cuba and Cyprus Crises in Four Norwegian Newspapers." *Journal of Peace Research* (1965); Dupree, J.D. "International Communication, View from a Window on the World." *Gazette*, Vol. 17 (1971), 224–235; Ahern, T.J. Jr. "Determinants of Foreign Coverage in Newspapers," In R.L. Stevenson and D.L. Shaw (eds.), *Foreign News and the New World Information Order*, Ames, IA: Iowa State University Press, 1984; and Wu, Dennis. "Investigating the Determinants of International News Flow." *International Communication Gazette*, Vol. 60, No. 6 (1998), 493–512. Additionally, the notion that media systems play a significant role in changing the way that information is presented to global constituents, which then affects policy responses, is a common one in the communications literature more broadly. For the most complete account of this theoretical tradition, see Gilboa, Eytan. "The CNN Effect: The Search for a Communication Theory of International Relations." *Political Communication*, Vol. 22, No. 1 (2005), 27–44.

88 See Allen-Robertson, J., and D. Beer. "Mobile Ideas: Tracking a Concept through Time and Space." *Mobilities*, Vol. 5, No. 4 (2010), 529–545; Quercia, D., L. Capra, and J. Crowcroft. "The Social World of Twitter: Topics, Geography, and Emotions." *Proceedings of the Sixth International Conference on Weblogs and Social Media*. Palo Alto, CA: AAAI Press, 2012; and Takhteyev, Y., A. Gruzd, and B. Wellman. "Geography of Twitter Networks." *Social Networks*, Vol. 34, No. 1 (2012), 73–81.

89 See Greenhow, C., and B. Robelia. "Informal Learning and Identity Formation in Online Social Networks." *Learning, Media and Technology*, Vol. 34, No. 2 (2009), 119–140; Bennett, Lance W. "The Personalization of Politics Political Identity, Social Media, and Changing Patterns of Participation." *The Annals of the American Academy of Political and Social Science*, Vol. 644, No. 1 (2012), 20–39; and Stefanidis, Anthony, Amy Cotnoir, Arie Croitoru, Andrew Crooks, Matthew Rice, and Jacek Radzikowski. "Demarcating New Boundaries: Mapping Virtual Polycentric Communities through Social Media Content." *Cartography and Geographic Information Science*, Vol. 40, No. 2 (2013), 116–129.

90 Nye, "Cyber Power" (2010); Betz and Stevens, *Cyberspace and the State* (2011).

Further readings

Borghard, Erica D., and Shawn W. Lonergan. "The Logic of Coercion in Cyberspace." *Security Studies*. Vol. 26, No. 3 (2017), 452–481.

Gartzke, Erik. "The Myth of Cyberwar: Bringing War in Cyberspace Back Down to Earth." *International Security*. Vol. 38, No. 2 (2013), 41–73.

Gartzke, Erik, and Jon R. Lindsay. "Weaving Tangled Webs: Offense, Defense, and Deception in Cyberspace." *Security Studies*. Vol. 24, No. 2 (2015), 316–348.

Gompert, David C., and Martin Libicki. "Cyber Warfare and Sino-American Crisis Instability." *Survival*. Vol. 56, No. 4 (2014), 7–22.

Maschmeyer. Lennart. "The Subversive Trilemma: Why Cyber Operations Fall Short of Expectations." *International Security*. Vol. 46, No. 2 (2021): 51–90.

Mazanec, Brian, and Samantha Ravich. "Lessons for the Cyber Battlefield From the Early Nuclear Era's Single Integrated Operating Plan." Foundation for Defense of Democracies. January 18, 2019. www.

fdd.org/analysis/2019/01/18/lessons-for-the-cyber-battlefield-from-the-early-nuclear-eras-single-integrated-operating-plan/

Nye Jr, Joseph S. "Deterrence and Dissuasion in Cyberspace." *International Security*. Vol. 41, No. 3 (2017), 44–71.

Rid, Thomas. *Cyber War Will Not Take Place*. New York: Oxford University Press, 2013.

Russell, Alison Lawlor. *Cyber Blockades*. Washington, DC: Georgetown University Press, 2014.

Valeriano, Brandon, Benjamin Jensen, and Ryan C. Maness. *Cyber Strategy: The Evolving Character of Power and Coercion*. Oxford, UK: Oxford University Press, 2018.

Whyte, Christopher. "Ending Cyber Coercion: Computer Network Attack, Exploitation and the Case of North Korea." *Comparative Strategy*. Vol. 35, No. 2 (2016), 93–102.

10 Cyber conflict as "grey zone" conflict

In 2012, U.S. Secretary of Defense Leon Panetta drew the attention of the world – or, at least, of Western defense communities – when he claimed in a speech that the one of the biggest threats faced by the United States was that of a "cyber Pearl Harbor."[1] In saying this, he implied that a catastrophic and unpredicted cyber assault on the country could dramatically harm national function, potentially as part of a broader set of conflict actions taken by a foreign adversary. According to his speech, designed to highlight shortcomings in Western approaches to cybersecurity at the national level, there is significant potential that a cyberwar scenario will play out in the real world in the foreseeable future.

As we discussed in Chapter 9, there is broad consensus among both scholars of international affairs and cybersecurity practitioners that such cyberwar scenarios are likely to remain the stuff of special circumstances in world politics. Where powerful states desire to interfere in the affairs of small ones or aim to punish political intransigence for coercive purposes, the implementation of a cyber blockade or of broad-scoped attacks against critical infrastructure seems possible. But where states of comparable strength and competing interests find themselves squared off, waging cyberwar makes little sense. Unless accompanied by broader aggression in other domains, cyber victories are simply too fleeting to be worth it.[2] Moreover, cyberattacks might worsen relations between states such that conflict in other domains is forced. Except for where broader conventional war is the intention, disruptive cyber campaigns might thus be disastrous.

Despite the clear logic of this position, the reality is that cyber conflict is a widespread and rising phenomenon in world affairs. As we discuss in this chapter, it is necessary for students of IR that we move the conversation on how cyber conflict manifests from the traditional framing of warfare between states as a dichotomous affair – i.e. states are either at war or not – to notions of contention increasingly common in IR's strategic and military studies subfields. Specifically, cybersecurity students would do well to think of information security as enabling forms of conflict that exist between war and peace.[3] Conflict outside of warfare – what many are calling "grey zone" conflict,[4] a phrase we unpack in detail later – is an increasingly common feature of IR in the 21st century. Moreover, conflict of this kind increasingly contains digital elements.

This chapter describes conflict that takes place between states beneath the threshold of conventional warfare and considers reasons as to why increased incidence of such contention might stem from the effects of the information revolution. Then, we discuss the specific rise in incidence of "grey zone" conflict as relatively unique to the 21st century and consider a series of arguments as to why states that otherwise would be able to effectively fight one another might increasingly opt for sub-optimal conflict strategies – including

DOI: 10.4324/9781003246398-10

cyber conflict and cyber-enabled information warfare and influence operations – as a matter of course.

Between war and peace

Conflict between states, or between states and significant, recognizable non-state entities, occurs outside the bounds of declared warfare more often than it does within.[5] In other words, though interstate warfighting is a relatively common feature of the landscape of international affairs, conflict that occurs between war and peace nevertheless presents as a sort of constant background noise of world politics. For many countries, deaths – both military and civilian – stemming from foreign aggression are not the result of formal military actions most years. State-sponsored militant groups might kill a handful of civilians by firing improvised rockets across a border, while state-linked terrorists might target a military checkpoint with homemade explosives, all without such actions escalating to the outbreak of conventional interstate warfighting.

Asymmetric warfare, in particular, is an age-old feature of conflict in IR.[6] Asymmetric warfare quite simply describes conflict that involves actors of disproportionate military capability (i.e. strong actors vs. weak actors). The term might be applied to characterize a broad range of conflict scenarios, from state military efforts against terrorist or insurgent forces to the clash of state proxies (i.e. non-state arms of state security forces). Such conflicts tend not to look much like traditional military conflicts in tactical or strategic terms. Given the relative indifference in levels of power and ability involved, combat often entails limited engagements chosen by the weaker force to maximize either military or terroristic effects (e.g. bombing a truckload of soldiers rather than engage them with comparable forces).

Asymmetric warfare – and conflict between states more broadly that, regardless of the relative strength of those involved, occurs below the threshold of warfighting – often takes non-violent forms where the point is to achieve favorable changes to political and security circumstances via manipulative and coercive means. In part, such contention takes the form of what is called **political warfare**.[7] Political warfare regularly occurs where asymmetric warfighting dynamics exist, but it is often prosecuted in situations beyond violent clashes between mismatched combatants. Political warfare is a form of coercion that doesn't rely on the use of force or threat of force so much as it seeks to compel changes in behavior by making conditions hostile for the target. Naturally, this *does* sometimes include low-level violence conducted by state security actors or, more often, state proxies against sub-state organizations. But as often it includes the manipulation of public opinion, the suppression of pro-target voices in society, the spread of propaganda, bribery and more, all to create an environment that is hostile to the interests of a given target.

Box 10.1: How else should we think of cyber threats?

Given the idea that cyber conflict does not necessarily constitute warfighting in some meaningful way, but rather a set of enabling dynamics and techniques that permit new forms of contention outside the scope of traditional warfare, it is worth considering the form that cyber threats to national security take beyond individual incidents. In short, if we drop the U.S. Department of Defense's assumption that cyberspace is best thought of as a domain of warfighting and instead assume that cyber conflict simply emerges from a broad-scoped "informationization" of global society, what kind of problem is the cyber phenomenon?

Traditionally, of course, public policy practitioners have thought about cyber threats to homeland and national security in terms of the technological and political sources of insecurity involved. Many think of cybersecurity as inherently technical and would likely argue that security threats to society as a whole are premised on technological challenges. As a result, solving technical problems should take precedence over other forms of problem solving, and the most critical actors in the construction of better national cyber "health" are technologists, researchers and developers. By contrast, others (many of those cited in the writing of this book on cyber conflict) are of the mind that cybersecurity is best understood in thinking about the threat actors that threaten society in this domain – i.e. criminals, foreign states, intelligence organizations and proxies of all of these. Combating cyber threats, if you think this way, inevitably means combating those specific threat actors and empowering the processes of national security that are tasked with doing so.

But is cybersecurity more than just a technical or political issue? Given the way we've talked about the information revolution in this book as having systematically rewired the world system and as evolving from complex societal processes, would it be better to think about cyber threats as environmental phenomena rather than man-made issues?

Two ideas in this vein are worth mentioning. One is quite simply the increasingly popular notion that cyber threats to global society present as pollution – a sort of ever-present force that occurs in the everyday operation of the international system. Thinking this way, cyber threats might be best thought of as emerging from the negative externalities of human social, economic and political systems. For instance, rapid growth of new information services motivates reduced focus on security practices so as to ensure efficiency and productivity. Therefore, the public policy challenge is less meaningfully about addressing specific cybercriminal or security threats – perhaps except for the major ones that loom on the horizon – and more about incentivizing good behavior and best practices across all societal sectors. The implication, of course, would subsequently be that the most important actors for achieving a broad-scoped condition of national cybersecurity for any country would be economic and social stakeholders in a position of influence over usage of network technologies.

Similarly, another idea about the nature of cyber threats is that they are best thought of as a public health issue. This idea makes some considerable sense if one thinks about the broad landscape of cybersecurity threats as emerging in pandemic-style patterns. Particularly where we're talking about major data breaches or global ransomware attacks, cyber challenges emerge patterned in a similar fashion to pandemic disease outbreaks. Cyber threats are hard to predict, can spread unpredictably and affect a diverse cross-section of societal sectors. Therefore, the public policy challenge is about making sure facilitators – i.e. stakeholders that can coordinate across sectors of society for rapid crisis response – are enabled to act in the public interest in the most effective manner possible.

More than just political warfare or asymmetric warfare involving relatively autonomous proxies of foreign competitors, however, much interstate contention between war and peace takes the form of what many have increasingly called **hybrid warfare**, where political warfare techniques are blended with limited applications of more traditional security techniques in order to more effectively compel a target in some way.[8] Though the definitional line between these terms is reasonably blurred, the difference between hybrid warfare and some combination of asymmetric and political conflict is the implication that hybrid warfighting is a highly directed effort. Wherein political disinformation and propaganda efforts might aim to reduce the resolve of foreign populations to stand up to aggression, limited kinetic security actions – such as the deployment of un-uniformed soldiers to contested territory, as happened in Crimea in 2014 – can be planned to coincide with such effects and maximize gains for an attacker without actually risking outright war.

Information warfare

In thinking about and analyzing conflict between war and peace, much focus is often placed on manifestations of **information warfare**.[9] This is a term whose meaning has morphed over time to imply more or less specific elements of information security that manifest in conflict processes. Unfortunately, the term remains under-defined. There are no real consensus positions on what is meant by information warfare beyond those that attribute most warfighting actions to be, in some sense, intrinsically about information dynamics. Warfare is an inherently human affair, and what is human is inevitably about how humans communicate, organize, institutionalize and behave in a social sense. It is, in essence, inevitably about information – the informational content of human societies, the *meaning* of such content and the strategic value of all of the above. As then U.S. Secretary of Defence Carter stated in the *2016 DOD Strategy for Operations in the Information Environment*, although the term information environment is relatively new, the concept of an "information battlefield" is not.

What constitutes the toolkit of information warfare depends on the era and the circumstances we are interested in analyzing. Information warfare has enduringly been said to include **propaganda** efforts, which involves the circulation of biased or misleading material to politicize a particular topic; **psychological operations**, which involve the manipulation of information specifically to affect the reasoning capacity of a target (such as a foreign state's military leadership); **military deception**, which is efforts that aim to mislead competitors about the extent of a state's military preparedness; and **internal security measures**, by which governments enact protocols designed to shield security functions from interference. Naturally, affecting the information environment in which politics, commerce and national security functions take place might also involve physical destruction and, at least from the midpoint of the 20th century onward, **electronic warfare** aimed at disrupting the infrastructure of military, government and societal communications.

But, particularly given the degree to which discussions of information warfare have come back into vogue in the digital age, it seems wise to introduce the topic in terms of outcomes rather than specific modes of conflict interactions. Regardless of the tools being used to achieve informational outcomes for political and security purposes, certain modes of interactions persistently characterize information warfare campaigns. Specifically, five modes of interaction and activity cover the gamut of tools and targets typically involved in such campaigns, regardless of era.

Much of what is involved with information warfare efforts is not actually what one might be tempted to label an attack. Rather, sophisticated and effective information warfare campaigns often involved significant efforts aimed at ensuring appropriate infrastructural conditions for success. **Information transportation** is the broad term used to describe any effort to construct such conditions. Information transport efforts might include any attempt to shape how adversaries transmit information, as well as defensive attempts to ensure redundancy in communications infrastructure. Significantly, information transport efforts often entail actions that ensure one of two things: first, that only certain kinds of information are being communicated or accessed by a target and, second, that information is as susceptible to interception and manipulation as possible. The U.S. bombings of Saddam Hussein's fiber optic cable tranche stations during the First Gulf War, which forced Iraqi forces to use legacy radio systems (which were easily tapped by U.S. intelligence) to communicate, is a good example of an operation that aimed to affect how information was transported.[10]

Similarly, information warfare efforts aimed at constructing appropriate infrastructural conditions for operation usually contain basic **information collection** activities. The idea here, quite simply, is to achieve enhanced situational awareness such that disruptive conflict actions can be better planned and put into effect. Again, it is worth mentioning here that what constitutes "information warfare" might be almost anything, as "information collection" essentially describes the function of state intelligence services.

Information protection or information security involves the use of any technique that minimizes the information collection activities of one's adversaries. If effective information collection is desirable because heightened situational awareness lends itself to better planning, then information protection is desirable simply because reducing an adversary's awareness diminishes the opportunities they'll have to gain by engaging in information warfare. This is a two-way street in that information protection activities can be both offensive and defensive. Defensively, information protection is constituted of security measures taken to blunt the effect of foreign interference. Offensively, information protection is any action that degrades the view of one's opponent.

Information disinformation and **information manipulation** involve altering the content of an information environment in an effort to shape how adversaries – governments, specific institutions or a broader population – view the world around them. With information manipulation, the goal is often twofold. First, disinformation is designed to inject false narrative and facts either into sociopolitical discourse or into the decision-making processes of specific institutions. The purpose of doing so is to muddy the water and prevent the effect function of systems that might otherwise result in prudent debate and analysis of unbiased information. Second, information tampering – particularly where information is altered but not entirely falsified – is often used to inject uncertainty about the credibility of information resources. Even where a target population reaches a consensus position about major issues or facts, the knowledge that information being received in news media coverage or in non-news reporting might have been manipulated often causes doubt over the reliability of traditional pillars of political and institutional information security.

Finally, **information disruption** refers to efforts that aim to prevent adversaries from receiving the full picture by directly attacking information and information systems. In reality, information disruption – as it pertains to information warfare – includes several different techniques, including *information denial*, *information degradation* and *information disturbance*. The point in each case is that an information warfare attacker reduces the ability of an enemy to receive or rely on information by directly tampering with the way in which information is made available thereto. Information denial simply entails the deletion of information, often strategically so as to ensure a skewed view of incoming information on the part of the target. Information disturbance more specifically aims at reducing the reliability of incoming information, often by introducing additional information to confuse the target. Such additional noise makes it hard to parse out relevant detail and meaning from incoming information. By contrast, information degradation involves those techniques that take aim not at the nature of incoming information or its reliability, but rather at the ability of a target to receive or analyze it. Degradation attacks might, for instance, include denial of service attacks that prevent the use of a given network connection or even the physical bombing of, say, a receiver station for radio transmissions.

The utility of cyber conflict for "grey zone" conflict

Many scholars and security practitioners place "cyber warfare" as a new part of the toolkit of information and political warfare.[11] Yet others describe cyber conflict as a mode of contention and disruption that is adjacent to and overlaps with, but is distinct from, political-security activities like intelligence gathering, counterintelligence efforts, terrorist campaigning and military operations. Regardless of the specific approach one takes to categorizing cyber conflict as a subset or corollary to traditional forms of political contention, it is clear that Internet technologies have had a profound impact on the conduct of warfighting. Moreover, there has been an explosion in the number of conflicts in the space between war and peace since the late 1990s, many of which prominently feature Internet technologies in enabling or disruptive roles. This section outlines several reasons as to why the Internet – or, more accurately, cyberspace – presents as an attractive new domain of operation and set of tools for those interested in engaging in contestation short of war.

The Internet as a global control system

In 1914, a few hours before the outbreak of World War I in Europe, the first offensive action of the British Empire was ordered and preparations were made.[12] Quite apart from being an attack against the military forces of the Central Powers, the purview of the orders given was the communications capabilities of the German Empire. Setting sail on a ship called the *Alert*, British agents traveled a handful of miles into the English Channel and – not long after the official declaration of hostilities – dredged up and severed a series of telegraph cables. Though wartime development of better radio systems and construction of new telegraph infrastructure would not allow the Allies an enduring upper hand in the communications war, the effect of the *Alert*'s action was rapid and remarkable. Great Britain was able to force significant delays in communication between different elements of the German and Austro-Hungarian armies over the opening months of the war. More importantly, Britain's broad-scoped control of global telegraph infrastructure – or, at least, access to British-operated infrastructure – would allow for the development of an unprecedented communications interception regime wherein agents of the Empire (called "censors") would sift global information transmissions at critical junctures to obtain valuable intelligence and to engage in information warfare.[13]

The British Empire's control of global communications infrastructure in 1914 was impressive and awarded major advantages in the conduct of the conflict (though, admittedly felt more beyond the European landmass than on the continent itself). More than simply allowing for British censors to intercept information about the Central Powers' military maneuvering, the setup allowed for broad-scoped analysis of data about global commerce during the war years. The result was the interdiction of companies from Germany, Austro-Hungary, the Ottoman Empire and even neutral countries acting to provide the Central Powers with the commodities, natural resources and capital needed to wage total warfare in Europe. In shutting down many such operations, the British Empire was able to gain strategic advantages that may have prevented an early end to war on the continent prior to U.S. intervention in 1917.

Box 10.2: Semi-state actors, the Internet and contested cyber sovereignty

The example of Britain's cutting of German telegraph cables here illustrates a significant point, which we previously brought up in another form in Chapter 4. Private corporations play a critical role in the function of infrastructure that underwrites the function not only of national societies but also of militaries. In Chapter 4, we grappled with the notion that governments need to protect infrastructure and sophisticated information technologies from export in order to secure certain national security imperatives. Here, though, it is worth considering another element of the issue, namely that private control of information infrastructure presents as a limitation on state sovereignty.

Sovereignty is often defined, in reference to Max Weber's discussions on the nature of the thing, as something that exists where there is a legitimate monopoly on the use of violence within a given territory and over a given population. Simply put, this means that an authority able to effectively police its borders and to ensure that no other domestic force is able to raise an armed force with impunity is the sovereign authority.

With cyber conflict, it is simply not true that states have a legitimate monopoly on the use of force online. There are two elements of this that we should break apart. The first is about the "monopoly" element of sovereignty. More than 95% of military and intelligence actions online occur on civilian networks owned and operated by various backbone operators and information product vendors. Though they tend not to do so in the West, these private operators could quite easily blunt or block offensive or exploitative cyber operations undertaken by state actors. The corollary for a more primitive era would be a landowner or aristocrat that legitimately operates infrastructure and has, as a result, a de facto ability to block the military operations of the government by, say, blocking off a major road.

Naturally, the crux of the matter here is what private actors might be compelled to do in support of state cyber operations. If the state can compel these actors not to interfere, then there is no violation of the *legitimate* monopoly on force held by the state. The state, in other words, retains the full claim on sovereign control of its territory and population.

And yet, it is not clear that the state – at least, the United States and similar democratic Western countries – have this power. Indeed, as incidents like the well-known Apple versus the Federal Bureau of Investigation (FBI) case demonstrate, there is at least some clear tendency among private actors to interpret their actions as legitimate via direct reference to public interests and not to government procedures. The question becomes, quite simply, one of the origins of legitimacy. If a private set of actors wanted to blunt the ability of their government to engage in cyber warfighting by simply not allowing actions of that kind to take place on their private systems, would they be in the right by referencing public interest that might be harmed by government action? In other words, even if the government is legitimate because of its election by the people, could private stakeholders claim like legitimacy in defending public interests against government wishes? If yes, the implications for state power and cyber conflict are immense.

The parallel to modern global information infrastructure here is obvious. The function of global commerce, society and politics is enabled and underwritten by an immense apparatus of physical information technology infrastructure. But the vignette about British efforts at the outset of World War I is incomplete. The reason for this has to do with the nature of the most recent information revolution as not fundamentally about connecting people directly, but rather about connecting people *via* their use of computerized systems. As intelligence organizations across the West and the Soviet bloc realized in the 1970s and 1980s, espionage in the world of the Internet is not merely about the encryption of information during transmission; it is about direct acts of espionage that can occur in the system itself (i.e. in the remote usage of networked computer systems). Thus, whereas a degree of control over global societal functions in aid of the Allies' cause came in the form of British interception of communications and subsequent actions taken in the real world, today it also occurs via direct interaction with global information systems. Past communications technology allowed global powers better ways to coordinate industrial and commercial systems; today, the Internet and the computers that connect to it have essentially become these control systems. Given this, it's clear to see why cyber operations have so organically become part-and-parcel of security strategies designed to interfere with, manipulate and contest the status quo.

The Internet as a societal subconscious

As the last chapter described in some detail in its discussion of the sources and nature of power in the digital age, another reason that the digital domain presents as an attractive operational prospect for those interested in contesting the status quo without going to war has to do with the transforming effect of the Internet on the substance of human society. In many ways, it's best to think about this effect as a sort of transformation of societal subconscious. The human subconscious is the part of the mind that is active and impactful on human operation but is not really something that we're aware of. The subconscious regulates our reactions to the world around us, how we digest and process information and those parts of our bodies that operate as a matter of routine (from blinking and breathing through automatically looking left and right without thinking about it while standing at a road crossing).

In societal terms, much of what we talk about in describing the global information environment qualifies as a component part of humankind's subconscious. Even where sizable tracts of society, industry, government or politics are aware of different elements of the information environment's functions, the function of the whole is nevertheless dependent on background systems and processes that fall below the conscious gaze of the average citizen. Specifically, the unique way in which people access information, are conditioned to analyze incoming information, have information framed for them, and then themselves communicate with others stems from a complex range of externalities related to the setup of industry, culture and politics. Exposure to differing kinds of bias in news reporting, for instance, often emerges from the interacting interests of those organizations that fund media companies. And knowledge of politics and social issues, which affects how people respond in public opinion polls and in political participation, often differs depending on the platform via which one receives their news (e.g. social media users tend to create echo chambers by subscribing to sources that offer agreeable perspectives at the expense of those offering more objective or competing ones). The result of such a dynamic is that the manipulation of informational content and of the underlying infrastructure of the global information environment (by, for instance, incentivizing talking heads to shill one perspective, or bribing ISPs with favorable legal status in exchange for adherence to censorship guidelines) is not merely attractive, as it has always been; in the Internet age, subversion and information manipulation are often directly attainable from cyber actions.

Muddied waters: the Internet's special characteristics at work

Finally, though most simply, another reason that cyberspace and cyber conflict techniques appear as an attractive option for those interested in contesting the status quo without going to war, is that operation in the digital domain allows one to take advantage of various special characteristics of the thing. We discuss many such characteristics in past chapters, but, briefly, those operating online to undertake disruptive cyberattacks, engage in espionage or information manipulation activities, or otherwise attempt to engage in information warfare particularly benefit from the enduring problem in linking online action to kinetic inputs. The attribution problem for defenders and investigators incentivizes cyber aggression because the risks of exposure are mitigated by (1) the challenge of gathering evidence about such actions and (2) the enduring mismatch between the possible scope of such evidence gathering and evidentiary standards for invoking international law. The incentive to take advantage of cyber options for disruption is further enhanced by the fact that few constraining political frameworks exist that might punish such action anyway (which we discuss in later chapters). Moreover, the costs of entry to operation in the domain are low, and state actors can enhance their deniability in cheap, effective ways by sponsoring the operations of spy agencies, mercenaries and "patriotic hackers."

Making sense of cyber conflict in the "grey zone"

Asymmetric warfare, political warfare and hybrid warfare. These forms of political contention are the regular background noise of IR and have been so for untold centuries. The Roman Empire fought insurgent forces in areas now within Germany, Great Britain, Syria, Israel and Jordan frequently across decades-long periods of expansion. Both Louis XIV and Ivan the Terrible employed, in their different respective eras, innumerable spies, mercenaries and agitators to work subversively in central Europe to pave the way for later military conquest. And much of what constituted the European imperial race to conquer African territory in the 1800s, where bribery of local elites and early versions of gunship diplomacy allowed for predatory colonization arrangements, is essentially the story of hybrid warfare employed in line with grand strategic ambitions.

In the 21st century, however, it is worth differentiating these forms of conflict short of war from what many are now calling **grey zone conflict**. Conflict in the "grey zone" between war and peace is often described as contention that involves two (or more) state actors but primarily avoids the direct use of anything more than limited military force. Of course, using the term in this way suggests little different than those others (hybrid warfare, etc.). But while "grey zone conflict" is colloquially used to describe any form of "not war" contention, the reality is that it indicates something relatively unique about the dynamic of "not war" conflicts in the 21st century.

Specifically, grey zone conflict refers to conflict under the threshold of declared or officially observed warfare or intensive competition that happens between two states of similar power, influence and capabilities.[14] This kind of conflict between war and peace is much more unusual in the landscape of recent world history, at least insofar as we're talking about the immediate interests, territory and resources of a country and not its extended colonial holdings. Most often, incidence of asymmetric warfare occurs as a function of the strategic choice by weak actors to, for obvious reasons (i.e. they wouldn't be able to win otherwise), engage stronger actors unconventionally. Likewise, though the Cold War saw well-documented, extensive political warfare campaigns between countries of the Western and Soviet blocs, the scope of most remained limited, arguably so as to avoid escalation toward "hot" conflict. Where hybrid warfare did occur during the Cold War (i.e. where information warfare was married to limited military confrontation, the use of proxy forces to irritate an entrenched target, or diplomatic pressure), the dynamic was again most often one of strong states targeting weaker ones.

So far in the opening years of the 21st century, we've seen the prosecution of hybrid warfare between states that, while not necessarily equal in military terms, are fully capable of engaging one another in open hostilities on relatively even ground. Russia, in 2014, employed pro-rebel propaganda and unmarked military assets – a body of soldiers, dubbed "little green men," operating in Eastern Ukraine without uniforms or other markings to identify them as Russian – to support the outbreak of civil war and the secession of the Crimean Peninsula from Ukraine.[15] China has regularly turned to the use of unconventional techniques, including the employment of merchant shipping vessels to hassle foreign naval assets and the construction of sand islands in the South China Sea, to expand influence and provoke limited confrontation with the Philippines, Vietnam, Japan and South Korea.[16] And Russian security services have engaged in yet broader-scoped campaigns to interfere with the political process of peer competitors in Europe and North America via a combination of cyber operations and information warfare efforts.[17]

Box 10.3: Beyond cyber: automated, informational and industrial conflict in the grey zone

Though it is not the main point of this book, it is important to note that grey zone conflict manifests beyond just the use of cyber tools. Briefly, a series of new technological developments have enhanced the ability of state actors to engage in conflict short of war with their peer competitors. Advances in automation have enabled state usage of drones to engage in low-intensity kinetic operations against adversaries and against non-state actors. In the future, there is great concern about **swarm warfare**, wherein large numbers of small drones coordinated by sophisticated machine learning processes might make grey zone conflict even more attractive given the extreme efficiencies in its execution. Likewise, there are obvious information advances related to the rise of the Internet and to emergent information technologies that offer states advantages. We discuss these more in later chapters. And finally, new techniques for construction and extraction of resources have made grey zone operations an attractive option for states looking to antagonize without engaging in conventional warfighting. The People's Republic of China, for instance, has employed island-constructing platforms to the South China Sea to build sand-based territory for the purposes of enhancing their claim to parts of the ocean.

The incidence of so much grey zone conflict involving strong state actors in the 21st century is interesting, not least because it presents as something of a theoretical puzzle for international security scholars and practitioners. It's not immediately clear as to why states of relative parity in economic, military and political power terms would push one another consistently in a highly aggressive fashion without indicating an interest in broader conflict. Certainly, states spy on one another constantly and engage in limited military contestation without triggering the outbreak of broader warfighting. As bargaining theorists and others in the IR field might explain, such actions make a great deal of sense as a means of trying to better secure state interests by testing a competitor's resolve. But it's not clear that this is what is happening with grey zone conflict. Aside from the fact that such contention often targets specific political or security processes while the attacker is openly involved in conciliatory efforts on other fronts, it's not clear why states would choose to leave many of their most potent assets at home. If the purpose of "not war" conflict is to bargain by demonstrating resolve and capability, wouldn't states be incentivized to indicate their willingness to employ their full strength?

The remainder of this section is dedicated to describing why grey zone conflict might be increasingly attractive to states interested in contesting the power and influence of their peer competitors in international affairs. We offer five distinct arguments, though it is worth noting that many are compatible with one another. In most cases, to the benefit of our interest here in cyber conflict, digital dynamics play a central or otherwise significant role in supporting the logic of states' emphasizing grey zone strategies.

Expanding space for limited contention in the digital age

One reason that grey zone conflict may be more prevalent in the 21st century than it has been in decades and centuries past may have to do directly with the most recent information revolution. In an age where international affairs is defined by industries, social trends, political institutions and more than have grown up around the Internet, there exists an expanded space for contention in the possibilities of network-supported communications and control infrastructure. As has been the theme of sections and chapters before this, the Internet and related technologies provide an immense set of new coordinative, disruptive and persuasive options for those interested in participation and contestation of all kinds. The next chapter goes into this in yet further detail in describing the manner in which

Internet technologies have revolutionized the activist enterprise. Simply put, however, cyber actions – from disruptive attack to manipulation of data and the use of broad-scoped, diverse types of information-sharing resources – constitute a toolkit of contention that is expansive and fits between the thresholds of legitimate political participation and violent protest. In terms of state efforts to interfere with the interests of their competitors, this translates to an expansion of the toolkit of hybrid warfare, broadly writ.

The problem with this explanation of why grey zone conflict is more readily apparent in the 21st century than it has been in the past is that it does little to explain the rising incidence of such conflict where it is not primarily or even at all characterized by digital actions. This, of course, makes some substantial sense as digital actions are limited in their ability to affect kinetic dynamics (i.e. to wage physical warfare or achieve a kinetic disruption). As such, it may be fairer to simply say that ICT and the special characteristics of the cyber domain have multiplied the incentives for states to use cyber techniques insofar as new information technologies pertain to the informational aspects of conflict.

Information affordances in the age of the Internet

Though it carries with it the same problem of not fundamentally being able to explain the broader trend toward engagement in grey zone conflict seen so far in the 21st century, an interesting corollary rationale for the thing lies in the argument that the Internet era has brought with it unique and uniquely manipulable affordances. An affordance is a dynamic that exists wherein particular environmental characteristics – objects, people, technologies, etc. – permit specific kinds of social actions at the expense (or at least at the non-occurrence) of others. The term was first used by social psychologists seeking to explain how human (and animal) action and thought is defined by environmental conditions.[18]

We discuss the idea of affordances with regard to the Internet in part in the last chapter, specifically describing the manner in which changes in how humans access, frame and consume information translate to behavioral – and subsequent institutional – changes in how people engage in society and politics. Here, the idea is no different, but is perhaps worth revisiting the case of the Gutenberg printing press – a seminal example in scholarship on comparative politics and sociology – as an example of how information revolutions have regularly resulted in micro-motivated changes to the global social and political environment.

Johannes Gutenberg's movable type printing press was developed in the mid-15th century (ca. 1439).[19] The original machine was a remarkable innovation that incorporated new technologies and materials into a novel design that allowed for the mass production of books (and records, proclamations, etc.). Prior to the invention of Gutenberg's printing press, the ability of governments and companies to mass produce text was limited. Existing methods for duplicating text were poor, and so the task of recreating original work fell to dedicated armies of bureaucrats and, in the case of books, clergymen.

Gutenberg's invention and its subsequent spread across Europe had a profound impact on the political and social fabric of the continent. Quite arguably, the printing press is one of the developments most responsible for the turn in Europe toward national polities in the 16th and 17th centuries. In large part, the reason for this has to do with the fact that the limited ability to systematically duplicate text supported the perpetuation of systems that separated the aristocracy from lower segments of society and failed to incentivize the emergence of non-local communities (i.e. people rarely thought about life beyond their town or local region). Benedict Anderson, in his now-famous work on "imagined communities," describes three broad-scoped changes in particular that came about in Europe as a result of the printing press.[20]

The first of these was that the ability to mass produce text prompted an unprecedented shift away from the primary usage of Latin as the written language of choice in Europe. Instead, the vernacular tongue – Italian, German, French, etc. – was embraced as new forms of publication were created and multiplied.[21] This, in turn, led to greater interest in literacy for ordinary folk, as much social and economic activity turned to revolve around new kinds of entertainment (i.e. newspapers, pamphlets) and reporting.[22] The result of all of this was the development of affordances coming from a sort of nationalization of the worldview of even the lowest of Europe's general population. The development of newspapers allowed for regular news from distant parts of the continent and encouraged a centralization of perspective on societal and political issues as being about national – rather than local or urban – identity. Much as the development of papyrus scrolls for record-keeping did in ancient Egypt, texts mass produced in the age of the printing press also encouraged the further standardization of concepts and measures of the world around Europe's citizenry, including of time and value.[23] And, as all of this encouraged the rise of commerce and a workforce more comfortable with the idea of prosperity emerging from non-local opportunities, governments were incentivized to organize and enable their populations around the goal of greater economic power. Thus, in the 16th and 17th centuries, Europe experienced the proliferation of national education programs, centralized banking and conscription-based militaries.[24]

Box 10.4: Is the Internet an imperial force?

Of the various affordances the Internet brings to activist and subversive efforts, none is perhaps quite as meaningful as the opportunity for actors to inject some means of ideational influence over far flung populations. However, in thinking about this, it is worth considering implications beyond the operational. That is to say that it is worth thinking about more than just the implications of the Internet for specific social movements or for information warfare. Clearly, information that is distributed and then proliferates online leads to broad-scoped societal effects over and above what is intended by those who attempt to influence. It has, in many ways, a mind of its own. This is not new in the information age, but it is certainly more noticeable.

The question we should then consider is whether or not the Internet is an "imperial" force. By "imperial," we mean to invoke the image of colonial European powers spreading their empires around the world and, as a result, influencing the development of non-European civilizations in line with European economic, political and social systems. We must ask if something similar is happening in the age of the Internet given the way in which the thing is structured. Certainly, usage of the web has become more evenly distributed across countries in the past 20 years as the developing world has come online. Whereas in the 1980s and even 1990s it might have been fair to say that the Internet was largely U.S. or Western, that is simply not informationally true anymore, at least in a basic sense. The bulk of content stored online is increasingly less and less in English; it is in Mandarin, Urdu, Hindi, Russian, Spanish and more.

However, it is certainly first movers in the Western world who dominate content production and distribution in meaningful ways. And more than just the fact that Disney, Netflix and Amazon are some of the most prolific content platforms across even the non-English speaking worlds, search engine giants like Google and Microsoft are behind the design of search and distribution algorithms that determine what information individuals find when they go online. Given this, is it possible that the Internet represents a new form of imperialism?

The information environment defines what affordances – i.e. what behaviors and resulting institutional developments – exist. The Internet age undoubtedly includes a series of affordances that are having a profound impact on the shape of global commerce, politics and society. A broad range of scholarly projects in recent years have identified many such affordances, including that individuals that use social media platforms extensively become inclined toward network-oriented problem solving techniques and that shared stylistic preferences across distinct cultural sub-groups vary directly in line with the prevalence of prominent examples with viral exposure.

The implication of all of this for incidence of grey zone conflict is, again, quite simple. The unprecedented, systematic emergence of affordances linked with the function of Internet infrastructure implies that there are immense opportunities to be had from interfering with and manipulating those underlying systems. We expand this in discussing constitutive power strategies, but it is worth considering more simply this argument; that the development of new affordances linked with information infrastructure in the digital age itself incentivizes interference – whether via digital or non-digital techniques – of those societal functions that now depend on the Internet.

The stability-instability paradox, deterrence and grey zone conflicts

A main element of the puzzle about grey zone conflict upheld in debate by security practitioners is that such conflict implies a failure of deterrence.[25] In other words, if one country is being successfully deterred from attacking another, then any new form of limited conflict beyond "normal" espionage and non-confrontational military developments would be considered a failure of the dynamic. After all, the purpose of deterrence is essentially a sort of temporary holding action (even when deterrence is affected over long periods of time) wherein an adversary is incentivized to sideline any plans (that wouldn't be considered "normal" diplomatic or economic behavior, that is) for bettering the strategic situation. Thus, grey zone conflict implies that deterrence has largely failed.

The issue is that prominent instances of grey zone conflict don't seem to exhibit such a loss of deterrent relationship characteristics. Where Ukraine was verbally offered some measure of commitment of protection from other European countries in 2014 and 2015, the Russian Federation appeared to take active steps to limit its military position in the eastern parts of the country while still continuing clandestine and asymmetric support for rebel forces. However, this was not the case in the 2022 Russian invasion of Ukraine where little restraint was shown in its conventional military operations. In this instance, the example is not apt as it was generally more of a conventional conflict than grey zone competition. Similar to the 2014 and 2015 Ukrainian example, China made various diplomatic concessions in the interest of better partnerships with regional stakeholders while supporting low-level efforts to advance territorial claims in the South and East China Seas. So what's happening?

Scholars have increasingly turned to the notion of the **stability-instability paradox** to explain grey zone conflict in the context of deterrent dynamics of interstate relationships.[26] The paradox is most often used in IR scholarship to explain the propensity that nuclear weapons' states have to engage in low-intensity conflict with one another, most often through the use of proxy non-state actors and allied governments. The logic here is fairly simple. States with nuclear weapons – and specifically with the ability to launch a retaliatory **second strike** in the event that they are nuked themselves – tend toward reasonably peaceful relationships, even where the countries involved are entrenched geopolitical or historical rivals. However, the strength of the deterrent force involved – i.e. the nuclear arsenal of the opposing side – actually offers a credible sense of security in that your adversary has limited incentive to use their capability short of the appearance of an existential threat. The result is a threshold past which military action would be unwise, but under which confrontation is excusable. So long as that threshold is not breached, major warfare will not materialize. Thus, successful nuclear deterrence leads to a paradox wherein high-level peace incentivizes low-intensity conflict as a relatively consequence-free way to incrementally improve one's strategic position.

With grey zone conflict, scholars have increasingly recognized (as the two sections above do) that an expansion of the space between war and peace in the digital age likely means that there are more opportunities for contention that don't violate the deterrent threshold. Particularly where nuclear weapons are involved, we might expect states to embrace the notion of broad-scoped grey zone conflict as a substitute for normal competition below the threshold of war. In this way, grey zone conflicts might be perceived as a net positive development in IR, as they emerge from strategic success in deterring a foreign adversary.[27] The challenge for states in coping with such conflict will be two-fold insofar as the aim is to avoid massive destabilization of a broadly peaceful status quo condition in the global system. First, states that are the targets of grey zone conflict must decide where and when to respond with major force or threat of force.[28] In essence, states must describe what kinds of hybrid warfare push too close to the threshold of what is permissible short of war and then act to raise the costs of adversaries emphasizing such approaches. If done effectively, this would again be a net positive for international affairs, as grey zone conflict could be normalized in relation to prevailing views on the balance of global power. Second, the same set of decisions must be made where grey zone conflict emerges between non-nuclear states (either without nuclear weapons or not under a nuclear umbrella). In those situations, the threat of nuclear retaliation does not exist and so the nature of such "not war" conflict plays more into the question of appropriate retaliation.[29]

A logic of circumstances

Apart from some prevailing condition of deterrence between states or the implications of the information revolution, a notable – if simple – reason for the rising incidence of grey zone conflict in the 21st century may have to do with circumstantial effectiveness.[30]

Scholars and security practitioners think of conflict between war and peace as sub-optimal, because states are essentially opting not to bring their best assets into the contest. The argument above posits that there is a strategic logic of certain situations where the established payoff structure of possible outcomes makes grey zone conflict a viable option, essentially by reducing the likely costs of retaliation. That said, it is worth noting that grey zone conflict may be an attractive option because of situational cost-benefit calculations. Though the argument might appear very similar, it differs from the idea that limited costs from potential retaliation systematically incentivizes grey zone conflict. Rather, states simply sometimes don't feel like risking the costs of lost military assets, personnel and resources; alternatively, some may be comfortable with risking and perhaps forfeiting military equipment because its value has circumstantially diminished.[31]

These types of cost-benefit analyses are most often a function of institutional or strategic variables that directly impact upon the ability of a military to fight. If an army has just been withdrawn from a decade-long conflict in another country – as occurred with the Soviet Union after Afghanistan, the United States following Korea, Vietnam, Afghanistan, Iraq, etc. – then it may be unwilling to launch into a conflict because personnel and equipment attrition has reduced the effectiveness of the fighting force in the near term. Here, grey zone conflict simply may present as the most efficient option for confrontation and contestation, just as it might if a political administration funded an unprecedented expansion of a state's intelligence or special forces apparatus. Conversely, a state five years removed from such a decade-long conflict may be more incentivized to engage in a conventional war, as the replenishment of personnel and equipment has left its military with a surplus of expendable, battle-worn hardware. In short, grey zone conflict – and particularly cyber and cyber-enabled conflict – may occur simply where a state or an element of a state's security services finds itself well suited to the task.

Grey zone strategies as constitutive power challenges

For most security practitioners, explaining grey zone conflict involves understanding why and when it might materialize as a form of contention in some interstate relationships. Most explanations of "not war" strategies, however, do a poor job speaking to the conduct of grey zone conflict beyond timing. In other words, most explanations do little to explain the construction of a given campaign and the targeting choices made by "not war" belligerents.

One set of explanations for the rising incidence of grey zone conflict addresses *both* the significance of the information revolution and the specific shape of how contention takes place between war and peace. In Chapter 9, we engaged with the question of what it means for a state to be cyber powerful and what power itself looks like in the information age. In doing so, we invoke Barnett and Duvall's famous taxonomy of the sources and facets of power (compulsory, structural, institutional and productive).[32] In this breakdown of what kinds of resources, processes, and strategies should matter to state actors, we locate yet another argument about the nature of grey zone conflict between peer competitors.

In discussing the competing incentives and capabilities of states vying for security and influence in world affairs, most IR scholars locate their understanding of what matters to states in the first three of Barnett and Duvall's taxonomy of power. Power is understood to emerge from the material military capabilities of states, the latent economic and societal resources that produce those capabilities, the institutional influence over rules of the road that governs much international behavior and the norms of behavior that exist in normal interstate relationships. These facets of power are addressed in most major IR theories in some sense and are variably – depending on the degree to which scholars argue that one part of the thing has more explanatory power than others – the prominent features of the field's main macro-theoretical schools of thought.

Often left out of direct consideration is the fourth and final facet of power, what Barnett and Duvall call the productive facet. To quote the last chapter, "productive power describes indirect constitutive control over others in the context and construction of those broader socio-technical relationships within which all states are embedded." In short, productive power involves the ability to shape and direct the constitutive processes that lead to manifestations of power in all the other categories.[33] In reference to the arguments addressed earlier, constitutive power might thus be said to include the ability to influence how social affordances are created and felt across broad populations. Constitutive power strategies manifest in some elements of information warfare. Whereas information operations designed to mislead people and institutions take aim at the integrity and viability of mechanisms of compulsory, structural and institutional power, those intended to affect how people interact with their environment – by, say, reducing the credibility of particular sources of information or forcing changes in how people think about the information they're receiving – organically address the constitutive power of an adversary. In the next chapter's description of non-state actors interested in affecting broad-scoped normative changes in prevailing societal sentiment, we describe such actions as being about subversion.

Subversion in the form of constitutive power strategies falls generally outside the scope of this discussion about deterrence and grey zone conflict between states operating on roughly equal footing. And yet, it provides a compelling reason as to why states might act to interfere with the machinations of foreign competitors' societies without directly addressing those competitors via directly contentious actions. In particular, taken in tandem with the argument that the information revolution has opened up new space for non-violent contestation and been the source of new global affordances that are shaping societal trends, it's easy to see why grey zone conflict has become a favored strategy for states on certain fronts. Even where the approach does not involve direct manipulation of information technologies for disruptive and manipulative purposes, the notion that states should be incentivized to engage others in limited fashion in order to affect constitutive processes, wherein the approach taken does not interfere with prevailing deterrent dynamics, explains why occasional "not war" conflicts between peers should be expected. The bulk of other incidents that constitute the 21st century's bulging number of grey zone conflicts thus might be explained by the advent of Internet-aided opportunities for manipulation of underlying constitutive conditions. Moreover, the idea that constitutive power strategies are not uncommon beyond the scope of deterrent relationships and are increasingly common in the form of grey zone campaigns, makes substantial sense when one considers the range of targets and methods employed in paradigmatic cases since 2010, including information operations conducted against Ukraine, Germany, France, the United States, the United Kingdom and more.

Future trends

The next chapter, in dealing with non-state actors and cyber conflict, extends much of what we've talked about here. This is particularly the case when it comes to our discussion of "patriotic hackers," non-state proxies and cyber mercenaries, and subversive non-state actors. The connection between this chapter and the next is significant, however, in that cyber conflict organically implies a blurring of the lines between state agency and the actions of non-state actors. The information revolution has affected both state and non-state actors, not only separately but also in terms of the mutual interactions of relevant security stakeholders – state militaries, intelligence organizations, organized criminal syndicates and self-motivated proxies of national interest – in international affairs. Thus, as we move to discussing non-state actors it is important to keep in mind the varying security calculations made by states. In doing so, the objectives and payoff structures of non-state actors, insofar as they matter to state security interests, might be better understood.

Notes

1 Ryan, Jason. "CIA Director Leon Panetta Warns of Possible Cyber-Pearl Harbor." *ABC News* 11 (2011).

2 Gartzke, Erik. "The Myth of Cyberwar: Bringing War in Cyberspace Back Down to Earth." *International Security*, Vol. 38, no. 2 (2013), 41–73. Also see Jensen, Benjamin, Ryan C. Maness, and Brandon Valeriano. "Cyber Victory: The Efficacy of Cyber Coercion." In *Annual Meeting of the International Studies Association*, 2016.

3 Such conceptualizations of conflict have been commonplace since at least the 1980s. See *inter alia* Schultz Jr, Richard H. "Low-Intensity Conflict: Future Challenges and Lessons from the Reagan Years." *Survival*, Vol. 31 (1989), 359; Kornbluh, Peter, and Joy Hackel. "Low-Intensity Conflict: Is It Live or Is It Memorex?." *NACLA Report on the Americas*, Vol. 20, No. 3 (1986), 8–11; Kober, Avi. "Low-Intensity Conflicts: Why the Gap Between Theory and Practise?." *Defense and Security Analysis*, Vol. 18, No. 1 (2002), 15–38; Hammond, Grant T. "Low-Intensity Conflict: War by Another Name." *Small Wars & Insurgencies*, Vol. 1, No. 3 (1990), 226–238; and Kinross, Stuart. "Clausewitz and Low-Intensity Conflict." *Journal of Strategic Studies*, Vol. 27, No. 1 (2004), 35–58.

4 Votel, Joseph, Charles Cleveland, Charles Connett, and Will Irwin. *Unconventional Warfare in the Gray Zone*. Technical Report. Washington, DC: National Defense University Press, 2016. See also Wright, Nicholas D. "From Control to Influence: Cognition in the Grey Zone." (2017). http://nsiteam.com/social/wp-content/uploads/2017/07/Wright2017_CognitionGreyZone_v2.pdf.

5 Bragg, Belinda. *Integration Report: Gray Zone Conflicts, Challenges, and Opportunities*. Technical Report. Arlington, VA, 2017.

6 Bachmann, Sascha-Dominik, Hakan Gunneriusson, K. Hickman, M. Weissman, and N. Nilsson. "Hybrid Threats and Asymmetric Warfare: What to Do?" *Conference Proceedings* February 2018, 1–41.

7 See *inter alia* Laqueur, Walter. *Guerrilla Warfare: A Historical and Critical Study*. Oxford, UK and New York: Routledge, 2017; Jensen, Benjamin. "The Cyber Character of Political Warfare." *Brown Journal World Affairs*, Vol. 24 (2017), 159; and Robinson, Linda, Todd C. Helmus, Raphael S. Cohen, Alireza Nader, Andrew Radin, Madeline Magnuson, and Katya Migacheva. *Modern Political Warfare*. Santa Monica, CA: RAND Corporation, 2018.

8 See *inter alia* Chivvis, Christopher S. *Understanding Russian "Hybrid Warfare."* Santa Monica, CA: The RAND Corporation, 2017, 2–4; Chivvis, Christopher S. "Hybrid War: Russian Contemporary Political Warfare." *Bulletin of the Atomic Scientists*, Vol. 73, No. 5 (2017), 316–321; and Fox, Amos C., and Andrew J. Rossow. *Making Sense of Russian Hybrid Warfare: A Brief Assessment of the Russo-Ukrainian War*. Arlington, VA: Institute of Land Warfare, Association of the United States Army, 2017.

9 For cyber-relevant discussions of information warfare, see *inter alia* Libicki, Martin C. "The Convergence of Information Warfare." *Strategic Studies Quarterly*, Vol. 11, No. 1 (2017); Nathan, Andrew J. "Cyber Dragon: Inside China's Information Warfare and Cyber Operations." *Foreign Affairs*, Vol. 96, No. 5 (2017), 193; Siegel, Alexandra A., and Joshua A. Tucker. "The Islamic State's Information Warfare." *Journal of Language and Politics*, Vol. 17, No. 2 (2017), 258–280; and U.S. Army. *Chinese People's Liberation Army and Information Warfare, Report on the PLA, Network-Centric Warfare, Electronic and Cyber Warfare, China Espionage, Implications for US, and Psychological Warfare*. Carlisle, PA: U.S. Army War College, 2017.

10 See *inter alia* Parker, Emily. "Hack Job: How America Invented Cyberwar." *Foreign Affairs*, Vol. 96 (2017), 133; Kaplan, Fred. *Dark Territory: The Secret History of Cyber War*. New York: Simon & Schuster, 2016; and Segal, Adam. *The Hacked World Order: How Nations Fight, Trade, Maneuver, and Manipulate in the Digital Age*. London: Hachette, 2016.
11 For instance, Libicki (2017).
12 For perhaps the best recounting of this story, see Corera, Gordon. *Cyberspies: The Secret History of Surveillance, Hacking, and Digital Espionage*. Oakland, CA: Pegasus Books, 2016.
13 Messinger, Gary S. *British Propaganda and the State in the First World War*. Manchester, UK: Manchester University Press, 1992.
14 See Gannon, J. Andres, Erik Gartzke, and Jon Lindsay. "After Deterrence: Explaining Conflict Short of War." *Working Paper* (2018).
15 Ibid. Also see Altman, Dan. "Advancing Without Attacking: The Strategic Game around the Use of Force." *Security Studies*, Vol. 27, No. 1 (2017), 58–88.
16 See Carter, Ash. 2016. "Remarks on 'The Future of the Rebalance: Enabling Security in the Vital & Dynamic Asia-Pacific.'" On board USS Carl Vinson, San Diego, CA; and Gady, Franz-Stefan. "'Little Blue Men:' Doing China's Dirty Work in the South China Sea." (2015). http://thediplomat.com/2015/11/little-blue-men-doing-chinas-dirty-work-inthe-south-china-sea/.
17 See *inter alia* Franzén, Simon. "Russian Information Warfare." (2017) http://fhs.diva-portal.org/smash/record.jsf?pid=diva2%3A1127531&dswid=-6992; Darczewska, Jolanta. "The Anatomy of Russian Information Warfare. The Crimean Operation, a Case Study." (2014). www.osw.waw.pl/en/publikacje/point-view/2014-05-22/anatomy-russian-information-warfare-crimean-operation-a-case-study; Thornton, Rod. "The Changing Nature of Modern Warfare: Responding to Russian Information Warfare." *The RUSI Journal*, Vol. 160, No. 4 (2015), 40–48; and Giles, Keir. *The Next Phase of Russian Information Warfare*. Vol. 20. NATO Strategic Communications Centre of Excellence, 2016.
18 With regards to digital age operations, this topic is more fully fleshed out in Earl, Jennifer, and Katrina Kimport. *Digitally Enabled Social Change: Activism in the Internet Age*. Cambridge, MA: MIT Press, 2011.
19 As noted again in later chapters, this example is the centerpiece of significant social science studies on the nature of emergent community identities that shaped sociopolitical context across European civilization following the Middle Ages. See Anderson, Benedict. *Imagined Communities: Reflections on the Origin and Spread of Nationalism*. London: Verso Books, 2006.
20 Ibid.
21 Ibid.
22 Ibid.
23 Ibid.
24 Ibid.
25 This is the main argument made by Gannon, Gartzke and Lindsay. "After Deterrence: Explaining Conflict Short of War." (2018).
26 For notable descriptions of the paradox, see Krepon, Michael. "The Stability-Instability Paradox, Misperception, and Escalation Control in South Asia." *Prospects for Peace in South Asia* (2003), 261–279; Cohen, Michael D. "How Nuclear South Asia Is Like Cold War Europe: The Stability-Instability Paradox Revisited." *The Nonproliferation Review*, Vol. 20, No. 3 (2013), 433–451; and Ganguly, Sumit. "Indo-Pakistani Nuclear Issues and the Stability/Instability Paradox." *Studies in Conflict & Terrorism*, Vol. 18, No. 4 (1995), 325–334.
27 Gannon, Gartzke and Lindsay. "After Deterrence: Explaining Conflict Short of War." (2018).
28 Ibid.
29 Ibid.
30 Ibid.
31 Ibid.
32 Barnett, Michael, and Raymond Duvall, eds. *Power in Global Governance*. Vol. 98. Cambridge, UK: Cambridge University Press, 2004.
33 Stevens applies this categorical breakdown in attempting to problematize cyber power. See Stevens, Tim. "Cyberweapons: Power and the Governance of the Invisible." *International Politics* (2017), 1–21.

Further readings

Lindsay, Jon R., and Erik Gartzke. "Coercion Through Cyberspace: The Stability-Instability Paradox Revisited." *The Power to Hurt: Coercion in Theory and in Practice*.http://deterrence.ucsd.edu/_files/LindsayGartzke_CoercionThroughCyberspace_DraftPublic1.pdf. Google Scholar, 2016.

Lindsay, Jon R., and Erik Gartzke. "Cross-Domain Deterrence as a Practical Problem and a Theoretical Concept." In *Cross-Domain Deterrence: Strategy in an Era of Complexity*. Gartzke, E. and Lindsay, J.R. (eds.). La Jolla, CA: Manuscript, 2016.

Lord, Kristin M., and Travis Sharp, eds. *America's Cyber Future: Security and Prosperity in the Information Age*. Vol. 1. Washington, DC: Center for a New American Security, 2011.

Thornton, Rod. "The Changing Nature of Modern Warfare: Responding to Russian Information Warfare." *The RUSI Journal*. Vol. 160, No. 4 (2015), 40–48.

U.S. Government Accountability Office. "Information Environment: DOD Operations Need Enhanced Leadership and Integration of Capabilities." GAO-21-525T. April 30, 2021.

U.S. Department of Defense. "Strategy for Operations in the Information Environment." Kune 2016.

11 Non-state actors

Terrorism, subversion and activism online

Though states have been behind many of the most visible and disruptive incidents involving the use of ICT, either directly or via proxy, comprehending the global cyber conflict ecosystem inherently means assessing and contextualizing the actions of non-state actors. From well-meaning developers to petty criminals and terrorist organizations, non-state threats to national security in the digital age are far more numerous and diverse in their potential for disruption than are those from foreign governments. Moreover, no understanding of interstate cyber conflict is complete without some greater understanding of the non-state corollaries of states' digital arsenals mentioned in past chapters – "patriot" hackers, criminal organizations and more. In many ways, it would be fair to say that non-state actors thematically dominate the global cyber threat ecosystem, as non-governmental entities – from hacker collectives to Internet Service Providers (ISPs) – all inevitably play some role in even interstate conflict episodes.

This chapter describes the various ways in which non-state actors use information technologies in world affairs. Naturally, as with most primers on non-state threats to national security in the digital age, this means engagement with well-known episodes in which the actions of individuals compromised the security of military or government systems. The Morris Worm incident in 1988, for instance, involved the release of a computer worm by a programmer that ended with disruption of network access affecting a significant percentage of all computers connected to the nascent Internet.[1] A decade later, the ILOVEYOU worm – investigation of which led to the arrest of two Filipino amateur programmers – attacked millions of Microsoft Windows computers across industry and governments around the world with similar effect.[2] And with Solar Sunrise, broad-scoped hacking of unsecured U.S. government systems ended up being the work of teenage hackers and an Israeli accomplice. However, understanding security threats and dynamics as they emerge from non-state actors' use of the web demands knowledge of more than just the adjunct role of individuals in state security crises. Likewise, no survey of digital age security topics would be complete if the focus were only on *malicious* non-state actors. As any researcher of democratization trends or counterculture will attest to, ICTs have also fundamentally shifted the balance of capabilities to disrupt in favor of non-state actors interested in social or political change. As such, this chapter describes a range of areas where ICT has altered the nature of operations for non-state actors, including social activism, terrorism and criminal enterprise.

Non-state actors and cybersecurity: some terminology

Non-state actors are more common in the digital domain than they are in any other. In any of the other traditional domains of conflict – land, air, sea or space – there are

DOI: 10.4324/9781003246398-11

significant barriers to entry for even the most highly organized and well-funded non-state groups (such as private companies or terrorist organizations). On land, non-state threats to national security take well-understood forms, from terrorist organizations to insurgent groups and organized criminal syndicates. At sea, non-state actors threaten national security in highly specific ways, such as through piracy and smuggling. In the air and in space, non-state actions that affect national security calculations are few and far between, largely because the costs of operating effectively in those domains is staggeringly high. With cyberspace, few restrictions on entry exist because of the common availability of pre-packaged tools, talent-for-hire and educational resources that impart the basic knowledge needed for misbehaving online.

As importantly, the architecture of the digital domain and the numerous strata of governance dynamics that relate to different parts of the online world is so complex as to *magnify* the incentives non-state actors have to be disobedient on the web. Whereas a naval pirate might be deterred by the high combined costs of ship upkeep and active counter-piracy measures in place in a given region, barriers to operation are less clear for hackers. The lack of physicality associated with entering the domain (i.e. acting online is not particularly shaped in any way by where a hacker is located) is particularly important in that, alongside low technical costs to operations, hackers have a broad range of options for avoiding the attention of authorities with viable criminal jurisdiction. The disconnect between physical location and cyber operation is also the source of magnified incentives to antagonistic digital actions because the international community has mostly yet to effectively apply international law to criminal acts involving persons and information interacting across international boundaries. In short, the fact that barriers to prevent or prosecute malicious action online is both technically and legally complicated makes it so that hacking is a pretty viable option for all manner of criminally motivated non-state actors.

But who are these non-state actors? The sections that follow describe different kinds of non-state actors that operate online in traditional terms – i.e. in terms of their political or economic distinguishing features. But more generally, what kinds of actors hack beyond the purview of state-directed cybersecurity efforts? And what roles do non-state actors play in the global cyber conflict ecosystem?

Generically, we might split malicious (or, at least, antagonistic) non-state cyber actors into two categories. First among these are (1) **hackers**. Though the term "hackers" is often used generically to describe individuals that act maliciously online, it actually describes a person with specialized knowledge of computer systems. In this way, hackers are distinct from those without technical knowledge of how their tools work, but rather engage in cyber conflict using pre-designed instruments and procedures. Hackers have programming skills and tend to fall into one of three sub-categories. **Black hat hackers** are persons that employ their hacking talents for personal or political gain. When a practitioner or journalist outlines steps taken by "hackers," they most often mean black hat hackers who are interested in causing disruption or stealing information that brings some economic or political benefit. By contrast, **white hat hackers** employ their skills for legitimate purposes, often the identification of security loopholes in existing systems. Companies around the world like Google, Facebook, Baidu and Weibo regularly hold competitions wherein participating hackers are encouraged to find unique vulnerabilities in existing IT infrastructure. In doing so (and in retaining white hat hackers to aid in design and development of products), governments and the private sector are able to shore up faults that might otherwise allow black hat hackers opportunities for gain. Finally, **grey hat hackers** are white hat hackers that sometimes undertake actions beyond the legal scope of pre-planned disruption. The term most often refers to digital vigilantes, such as individuals or groups that have been linked with the Anonymous hacking collective. Such hackers tend to be well intentioned but are not against resorting to technically illegal methods, such as the disruption of offensive websites or the theft of information from extremist organizations, to achieve their goals. Examples of grey hat hacker actions might include Anonymous-sponsored vandalism of Israeli government websites to protest against settlement construction in the West Bank, and left-wing vigilantes in 2005 in Germany that hacked the mailing list of a neo-Nazi organization in an attempt to expose the group's extremist supporter base.

By contrast with hackers, (2) **script kiddies** are people who use scripts and programs *developed* by hackers to prosecute cyberattacks. Script kiddies – a term emerging from the often-inaccurate assumption that teenagers are the primary demographic interested in hacking by lacking true technical skills – are generally uneducated in rudimentary programming skills. However, the proliferation of pre-packaged tools for malicious behavior since the late 1990s has meant that script kiddies are certainly capable of inflicting large-scale damage against government, industry and civil society. In particular, hacker development of toolkits that streamline the process of developing unique malicious code means that script kiddies are able to be specific in who they target and how they approach their objective. Thus, while it is true that the use of pre-packaged design tools means that detection and interdiction by authorities is often easier than it might be with true hackers, script kiddies represent an enormous potential threat to national security and prosperity.

Moving beyond this distinction, it is common to think of non-state actors in cyberspace in terms of traditional social, political or economic classifications. "Hackers" and "script kiddies" might obviously be grouped given any number of classifying factors, but the most common categories we might consider include (3) **cyberterrorists**, (4) state spies (or **cyberspies**) and **proxies**, (5) **hacktivists**, (6) **subversives**, (7) criminals and (8) **vicious employees**. As will become clear in the remaining sections of this chapter, the lines between these categories can be extremely fuzzy. State proxies and spies, which are discussed in past chapters, are worthy of mention as non-state actors largely because of the attribution problem. States often hire mercenary hackers or incorporate technical practices designed to distance official functions from malicious actions taken online, thus ensuring a degree of plausible deniability in the event of an outcry. Hacktivists are social activists that employ cyberattacks and digital circumvention to further political objectives. And yet, social activists (often activist "script kiddies") that don't particularly meet a threshold standard for hacking capabilities are not uncommon elements of localized cyber conflict episodes, particularly in authoritarian states. Likewise, subversive organizations might look remarkably like traditional interest groups or political parties but tend toward online actions that are antagonistic *explicitly because* the group or cause faces widespread opposition. And the difference between criminals and vicious employees is only meaningful insofar as the latter term describes a specific instance where crime is enacted by individuals with pre-existing access to systems. This often changes the nature of a given cyberattack episode, but the outcome is the same – crime involving either "property" damage or personal gain from information theft.

More specifically, these categories are useful for thinking about the substance of cyber conflict involving non-state actors and the preventative efforts of governments, but they link actors to digital actions remarkably poorly. Some hacktivism, for instance, is malicious. However, given that cyber actions tend not to be inherently violent even when they are aggressive, it is not necessarily fair to describe all forms of digital protest as such. From LGBT rights groups in Africa that employ off-the-shelf encryption to mask group communications to far left groups in Germany that vandalize the websites of their social adversaries, many activists might better be thought of as disobedient than criminal. Some literature on cyberspace and international security differentiates between such actors by pointing out that some undertake attacks focused on achieving technical disruption (i.e. **syntactic objectives**) and others take actions focused on the informational content of a given system (i.e. **semantic objectives**). However, even that characterization leads to imprecision when trying to understand different kinds of non-state actors. For instance, the lack of potential for violence *also* means that "cyberterrorism" only partially describes the traditional toolkit and set of objectives held by terrorists. In reality, cyberterrorism largely describes the use of the Internet to recruit, communicate and organize in preparation for traditional kinetic efforts. And though the end goals are different, the same might be said of subversives and non-governmental organizations (NGOs) that have an interest in avoiding government gaze.

In the remaining sections of this chapter, we discuss non-state uses of ICT in terms of the substantive categories outlined earlier. Though there will always be problems with thinking about non-state actors in cyberspace in terms of traditional definitions, the fact of the matter is that any attempt to categorize in extremely general terms can only take us so far. Describing non-state actors as responders or hosts or targets is valuable insofar as such designations let analysts quickly understand the contours of a given attack episode. But the truth is that there are no uniform sets of techniques or tactics unique to the different kinds of non-state actors outlined. Inevitably, it is necessary to think about different forms of non-state cyber conflict in the context of different political, economic and social actors, regardless of the analytic difficulties that might emerge from their many shared features.

Social activism in the digital age

For anyone thinking about non-state actors involved in conflict in the digital age, the first thought is likely to be of organized hacktivists like the Anonymous hacking collective or criminal hackers aimed at the theft of intellectual property. However, as previous chapters argued, it would be inappropriate to think of cyber conflict only in terms of aggression prosecuted online. Rather, we need to think about conflict and security *that has been impacted by the information revolution* – even if said conflict does not necessarily include malicious hacking for the most part – if we are to understand the degree to which ICTs have magnified the potential of non-state to affect political and economic systems.

Social activism is perhaps the most widespread and easy-to-understand category of activity in which non-state actors have felt augmented effects of the information revolution. Social contention and protest are also inherently conflictual but are rarely violent in an organized sense. Likewise, the shape of protest and activist efforts around the world has transformed in line with new abilities wrought from the rise of cyberspace, but they seldom involve sophisticated cyberattacks. Rather, ICT presents as a powerful toolkit for coordinating activities, for mobilizing resources and for reaching global audiences in unprecedented ways. This section describes the ways in which activists have benefited from the use of the web and additional problems that activists now face as a result of the information revolution.

Who are digital activists?

To be clear, "activism" refers to all efforts to advocate a position or a course of action. By talking about ICT and *social* activism here, we are making the argument that new advocacy abilities brought about by the information revolution are most meaningful for national and international security in the context of those seeking social (and requisite political) change. Though businesses have clearly enjoyed massive success directly from new abilities to advertise to audiences online and to use a remarkable number of diverse IT services to highlight products for consumption, it is political interest groups, NGOs, student movements and other similar social fronts that have, since the late 1990s, increasingly been able to affect real political change through the use of the Internet.

Social activism augmented by use of the web has many names – "**digital activism**," "cyber dissidence," "digital disobedience," etc. – and has been a relevant feature of world politics since the early 1990s.[3] Throughout the 1990s, a range of dissidents and political activists around the world set about constructing websites and sponsoring email newsletters to more efficiently reach their supporter base. Daniel Mengara, a Gabonese dissident living in exile in the United States, is well known for his early efforts to create a structured space on the web for advancing his perspectives on the national regime in Gabon. Mengara created a series of websites that were strongly inter-connected and could be used to coordinate supporters coming to the cause from a wide range of places and perspectives. In essence, Mengara created the first blogosphere and demonstrated that a distributed web presence could go a long way to cost-efficiently centralizing the efforts of diverse social movements.

Another secondary feature of digital activism is often the use of old and modified telecommunication technologies by dissidents – even some labeled insurgents by the governments of countries like Chad, Botswana, Mexico and Peru – to connect to the realm of websites, blogs and social media-supported networks often thought of as the domain of true cyber dissidents.[4] In China, for instance, Falun Gong practitioners have developed sophisticated programs to allow the banned community to directly utilize the uncensored Internet against the wishes of the Chinese government. Before such capabilities were developed, however, the group used cell phones, telegrams and even pagers to communicate ideas and content that could then be published by a foreign-based web of online repositories and mouthpieces. The Zapatista Army of National Liberation in Mexico (EZLN) and the Earth Liberation Front founded in the United Kingdom have had similar experiences in using basic telecommunications technology as a tool to support more sophisticated global web-based activist efforts.[5]

Of course, perhaps the best-known examples of digital activism enabling highly impactful social and political change are those social movements whose digital efforts have either caused or significantly contributed to the coalescing of a globally visible protest episode. Three of the most well known are the events of the Arab Spring in 2011, the Occupy Movement (also in 2011) and, emerging as an offshoot of Occupy in Hong Kong, the Umbrella Movement in 2014. Different from digital activism throughout the 1990s and early 2000s, these protest movements exemplify the organizing potential of messaging enabled through social networking services like Facebook and Twitter. These services support tens of millions of users around the world and, more importantly, function in such a way that information can become globally visible (1) very quickly and (2) without reference to traditional "gatekeepers" of global news (like traditional news media outlets and, in authoritarian countries, state-run information services).[6]

In each of these cases, the activism of a veritable multitude of dissident organizations and interest groups coalesced around responses to specific developments on social media into real-world protest incidents. With the Arab Spring and Umbrella specifically, social media was effectively leveraged to produce weeks-long events that saw real political concessions from a range of state entities. With the Arab Spring, pro-democracy protests in Tunisia and Egypt in particular gained global attention and, though researchers debate the causal role of social media, were certainly fueled by a tidal wave of citizen journalism that saw human rights information distributed via Facebook and organizing information published on Twitter. In Libya, where government forces fought a bloody campaign against revolutionary groups attempting to overthrow Muammar Ghaddafi, social media was uniquely useful in preventing harm to civilians caught in war torn cities in the country's eastern regions. Famously, dissidents tweeted at NATO's official Twitter account about the advance of armored government vehicles on rebel-held check points and, within an hour, were the beneficiaries of an airstrike by French planes against the approaching column. And in Hong Kong, Umbrella protesters made extensive use of (often banned) programs to bypass government censorship efforts and publish information about police interactions with those marching in the streets. Again, though causality is hotly debated here, a range of analysts argue that such publication demonstrably restrained China's crackdown on protests across the period of the movement. In short, ICT has emerged as a tool for social activists not only to reach global audiences with rhetoric and messaging but has also seen the development of a digital toolkit that allows for an unprecedented degree of influence over interactions with governments.

How do activists use the Internet?

Activists use the Internet – and, more broadly, web technologies – in a number of ways. Generically, we might think about digitally aided actions taken by activists as falling into one of three categories that line up with the general campaign goals of such actors (see Figure 11.1). First, web technologies help activists *mobilize*. Through the use of email and simple online messaging systems, activists are able to reach members and potential

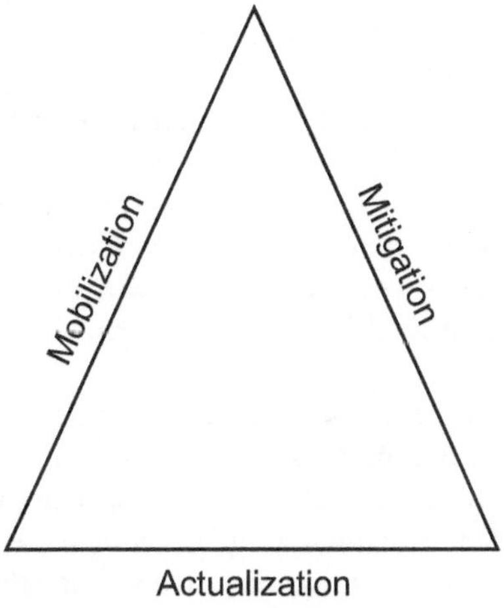

Figure 11.1 Operational elements of subversive campaigns

supporters in a way that is inexpensive and efficient. Falun Gong, for instance, succeeded in organizing a march on Beijing by 10,000 people in protest at the quiet exile of Li Hongzhi (the group's founder) in 1999 on only a few days' notice. Up until that point, such coordination would have taken much longer and would have been significantly more complicated. Removing the need to use telephones or physical messaging systems cuts down on restraints that limit the ability of activists to quickly organize and take action. Moreover, use of social media and more advanced messaging platforms allows activists to reach audiences beyond those already connected to an organization. These tools provide a forum for expansive information sharing and the mechanisms needed to link activists with broader domestic or global populations for the purpose of, for instance, finding new opportunities for financing and support.

Second, the Internet cedes activists new tools to *mitigate* threats to their cause. In some cases, this might mean the ability to disrupt the operations of governments or social groups opposed to a given activist outfit through denial of service attacks or vandalism. More commonly, for activists this can simply mean the means to counter disinformation and oppressive narrative with an alternative perspective.

Finally, web technologies give activists new abilities to *actualize* their campaign objectives. Though the line between mobilization and actualization is fuzzy insofar as both can involve persuading audiences, this is an important conceptual difference. Where activists are often trying to persuade a broader population to see their point of view and taken relevant action, it is the *way in which the digital revolution has changed the information environment* that activists operate in which dictates the impact of ICT on efforts to actualize objectives (i.e. not simply the mechanics of operation). The web not only logistically improves the ability of activists to mobilize, but it has also changed the nature of the task involved in influencing broader populations. The average global citizen produces, consumes and accesses information in new and unique ways relative to in the 1980s. Activists, in attempting to actualize their campaign objectives, are simultaneously able to reach large audiences and face new barriers to those audiences' attention.

More specifically, two new modes of using the Internet stand out as having dramatically altered the prospects of activists that choose to employ ICT. First, as Dorothy Denning notes, the development of the web has meant the construction of unique collection systems and procedures that make the job of the activist easier than it has ever historically been.[7] Denning asks us to think about the Internet as a vast library filled with information of every kind and organized in relation to the multitude of distinct sociopolitical and economic entities (governments, companies, social organizations, etc.) that use the web. Immediately, it is easy to see how great volumes of information on all manner of subjects is of particular use to those with a policy grievance or to actors interested in modifying some element of contemporary society. In short, the construction of an easy-to-access digital repository of human knowledge reduces or destroys many of the traditional educational barriers to informed sociopolitical operation that can stymie activist efforts.[8]

But the nature of the Internet as a kind of massive information repository is actually broadly significant for activists because of a secondary feature – the *distributed* nature of the repository itself.[9] In the digital age, it is almost inappropriate to talk of information access and availability without considering the role of governments as censors of information. In authoritarian regimes particularly (but not exclusively), the relationship between activism and sociopolitical change – in terms of the potential for success in activist efforts – has often been determined by the capacity of the government to block or redirect disobedient efforts into something harmless to the prevailing system. Since the late 1980s, numerous authoritarian governments have constructed the apparatus of digital censorship wherein access to certain kinds of information or the use of the Internet for particular types of political activity are limited by government intervention.[10] This is discussed in the following section. However, it is broadly necessary to note that even the most effective censor is constrained by the distributed and technologically diverse nature of web technologies. Beyond even common search engines and email distribution lists, web technologies allow groups to develop their own ways to store and share information. Against the efforts of censors, activists benefit from the ability to store information in servers abroad beyond the reach of a particular government. They gain from the ability of unique programs to bypass traditional, controlled methods of accessing the Internet. And they are substantially advantaged by the existence of a global marketplace for innovation of new techniques that enable individuals to hide from authoritarian forces, bypass censors and de facto assert freedoms of expression.

Alongside the construction of unique collection systems, of course, the rise of cyberspace has also brought with it new formats and opportunities for publication of information.[11] More than just providing yet another new medium for publishing opinions and data about goings on in the world, the disaggregated design of the digital domain means flexibility for activists looking to publish specific kinds of information to be seen by particular audiences. Using blogs, vlogs, chat platforms, email and much more, activists can publish information in a cost-effective and controlled fashion. There are few barriers to publishing in the digital age, and the toolkit for tailoring information presentation is great. Naturally, this also provides activists with unprecedented potential for publishing disinformation and for taking advantage of the fragmentation of a global information environment where consumers of information are asked to pass judgment on sources based on fewer credentials than ever before.

The experience of activists in the information age

Activists are simultaneously aided and constrained by the dynamics of operation in the digital age. As described earlier, the potential for organization, mobilization and actualization of desired sociopolitical changes through the use of ICT is enormous. And, over and above the immediate benefits of online operation for activists, ICT undoubtedly helps solve the kind of issues of global accountability that Anne Marie Slaughter has talked about. Greater connectivity, in short, allows non-state actors to observe elite activities, publish information on global governance processes and call global attention to distinct problems. At the same time, however, the toolkit of repression held by authoritarian states has been enhanced by both the development of specific technologies and, more broadly, by the emergence of a complex informational ecosystem that itself dictates new challenges for operation by dissidents.

The literature on democratization and democratic governance has broadly labeled ICT that aids activists in expanding their normative footprint as **liberation technologies**.[12] Jared Diamond first described this category of new technologies as any network system, service or piece of hardware that allows activists to bypass the traditional shackles of authoritarian rule and affect meaningful change. Liberation technologies are those that particularly allow for coordination in restrictive digital environments and coordination across national borders without reference to the controlling role of state-governed institutions (immigration control, news media organizations, etc.). The inspiration for this categorization of ICT lies with the role played by web technologies in the Arab Spring and, prior to the events of 2011, the popular revolutions in Central Asia, Ukraine, Iran and elsewhere from the mid-2000s onward. What is not clear about this categorization of ICT usage is whether or not liberation technologies include the broad repertoire of digital disobedience, from malware to denial of service attacks, not tactically focused on the traditional aims of activists (i.e. protest and persuasion). This is discussed further in the next section on hacktivism.

For activists, information technologies present two clear sets of new challenges to operation. One of the key advantages of usage of social media networking platforms and websites is the way in which it is possible to speak to a wide audience with incredible speed. However, the digital world is not exactly uncluttered. In reality, activists face issues with both mobilization and actualization that stem from the complexity of the global information environment. With regard to mobilization, the publication of information does not necessarily mean that said information will reach the intended audience. It certainly does not mean that information will go viral simply because the message is clearly important. The amount of information shared by billions of users of the web each day present as a sea of competing data points that activists must contend with when crafting their publications. Publishing at the wrong time, in the wrong medium, or (most often) without enough pre-planned connectivity means that information critical to activist success can easily be "lost in the crowd." Even if attention is paid to an activist's messaging, the half-life of that attention is likely to be exceedingly short as other stories/data quickly appear on a given user's digital horizon.

This limited window within which attention might be paid to a particular issue or organization presents a range of problems beyond simply the inability to get a message heard. Advertisers, news agencies and other activist organizations often adopt "clicktivism" approaches to staying relevant, abandoning quality of publication in favor of rapid-fire posting of content designed to consistently invite temporary attention. This competition for attention makes genuine activism difficult; it also means that co-option of messaging and the production of inaccurate information can occur as third parties attempt to capitalize on sudden visibility. This diversity of the global information environment also bring issues of coordination and actualization. It blunts the power of activist messaging in that potential supporters of an activist cause must self-adjudicate on the spectrum of information before them, rather than trust a single voice. Moreover, it makes the translation from digital interest to physical activism (i.e. protest activism) much more difficult, as communities of supporters are prone to moving from one cause to the next based on the short cycle of new information being made available online.

The other challenge activists face is less a natural development of the digital age and more a customized set of capabilities developed by those interested in stymying sociopolitical dissent. In the literature on authoritarianism and democratization efforts, it is often noted that, since around 2005 to 2007, governments across the Middle East and Asia have developed sophisticated censorship regimes that pose significant challenges for attempts to subvert non-democratic authority. Typically, this has been referred to as "upgraded authoritarianism" but, in actual fact, is not as overtly oppressive as the term suggests.[13] Yes, influence over ISPs and the use of massive government resources to develop tools of censorship have allowed governments like that of the People's Republic of China to directly mute the voices of pro-democracy activists. However, the use of ICT for control by authoritarian governments is increasingly far more insidious and focused on misdirection than it is on clear repression.

Box 11.1: Twitter, Facebook, Apple . . . is corporate policy foreign policy?

The digital world – and the prospects for contention and conflict therein – is based on private enterprise and technological innovation beyond government control. This is true regardless of whether the focus of one's attention is authoritarian or liberal democratic states. Even where authoritarian governments have direct oversight authority over technology companies and ISPs, the fact of the matter is still that cyberspace is functionally shaped by the actions of corporate entities.

Given this, it seems prudent to consider the question of whether or not corporate policy constitutes foreign policy of a sort. After all, the way that companies opt to operate can reflect foreign policy preferences. When Google took a position against the Chinese government and stated that they would no longer censor searches in Mainland China, they took a stand based on values intrinsic to Western political systems. They acted, in other words, to defend the principle of freedom even if, as was inevitable, they lost the favor of the Chinese government. Apple similarly took a stand against the U.S. government when it objected to an FBI writ demanding help in cracking open an iOS device found after a terrorist attack. Apple did so because cracking one phone would mean redesigning the security architecture involved such that compromise would be more broadly possible. In doing so, their corporate policy essentially became a public policy of sorts, reflecting the core values of liberal democracy and backing up their expression of a particular point of view with a credible stay on government power.

Perhaps more importantly, the way that a private firm chooses to approach the design and functionality of their network systems and other information products essentially determines – in admittedly greater or lesser ways depending on the context – the nature of the environment in which interstate and inter-societal interactions occur. How YouTube implements sharing and viewing algorithms massively determines the degree to which some content goes viral across international boundaries and some doesn't, therefore determining the shape of ideational diffusion around the world. More directly, the design of Facebook's platform and the rules of conduct around how Twitter users can use their account are key considerations for anyone wanting to understand the manner in which revolutionaries across the Arab world organized despite attempts at government repression during the Arab Spring. Likewise, the same design and operational considerations are critical variables in any attempt to understand the success of Russian-based information warfare efforts aimed at Western democracies since at least 2014.[14]

In essence, ICTs have increasingly allowed authoritarian (and even semi-democratic) governments to appropriate civil society. In addition to new regulations on the kinds of civil society actors that can legally operate – a step that allows governments to stamp out more informal activist efforts while maintaining the illusion of permitting contestation – authoritarian regimes regularly act to manipulate and direct civic dialogue online. By censoring calls to assembly but allowing anti-government speech in general, such governments can prevent the coalescing of protest movements that capture global attention while at the same time permitting a "pressure valve" for dissent. Such allowance of free speech without the opportunity for mobilization prevents instantaneous revolution, such as the continental revolutions that led to the dissolution of the Soviet Union in 1989, and simultaneously functions to mitigate the potential for more organized protest. At the same time, censorship of legitimate criticism following clear democratic failures in authoritarian or semi-authoritarian states (i.e. allowing the publication of accusations regarding vote fixing) has allowed countries like Russia, Belarus and China to direct national conversations about the state of current political systems. Specifically, such narrative manipulation can actually reduce faith in democratic processes while still allowing for the illusion that opportunities for contestation in contemporary civil society are relatively robust. In short, governments' use of ICT to counter the advocacy of non-state actors is, in many countries, sophisticated and effective in preventing activist efforts to mobilize and actualize change.

Hacktivism

By contrast with simple digital activism wherein advocacy groups and individuals employ ICT to magnify the reach of their message, "hacktivism" is a term often used to describe a set of activities and tactics beyond the traditional focus of activists on persuasion and citizen mobilization. Again, the line between activism and hacktivism – and subversion, discussed later, for that matter – is extremely fuzzy. Nevertheless, hacktivism most often denotes protest actions that are antagonistic and unlawful, even in the most liberal democracy where speech and assembly are protected rights. In other words, hacktivism describes civil disobedience – from website vandalism to more sophisticated cyberattacks – powered by web technologies.

The process and problem of hacktivism

Hacktivism is the process of malicious hacking for the purposes of civic activism.[15] Hacktivists are the archetypical grey hat hackers in that they break into computer systems and disrupt the function of services in order to advocate a particular point of view. And yet, hacktivists do not undertake advocacy operations in the same sense as do traditional activists. Hacktivists seldom publish persuasive arguments about specific issues and almost never aim to actualize their advocacy in the form of civic protest. So what is the logic of hacktivism?

The fuzziest part of the line between hacktivism and digital activism surrounds the development of tools to circumvent what is seen to be unfair regulation and censorship by governments. The development of programs and platforms like Freenet.org are, in many ways, the plainest kind of digital antagonism practiced by hacktivists aimed at affecting social change. But hacktivists also engage in cyberattacks on government websites, take down state and private sector services and more. In doing so, hacktivists do not seek to communicate a nuanced perspective on a social or political wrong. Rather, the basic logic of hacktivism is that overt antagonism costs targeted entities in several distinct ways. First, digital interference causes companies, governments and individuals to divert resources from other areas of operation to protect against further attack. Second, antagonistic hacks draw attention to a hacktivist's cause via the logic of serious disruption. Media organizations and social networks are far more likely to pay attention to unique conflict episodes than they are to the average activist message that contends with the multitude of alternative information in circulation in the modern global information environment. In many ways, hacktivists follow a similar logic to that of terrorist organizations, though physical harm from disruptive actions is not intended or particularly likely. Finally, hacktivism is demonstrative. Though the barriers to operation in the cyber domain are exceedingly low insofar as access to both talent and techniques is remarkably unlimited by market or government dynamics, hacking as social or political advocacy is still a relatively rare occurrence in world politics. Hacks prosecuted against government services or the websites of specific political elites, for instance, is exemplary in that observers are able to draw parallels with existing grievances and technical capabilities beyond the immediate situation.

Box 11.2: What is Anonymous?

Anonymous is a collective of online activists and hackers that have been responsible for a range of high-profile digital attacks since the early 2000s. It would be imprecise to call Anonymous a group; rather, Anonymous lacks real leadership, though there are certainly individuals more responsible for driving the movement toward particular actions than others. Nevertheless, the collective takes the form of a community that is sometimes motivated toward collective action usually aimed at addressing some sort of societal wrong or excess. Though Anonymous was not the first "hacktivist" outfit, it has arguably been the most impactful on global society in the first two decades of the 21st century.[16]

Anonymous has its roots in the online discussion board site 4chan and can be traced back to a series of postings (and subsequent discussions) in 2003. The motivation for Anonymous's vigilante antagonism – which most often takes the form of denial of service attacks against targets, vandalism of websites or the theft and release of private information – is not always clear. Like many hacktivist outfits, the group has variously attacked both state and non-state actors in support of social justice and "for the lulz" (meaning for the fun involved in doing so). The group has been behind attacks on the Church of Scientology, the Tunisian and Egyptian governments during the Arab Spring, the Israeli government, the Islamic State, PayPal, Sony Pictures, Inc., the Westboro Baptist Church and even government agencies in the United States. They have regularly espoused support for WikiLeaks and for protesters involved in the Occupy, Umbrella and Pussy Riot movements.

The question of whether or not Anonymous is either a positive or a negative force in the global cyber ecosystem is up to the individual. There is little doubt that Anonymous has, at times, acted in support of status quo forces fighting extremism around the world. The "Operation Paris" attacks on Islamic State following the Paris attacks in 2015 are an example of such activities. Likewise, Anonymous answered the call to resist the 2022 Russian invasion of Ukraine in force, with hundreds of hackers from across the globe launching an unprecedented and multifaceted campaign to "crack" Russian society, inject information about the war into the Russian media environment and expose the hypocrisy of Russia's ruling elite. At the same time, individuals associated with the collective, many of whom have been arrested, have attacked conservative civil society organizations and elected politicians. While there is a clear argument in most cases that victims violated modern social norms in an egregious fashion, Anonymous obviously operates outside of and against the rule of law. Thus, opinion is almost enduringly split between calling the collective modern-day "Robin Hoods" and labeling them merely as a sort of digital angry mob determined to brutalize those it is offended by.

Whereas digital activism is often applied to describe – even where the scope of protests, like with the Arab Spring, is transnational – the actions of groups operating in a specific domestic context, hacktivism has no such connotation. In fact, most problems that governments and inter-governmental agencies have in investigating and cracking down on criminal acts by hacktivists emerge from the fact that participants in such efforts are commonly distributed around the world. Such a dynamic makes crackdowns tricky for a number of reasons. Most significantly, authorities face the extended attribution problem acutely with transnational hacktivists that often don't exist with simple digital activism. The technical ability to disguise agency in online actions is enhanced by the distributed footprint of hacktivists in that government influence over ISPs and domestic civil society factions is much less complete. And authorities face unique legal issues in dealing with the threat posed by hacktivists insofar as legal standards for prosecution are highly variable across global jurisdictions. In many countries and localities, prosecution of particular types of cyberattack is not forbidden or evidentiary standards for prosecution are inappropriately extensive. In short, even where hacktivists are locatable, they benefit from the fact that legal systems around the world are perpetually playing catch-up in regulatory terms.

The tools and tactics of hacktivism

With hacktivists, tools and tactics might be broken into three categories: tools of (1) advocacy, (2) disruption and (3) information redistribution. Tools of *advocacy* in some ways differ little from those used by social activists more broadly to enact meaningful change. From the use of blogs to social media, the Internet provides activists with a number of ways to reach out and shape how large numbers of people make decisions about pressing issues. Specific to hacktivists, however, is the use of advanced spamming techniques to reach large audiences beyond the scope of legitimate services and platforms for Internet communication. Spamming programs are a common feature of world affairs and can support the ability to send messages via email, common instant messaging applications and more at very little cost to the user. Spamming is illegal around the world, as unauthorized sending of content to entire online communities is commonly against the terms of use of different web platforms and is often tied to service disruptions.

Tools of *disruption* are more diverse than are tools of advocacy and might be understood in two ways. First, we might think about tools of disruption in terms of the actual vehicles of attack used by hacktivists. As noted in Chapter 2, these might include Trojans, worms, viruses, logic bombs or the various methods of disruption (DDoS) attacks, which include *inter alia* SQL injection and cross-site scripting. Perhaps more usefully, we might think about disruption in terms of the strategic shape of different kinds of assaults. Generally, these fall into two categories. *Vandalism* refers to simple attacks that involve either the deletion or the manipulation of content on a website. With defacements, site content can be altered to reflect a nuanced manipulation of existing information, to portray alternative content introduced by the hacker or to show nonsense content, such as randomized links or links to obscene material. *Blockades*, also known as "cyber sit-ins," are a form of denial of service attack that overwhelms legitimate systems with traffic over a defined period of time. The distinction to be made here between DDoS as a technique and blockades as a strategy is that the latter are designed to maximize economic or reputational harm to the target through either repeated attacks or attacks timed to affect the most users of target services.

Finally, tools of *information redistribution* are really the combined ability hacktivists have to steal and then strategically publish information, an act commonly known as "doxxing." In part, intrusion is accomplished in no different manner than it might be if the end goal is systems' disruption. However, with the exfiltration of data, hacktivists are able to undermine the efforts of oppressive (or simply opposition) elements of global society in a unique fashion. Whereas the conventional wisdom is that activists rarely publish information not broadly available in the public sphere, hacktivists often aim to publish private details that will incriminate targets in illicit or immoral activities. In this way, much as the more general logic of hacktivism is to draw attention to societal opponents through severe actions, hacktivist redistribution of private information to the public sphere – an activity popularized by organizations like Wikileaks – aims to change the course of contemporary politics by presenting large audiences with shocking or salacious private information.

In some sources, this toolkit of hacktivism is even more broadly divvied up into strategic categories of approach to include **political cracking**, **performative** activities and **political rewiring**. Political cracking, simply put, involves methods employed to hack social or political systems and cause them to cease normal operation. This might involve website vandalism or the doxing of elites to bring about tension and unrest in society. Performative activities involve those designed simply to draw attention to a cause or narrative. And political rewiring references the development of circumvention tools and the provision of resources to empower dissidence and make real sociopolitical change possible.

Terrorism and information technologies

For terrorist groups around the world, web technologies have become part and parcel of upgraded efforts to more effectively prosecute violent campaigns.[17] Information technologies allow terrorist groups to undertake specialized recruitment, to distribute information both widely and secretly, to publicize terror activities and – though few such incidents have yet taken place – to terrorize via the prosecution of disruptive cyberattacks. In short, much as is the case with other kinds of non-state actors, ICTs present as a powerful tool for terrorists to improve the effectiveness of their campaign efforts.

The shape and causes of terrorism in the digital age

Much in the same way that cyberspace has enabled activism of all kinds by deconstructing barriers to non-violent sociopolitical advocacy, so too has it amplified the efforts of terrorists operating both locally and transnationally. The next section outlines the ways in which terrorists have generally turned to using the Internet (and web technologies broadly writ). However, it is first worthwhile thinking about the ways in which the information revolution has fundamentally affected terrorism today.

If one enters a phrase like "digital terrorism" or "cyberterrorists" into a search engine, they are likely to come across news articles, websites and social media focused on the threat posed to the international community by transnational Islamic terrorism. The reasons for this are numerous, not least the fact that the Western world has engaged groups like Boko Haram, al Qaeda and the Islamic State prominently since the late 1990s. Though it would be entirely disingenuous to say that fanatical Islamic-motivated terrorist outfits are the primary terrorist threat facing Western societies (indeed, at the time of writing more than 70% of all violent terrorist incidents that took place in the United States or Western Europe were prosecuted by neo-fascist or racial supremacy groups), it is certainly the case that the structure of such organizations is archetypical examples of a form of transnational, religiously motivated "fourth wave" of terrorism that has only coalesced over the past few decades.

Likewise, while it would be disingenuous to say that ICTs are causally linked to the emergence of such groups, it is certainly the case that cyberspace has augmented and greatly enhanced the ability of such groups to function effectively around the world in efforts to mobilize, recruit and terrorize. From Islamic State's "Digital Islamic Caliphate" to al Qaeda's trademark use of encryption for specialized recruitment and knowledge dissemination, such groups reflect a modern version of traditional terrorist efforts wherein cost efficiencies and coordinative boons from ICTs enable expanded outreach and hamper counterterrorism efforts. Of particular note, the information revolution has had distinct effects on the underlying causes of terrorism. Again, though we can't go so far as to say that information technologies have specifically, *causally* prompted the emergence of transnational terrorist threats over the past few decades, we do know that the online world and web technologies exacerbate the underlying development of the phenomenon.

Scholars of terrorism tend to consider three "levels" or categories of driving factors that produce terrorist campaigns. We might label types of explanations included in these categories as (1) *system-level*, (2) *state-level* or (3) *individual-level*.[18] Students of IR will recognize these categories as synonymous with the "levels of analysis" commonly outlined in introductory coursework on world affairs. Here, we simply extend to terrorism the general notion that political behavior emerges from the setup of international affairs, the structure of national systems or the psychology of individuals.

System-level explanations about terrorism rest on the notion that the structure of macro world events provides the motivation for the development of groups and campaigns. One popular meme that links the emergence of new terrorist groups to the world system – only partially an academic theory – is that of the "clash of civilizations."[19] First suggested by Samuel Huntington, the theory suggests that the "victory" of liberal capitalism over communism at the end of the Cold War has produced a world devoid of the political ideological conflict (in the global sense) for the first time in more than a century. As a result, fault lines for global conflict are increasingly likely to emerge on the boundaries between different "civilizations" – Western states, the Islamic world, Russia, etc. Here, terrorism emerges as the product of the encroachment of one set of civilizational values and influences into others. The actions of al Qaeda, for instance, can be understood as a rebellion against the encroachment of Western values into traditional Islamic cultures.

State-level explanations for terrorism emphasize national characteristics. For instance, failed states that boast no economy and perpetual poverty for segments of the population might be more likely to produce terrorist groups. As one vein of thought has it, this is particularly likely for a number of reasons. Failed states imply no ability on the part of the government to govern, and so terrorist recruitment and training are relatively unhindered. For the same reason, transnational terrorist organizations are likely to invest in such countries as a base for more easily moving material and personnel around without being caught. Likewise, failed states provide individual members of the population with both economic and social incentives to join extremist groups.

Finally, individual-level explanations for terrorism hold that membership of such organizations emerges from psychological conditions. On the one hand, membership of a terrorist organization might be of unique value to particular kinds of individuals that seek economic benefit or social status. On the other hand, terrorism often presents as an attractive choice for those who feel themselves distant and detached from mainstream society. Feelings of isolation, oppression and abandonment often make for ease in recruitment by terrorist organizations.

Naturally, these various categories of analysis explain some elements of the terrorist enterprise better than others. Typically, individual-level explanations focus on membership dynamics and recruitment in terrorist organizations, while the others attempt to explain strategic developments. Regardless of their individual utility, explanations across each category are affected by the dynamics of the information revolution. Though the "clash of civilizations" is not a direct result of the information revolution, for instance, it is certainly magnified by changes in the way that information is accessed, presented, and framed in international affairs. Development of new information infrastructure allows the "degradation" of traditional societies at an accelerated rate, as individuals (1) are more easily able to find information about other cultures, (2) are more readily influenced by content produced from alternative cultural perspectives and (3) are incentivized to engage in parts of the global system to thrive. Likewise, ICTs allow terrorist groups to more easily target ostracized communities and individuals to expand their membership, and to articulate an alternative vision of contemporary society that is easily accessible by those vulnerable to terrorists' influence.

How terrorists use the Internet

Beyond the broad effect that the information revolution has had on the underlying causes of terrorism, of course, web technologies have significantly enhanced the ability of terrorist groups to achieve campaign objectives.[20] But how do terrorists specifically use the Internet to their advantage? A great deal has been written about the ways in which terrorist causes have been enhanced and made more efficient by the Internet. Such studies extoll several specific characteristics of cyberspace and of network technologies as being of particular utility to terrorist groups. Besides the low barriers of cost to enter the domain, there has traditionally been little to no regulation of cyberspace by comparison with other domains. Likewise, the potential for anonymity protects terrorists online to a degree, and the potential ability to rapidly reach large audiences and to manipulate a complex media environment through low-cost methods incentivizes risk-taking in integrating web activities into terrorist campaigns.[21] More specifically, however, we might think of eight specific types of activities that terrorists are able to undertake online.

Much of what terrorists do online is logistical. Information technologies enhance the ability of terrorist groups to plan operations and to develop the resources needed for a campaign. Specifically, just as noted earlier in the case of activists, terrorists (1) *data mine* – essentially taking advantage of the repository nature of the Internet to gather intelligence that can be used to strategically and tactically plan operations.[22] More than ever before, terrorist cells are able to compile and organize massive amounts of information off the web for use in planning. Databases available to terrorist groups function much like an information list or product database might for a private company. They allow individual members (employees, in the case of private companies) to more quickly adapt their behavior while doing their job. Terrorist cells can more effectively identify weak spots in security procedures or optimal times for a bombing attack based on information freely available on the web, such as traffic metrics or government-published schedules for local government operations. The nature of the Internet as a large library also allows terrorist groups to introduce operation insurance in a way that has rarely been possible before. Where bomb attacks might traditionally have stood a high chance of failure due to equipment malfunction, for instance, standardized procedures for communicating design instructions improves the eventual effectiveness of the tools that terrorists make use of.

Terrorists also use the Internet to reach out to specialized audiences. This takes one of two formats. On the one hand, the web allows terrorists to more effectively (2) *fundraise* in two ways. First, easy access to off-the-shelf methods of encrypting communications and coordinating with foreign counterparts ensures a degree of secrecy in financial interactions between terrorist outfits. Likewise, data mining efforts provide more accurate demographic information about potential sympathizers and citizen funders, which are then relatively easy to contact in discrete ways via the use of the Internet (and particularly through darknets, off-the-shelf encrypted messaging applications, etc.). On the other hand, terrorists are able to use the Internet to more effectively (3) *recruit and mobilize* audiences sympathetic to their cause.[23] In some cases, the Internet facilitates recruitment of new members simply by allowing for quick communication across long distances. Would-be adherents are more capable than ever before when it comes to making contact with active terrorist groups. However, the web has also made targeted recruitment across an unbounded geographic area – essentially worldwide – quite simple. Again, data mining efforts allow terrorist groups not only to solicit donations from potential sympathizers but also to shape sympathizers' worldviews and invite different kinds of participation with the cause. Terrorists that operate on social media, for instance, can customize their informational offering such that those with minimal sympathies become inured in the narrative of struggle and political change forwarded by the group. Moreover, terrorist groups can promote their own narrative to counter that of governments, thus protecting their access to the demographic that constitutes their potential supporters (and, in some cases, future members).

Box 11.3: The dark web, cryptocurrencies . . . what are they?

Today, much online activity is enabled by cryptographic protections embedded in information transmission mechanisms. The Dark Web and cryptocurrencies are two examples of this that are invariably brought up in the context of cybercrime and the actions of terrorists online. This is because these developments award netizens abilities to hide online actions – particularly transactions and speech – in ways that are naturally attractive to those interested in avoiding the scrutiny of governments and intergovernmental actors.

The Dark Web, which is described in brief in Chapter 2, is easy enough to understand. Darknet sites are no different from other websites except that they cannot be accessed in the way that other sites on the Internet can be. They are not indexable by web crawlers and so cannot be found using traditional search engines. Instead, to reach a darknet site, one must use specialized software that is designed to anonymize user actions online. One popular piece of software, **The Onion Browser (TOR)**,[24] accomplishes this by creating a peer-to-peer network of nodes that a user's traffic is redirected through. Within this network, nodes online know the address of nodes on either side of it – in essence, they do not know the full shape of the network. The result is that, with enough redirections, it is basically impossible for an investigator to track web traffic to the point of originations. This leaves users free to engage in activities absent concerns about identification by authorities.

The Dark Web, which, because it is not indexed by web crawlers is not really a unified web of sites, is naturally of interest to those wanting to avoid government oversight. From dissident citizens to criminals and terrorists, the ability to operate without worry about compromising one's identity enables a broad range of illicit activities. The Dark Web is full of sites dedicated to the sale of narcotics, child pornography, weapons, malware exploits and more.

On the Dark Web, a common means of transaction is **cryptocurrencies**, particularly Bitcoin (the first cryptocurrency). To some degree, it is immensely unfair to suggest that cryptocurrencies intrinsically have something to do with terrorism or criminal enterprise. They were developed to solve an underlying problem with global financial institutions, namely that much financial transaction involves trusting gatekeeping entities like PayPal, often to the average citizen's loss. Cryptocurrencies are based on **blockchain** technology (and variations thereon). Blockchain technologies are a peer-to-peer method for ensuring record keeping of transactions. The heart of the thing is a database of transactions *held by every member of a cryptocurrency network*. Transactions are verified by multiple "peers" so that there is almost no opportunity for fraudulent or mistaken outcomes in financial exchanges. The *distributed ledger* of all transactions is then updated for everybody in the network so that cheating the system is impossible.[25]

Cryptocurrencies are of obvious utility for illicit transactions, the type of which are commonplace on the Dark Web. Part of the point of cryptocurrencies is that they do away with gatekeepers of traditional value exchange – banks, governments and similar financial institutions. Again, though there are a massive number of legitimate purposes to which blockchain-based technologies might be (and are being) applied, the lack of required oversight is a boon to those seeking to fly under the radar.

Staying with the logistical advantages of using the Internet to enhance a terrorist campaign, terror operations and the units that undertake them directly benefit. At the tactical level, new and advanced communications technologies allow for unprecedented (4) *coordination* on the battlefield. Terrorists can plan attacks with greater precision in timing than ever before and can furthermore more effectively adapt plans to evolving circumstances (such as the unexpected presence of military forces or the move a crowd might make away from a hidden bomb). In direct support of operations, terrorist groups benefit from new abilities to (5) *network* and (6) *share information* with counterparts around the world.[26] Networking means not only that groups are more easily able to talk with one another but also that it is increasingly easy for terrorist groups to fashion themselves as highly decentralized cell-based organizations. Here, the point is that terrorists are more easily able to coordinate various elements of their campaign *beyond* individual attacks to optimize strategic gains. By sharing information, terrorist groups that are using the Internet to their advantage are able to overcome many of the traditional problems of coordination with other extremist groups – inequalities of capability and talent, uneven access to resources, etc. Terrorists that effectively mine data relevant to their particular domestic situation, for instance, are able to extend the effectiveness of *national* opposition to the sitting government by sharing that information with other resistance groups easily and relatively risk-free over the Internet.

Finally, terrorists are able to more effectively engage with national and broader global information environments for the purposes of (7) *propaganda* and (8) engaging in *psychological warfare*. Much as is the case with activists, terrorist use of the Internet naturally involves an enhanced ability to speak to audiences distributed both locally and around the world. This allows terrorists to expose relatively large populations to information that they might rarely have been presented with before. Use of social media, email, wiki sites and more allows terrorists to apply their own frame and method of presentation to information about their cause, hopefully bypassing traditional government and media framing of terrorist efforts to reach sympathetic audiences. Moreover, messaging about terrorist causes need not come directly from known terrorist outlets. The Internet both allows for anonymity in publication of information and, quite commonly, the use of proxy mouthpieces for such a purpose. Diffuse networks of outlets can act to advocate and debate pro-terrorist positions while terrorists can more directly attempt to psychologically influence populations. Images of destruction, hostage executions and more receive close scrutiny and attention by media outlets around the world. This can serve the dual purpose of inflicting psychological harm on opposition audiences – for instance, executing soldiers to convince a Western public that the costs of intervention are too high – or convincing sympathetic audiences that a particular action has succeeded in repelling the forces against which terrorists are arrayed.

Box 11.4: Terrorism and the threat to critical infrastructure

Countries are, from a security perspective, incredibly complex. While scholars often oversimplify their descriptions of warfighting capabilities or national power by pointing to the strength of military forces, the attractiveness of economic or social mechanisms or the degree to which governments hold sway over the rules of international regimes, it is hard to escape the reality that state security is fundamentally bound up in those functional systems that enable these things. From the educational preparedness of human resources to access to natural resources, national security planners invariably have to take into consideration *infrastructural* vulnerabilities when assessing the scope of threats facing a given country.

In the United States, high-level investigations following the Oklahoma City Bombings and the September 11 attacks outlined a range of vulnerabilities that exist in the regular operations of **critical infrastructure (CI)**. Critical infrastructure includes those elements of national industry that are significant to the ordinary functioning of traditional measures of national power – the economy, military capacity, etc. The Clinton administration outlined, in Presidential Decision Directive 63 (PDD-63), 16 sectors of CI in 1998. These include the energy sector, healthcare and public safety, nuclear reactors and facilities and financial services.

As discussed in Chapter 8, critical infrastructure pose a particular security concern for national governments. The management of CI is addressed in more detail elsewhere here. Nevertheless, it is worth mentioning here that primary concern about cyber-based assaults on CI pivot on potential terrorist motivations for doing so. As Chapter 8 outlines, major state-sponsored attacks on CI hold little in the way of strategic value outside of broader conflict scenarios. Though there may be reasons for states to target the infrastructure of competitors in an irregular conflict scenario, it seems unlikely that major powers would sponsor such attacks.

Terrorists, on the other hand, may seize upon the ability to attack CI remotely as a potential method of coercing changes to government policy or drawing attention to a cause. Though there has yet to be such an attack, the FBI famously outlined evidence of al Qaeda cyber reconnaissance efforts aimed at mapping CI vulnerabilities in 2001. While traditional means of instilling fear in target populations might be cheaper and simpler to affect, there is little doubt that there is strategic value in terrorists' cyber attacking CI that does not exist for states.

Not among these eight approaches to Internet usage by terrorists is the use of cyber "weapons" to disrupt or inflict damage on a target. Much as is discussed in Chapter 8, the reason for this largely lies with the nature of cyberattacks as aggressive without being violent. As the case of Stuxnet demonstrates, of course, it is not entirely out of the realm of possibility that physical damage can result from a cyberattack. Moreover, a core concern of analysts interested in terrorists' use of the Internet is the potential for attacks on CI. By disrupting industrial control systems in the energy sector, for instance, cyber terrorists could feasibly knock out a national electrical grid over a sizable geographic area. The resulting disruption of social and industrial services could well result in death – indirectly, in all likelihood, from a sudden inability to call an ambulance or from traffic lights suddenly going out. But, just as is the case with state-instigated cyberattack, any "victory" gained from such an attack is temporary in that systems can be restored in a relatively short amount of time. And unlike foreign governments, terrorists are inherently limited in their ability to mount a meaningful attack to coincide with cyber disruption. Certainly terrorists might engage in disruptive cyberattacks in the future, though few have yet to do so, but the reality of the possibility is that such actions would be motivated by a desire to cause harm beyond the physical. Targets of successful cyber terror attacks could suffer from loss of reputation for effective information security and would necessarily start paying more to deter such efforts. Likewise, cyberattacks by terrorists could be used to enhance visibility among a particular demographic or achieve that rare physical disruption of a politically meaningful target, such as a nuclear power plant or enrichment facility. That said, the bottom line is that terrorists' use of the Internet largely remains constrained to logistical and coordinative considerations.

Subversion

Whereas the distinction between digital activists and hacktivists largely lies with the portfolios of cyber contention each utilizes, the line between activists and subversives instead lies with the scope of the outcomes desired. The same is true with the difference between terrorists and actors that primarily focus on **subversion**. Understanding the nature of subversive non-state actors and their approach to the use of information technologies is important for two reasons. First, subversives represent an intermediate step on the spectrum of advocacy that stretches from general political activism to militant terrorism. Second, and relatedly, the subversive enterprise has benefited from the developments of the information revolution to such a degree that the number of countercultural and broadly seditious *non-violent* organizations operating in world politics is on the rise for the first time since the 1960s. From cult organizations like Eastern Lightning in China to religious social groups like the Muslim Brotherhood and liberal action groups in authoritarian states, the world is full of non-state actors that don't fit neatly into the activist or terrorist (or insurgent or criminal group) categorizations outlined earlier. Most abhor the notion of violent revolution but are occasionally violent. Almost all aim for normative change, but only engage in advocacy under specific conditions. Thus, subversion is worthy of understanding as a distinct category of non-state actor beyond the traditional dichotomous view of civil disobedience as either violent or peaceful.

How subversion happens

In previous chapters, we discussed what subversion looks like as a strategy for contesting sociopolitical power. Here, before we focus in on non-state manifestations of subversive activity, it is worthwhile thinking first about how subversion might mechanically take place.

Subversion takes place in a range of different formats. Scholars who have studied government-sponsored subversion scholars break the subversive enterprise out into two broad categories. **Internal subversion** involves attempts to affect the conditions necessary for subversive transformation by dissidents residing within a country,[27] while **external subversion** describes the actions of states in attempting to influence conditions abroad.[28] External subversion is a common tool of statecraft and is often used to achieve ancillary aims for states (or specific rulers) interested in affecting political change abroad through more traditional means, including conquest and the securing of favorable treaty arrangements. Louis XIV, for instance, employed subversion via the encouragement of corruption and the manipulation of cultural practices for years in advance of his military campaigns in central Europe. Centuries before, the competing leaderships of the fragmented Eastern and Western Roman Empires did much the same, extending influence into less well-connected parts of the European continent in an attempt to subvert both cultural and formal political loyalties along the frontier. Ivan III encouraged sedition in Russia in the 16th century from abroad as a preparation for the internal campaign to throw off the Mongol yoke, as would the Habsburgs, the English, the British, the Nazis, the Bolsheviks and others at various times over the past several hundred years as an aid to broader strategies of domination. The logic, in each case, was fairly simple – conquest and/or superior positions in international relations is made much easier by the acquiescence of a target's population and ruling elites. And the employment of subversive tactics by governments is not merely an artifact of the pre-modern international system. Forcible regime promotion through subversive (among other) techniques has received some recent attention by scholars[29] inspired by events in, *inter alia*, Iraq (2003), Afghanistan (in both 1979 and 2001), Panama (1989), Angola (1975), Lebanon (1975–76) and Cambodia (1970).[30] External subversion is a common feature of the modern international system and, in the context of this chapter's focus on non-state participants in cyber conflicts, yet another source of state sponsorship of non-state belligerents that employ ICT.[31]

What does subversion look like in action? Work on subversion in the context of terrorism, insurgency and militant activism – particularly Kitson's famous treatise on irregular and information warfare, Rosenau's discussion of modern sedition[32] and Rid's summation of modern hacktivism[33] – does well in describing the various modes of activities undertaken by subversive campaigns in propagandizing, persuading and corroding the legitimacy of status quo symbols and institutions. Rosenau, in particular, takes cues from a range of past works in summarizing three different kinds of subversive activity in line with distinct categories of strategic function.[34]

First, the subversive enterprise is commonly composed of *front operations*. Subversion is countercultural and naturally originates from a position set apart from mainstream norms and expectations of political behavior. Subversive groups require arms that appear unattached to the countercultural core in order to achieve both logistical and activist goals. In general, there are two types of front organization – (1) those knowingly linked to the subversive group and (2) those unwittingly or only informally operating as an agent of counterculture. The redirection of resources by pro-LGBT groups to religious organizations and education programs in countries like Chad, Burkino Faso, Iran and Sudan serves as good examples of the latter type of front group, where broad advocacy for one position is masked in the charitable operations of other, more permissible activities. By contrast, the function of entities like the Holy Land Foundation for Relief and Development, Union of Good or North American Islamic Trust by affiliated members of branch elements of the Muslim Brotherhood movement – which, in some countries, might be characterized as subversive – provides a good example of the former type of group, in which representation of more extreme perspectives is knowingly maintained through informal and interpersonal connections.

Second, subversion often involves *infiltration* and espionage-like activities to place sources of influence within the institutions of the prevailing status quo position. This means the placement of individuals either belonging to or sympathetic to the cause of a subversive organization in government, opposition or civil society institutions. The role of such agents is twofold. First, it is often the responsibility of such an operative to sabotage or divert organization processes that would otherwise hamper the subversive cause. Second, it is occasionally the role of the agent to affect institutional subversion in changing the shape and nature of an organization such that conflict with the subversive cause is reduced. For situations where the organization or community is not directly opposed to the function of the subversive enterprise, infiltration is often about persuasion and recruitment. This type of activity is not unique to subversion, of course, insofar as violent and legitimate political actors place operatives in locations of opportunity as commonplace practice. There exists an extensive set of cases where al Qaeda and affiliate groups have placed operatives in Muslim communities, organizations and mosques across the West in an effort to either mobilize support or target specific recruitment needs,[35] as did the IRA, Nepal's Maoist insurgency, Aum Shinrikyo and more in decades past. Islamic State agents likewise filled the ranks of Iraqi security forces in limited numbers prior to the initial push against Baghdad in 2014–2016,[36] much as had happened in 2003–2004[37] and much as did the Viet Cong in the 1960s and 1970s in South Vietnam.[38]

Finally, subversive groups functionally act to frame the contentious issue or broader normative conflict that motivates their campaign through active efforts to *generate public upheaval*. Civil unrest provides an important role for subversive organizations in setting the stage for normative contention in the public limelight and not entirely because civil incidents accurately reflect a tension between the mainstream and counterculture. Indeed, civil protests and unrest largely pivot on secondary issues bound up in the construction of the current status quo rather than on the main platform advocated by the subversive movement. Causing civil unrest can be beneficial for subversive organizations for a number of reasons. First of all, large-scale disruptions can consume valuable state and non-state opposition resources. Second, the side effects of upheaval can exacerbate the exact society-government relations that subversive groups necessarily need to weaken in order to bring about a sea change in perspective on a given issue. Third, civil unrest is a source of new allies valuable to the subversive enterprise. Though often uncompromising in the integrity of the subversive cause, countercultural organizations have regularly benefited from the patronage or partnership of sympathetic actors motivated by related concerns (such as the alliance between elements linked to Hamas and branch organizations of the Muslim Brotherhood in Europe). Public upheaval and disruption produces a crucible from which such relationships can emerge. Finally, encouragement of civil unrest is one way to shut down a national system that does not revert to violence as a tool for structural transformation.[39] Much as might be the case with an old computer system, disruption to key functional processes can cause a national system to freeze up. This creates temporary political space in which subversive transformation of fundamental policy, process or system norms might be affected.

How subversives use the Internet?

In some ways, the narrative of how subversives use the Internet is entirely similar to the various ways that activists, hacktivists and terrorists do. The web is a powerful tool for those aiming to subvert – i.e. for those that aim to advocate and affect normative transformation in extreme, but unusual ways. However, it seems fair to say that the information revolution is relevant for subversives more because of the way in which it has changed the global information environment than because of the development of new technologies for enhancing *logistical* functions. Subversion is the manipulation of hearts and minds, and the fragmentation and complexification of the modern media environment presents subversives with both new opportunities and new challenges to bring countercultural perspectives into the societal mainstream.

Subversives' use of the Internet is remarkably low-intensity. Though there are certainly elements in common with the online strategies of terrorist groups and activist organizations, subversives' web usage is characterized by minimal willingness to provoke either countersubversive forces (i.e. governments) or the broader population. While subversive organizations *have* been known to undertake disruptive cyberattacks, they almost always do so against non-governmental targets, such as opposition civil society groups. In Germany, for instance, far right activists have attacked the servers and websites of far left groups several times since 2004. Likewise, while subversive actors do steal information and illegally publish it, the clear aim of such acts is often to appear to be performing a public service. Data theft and publication in various instances in Bangladesh, Pakistan, Ukraine and Italy between 2010 and 2014, for instance, sought to expose elite corruption in the lead up to state elections.

This approach to the use of the Internet is one that has received a number of names in recent scholarship on non-violent extremism. From "twilight tactics" to operations undertaken in the "grey zone" of contention, the purpose of such light-footed digital antagonism in strategic terms is the maximization of gains for a subversive cause in ways that take minimal risks. Indeed, research has shown that subversive groups have flocked to the Internet in unprecedented fashion particularly because of the utility of web technologies for such low-risk/high-potential gain activities. While attempting to engage a broader population to promote a counterculture cause, subversives are simultaneously able to engage in low-level civil disobedience that is unlikely to turn public opinion against them. Quite divorced from the way in which activists, hacktivists or terrorists might choose to publicize controversial information and frame discourse accordingly, subversives aim to enrich the informational environment they function within so that their ideas and actions become acceptable over time. In this way, we might think of subversives' use of the Internet as being the employment of *information enrichment techniques* that aim, among other things, to create echo chambers and question established norms such that countercultural perspectives can gradually enter the mainstream.

Criminal and political hacking

Though criminals and agents of countries are not always thought of as belonging in the same category, the context of the information revolution makes such a grouping appropriate insofar as new dynamics of behavior with both kinds of non-state actors emerge from the development of new markets for digital crime. In short, the information revolution has augured massive changes in the infrastructure of global finance, public sector functions and more. As a result, there today exists great potential for the redistribution of resources (money, intellectual property, etc.) through disruptive cyber actions for all manner of self-interested actor. Self-motivated non-state criminals or those acting on behalf of national governments constitute an extensive emergent ecosystem of threat to national prosperity and security.[40] The sections below discuss different elements of the threat, including simple cybercriminals, malware developers, cyber mercenaries "patriot hackers" and state spies.

Cybercrime, cybercriminals and why they matter

Cybercriminal activity is an enormous problem for global society. Though in many ways hard to measure because victims have strong incentives to not reveal information about attacks so as not to suffer further reputation and income losses, cybercrime is demonstrably bad for national economies on a number of fronts. At the lowest end, where individual citizens are targets, **cybercrime** is responsible for hundreds of millions of dollars lost from economic circulation every year.[41] Ransomware attacks where hackers encrypt user information and demand payment for decryption, for instance, succeeds in earning malicious cyber actors tens of millions of dollars per annum just in the United States. At the higher end, where private firms are the targets of criminal activity, national economies suffer from *loss of intellectual property* (IP) and confidential business information. This direct cost then incurs various indirect costs and presents extended challenges for national cybersecurity. Lost business information increases opportunities for *stock market manipulation* and further business-oriented hacking. Likewise, successful data theft attacks – or even knowledge of such attacks – forces firms to *redistribute resources to securing information*. In doing so, the economy pays *opportunity costs* as firms spend less on innovation and more on the cost of operation. The same is true when firms, aware of the possibility of IP theft, select not to innovate. And national economies also suffer when companies' *reputations suffer*, as consumer confidence in one firm tends to reflect sectoral measures of consumer desire to spend money and invest.

Direct economic loss from cybercrime occurs in a number of ways. A business knocked offline for a day will loss tens of thousands of dollars (or more). A hack against another business that is made public might result in significantly reduced sales due to lack of trust, particularly if the business in question is in the technology sector. Again, however, the history of cybercrime and the losses incurred by national economies are hard to outline with any degree of accuracy. Nevertheless, there exist some estimates about the degree to which malicious online activity affects national productivity. In 2014, estimated monetary losses to the global economy were anywhere between \$300 billion and \$1 trillion. For the United States in the same year, losses ranged from \$24 billion to \$120 billion.[42] In reality, the U.S. number was likely somewhere in the middle of that range. However, at time of writing, consensus holds that cybercrime losses in the United States have certainly hit \$100 billion. This represents nearly 1% of the country's Gross Domestic Product and is the most severe criminal problem by impact behind drug trafficking and non-narcotics smuggling.

Fortunately, the dollar figure of losses is actually of secondary importance. The real threat to national prosperity *and to national security* comes from second order effects, namely the loss of a competitive edge from criminal activity. For instance, where a firm suffers the loss of valuable IP, it increasingly finds itself forced to compete in an international marketplace where, at best, competitors are able to take advantage of the firm's disrupted work. At worst, competitors may benefit directly (from the theft of IP) or indirectly (as stolen IP dilutes the marketplace for innovation in a certain sector). For countries, this is a particularly worrisome dynamic because national economic potential tends to undergird more concrete machinations of diplomacy and national power. The ability to attract favorable trade relationships, to sanction rogue states, to promise to support an international coalition or to coerce opponents extends from a country's ability to innovate and sustain a robust economy. Strong economies provide the hard means of national power – an ability to spend on the military, for instance – and underwrite the "soft" power a country has (i.e. the attractiveness of a country to foreign powers). Thus, both economic losses from cybercriminal activity and ineffectiveness in dealing with it – or even the perception that government regulation and attention to the health of public–private relationships is ineffective – stand to harm national security on a number of fronts.

But who are cybercriminals? Just as is the case with hacktivism, one way to divide up the population of non-state actors involved in the cybercrime ecosystem is by differentiating between hackers and script kiddies. Again, hackers are those with specialized skills that break into computer systems. Black hat hackers do so for malicious purposes, while white hat hackers are the employees (often freelance) of actors that want to improve their own information security. In the middle, grey hat hackers – a term less meaningful when economic crime is the subject of focus – tend to be vigilante actors that break laws for ostensibly laudable purposes (i.e. exposing corruption or drawing attention to objectionable political perspectives). By contrast, script kiddies often hack for the thrill of the thing and have no specialized skillset. Instead, script kiddies use pre-designed and implemented instruments to prosecute cyberattacks.

With regard to economic cybercrime, however, three additional kinds of actors are worthy of note. We might use the term 'hacker' or 'script kiddy' to simply describe those involved in relative unorganized criminal actions. That said, several additional terms are commonly used to denote specialized roles within the cybercrime ecosystem. **Cyber spammers** seek financial gain specifically through acts of social engineering online. Using email (or, occasionally, SMS, IM, web forums, etc.) spammers send fake information to large numbers of web users. In many cases, spamming is demographically targeted at vulnerable demographics, such as older users. The purposes of spam messages is to start a dialogue wherein the spammer can manipulate a user into giving up valuable assets – money, user information for email accounts, bank login information, etc. Spammers either directly make money from such efforts or sell information gained from spam attacks onto other criminals.

Just as spammers are a specialized form of hacker, so too are **malware** developers. Where spammers might be differentiated from other black hat hackers by their tools and methods of choice, malware developers are different insofar as they are focused on producing hacking tools. Likewise, whereas spammers tend to have limited technical knowledge in line with the simplicity of their chosen approach, malware developers tend to be highly skilled. Developers produce programs that allow other hackers and script kiddies to evade detection when operating, to more effectively intrude into target systems and to fool victims of spamming attempts more readily. They also produce "creation kits," which are themselves programs designed to let unskilled criminals customize their tools prior to the launch of an attack.

Organized cybercriminals are yet another category of cybercriminal. Organized networks in this vein tend to mirror their real-world counterparts in that members are geographically distributed around the globe. In a great number of cases, traditional organized crime groups directly control transnational cybercrime networks. According to a 2014 report, organized cybercriminals are behind the bulk (~80% globally) of serious economic loss from data theft. This is, in many ways, unsurprising. This statistic conjures an image of mafia organizations running massive online criminal empires and it *is* the case that large criminal efforts exist that have great impact on the global economy (and on international security). Famously, Russian organizations controlling massive botnets were hired for the purposes of creating massive disruption in Estonia in 2007. But there is no clear evidence that this kind of organized criminal network with global reach is the norm. Rather, the nature of the cybercrime ecosystem simply links a massive number of otherwise unaffiliated hackers to organized crime. These networks are a primary source of tools that can be bought for use by petty criminals to attack bank accounts, company servers and more. Likewise, organized criminal networks can be quite small. Some are no more than a dozen people and essentially constitute the efforts of petty criminals to pool resources and amplify revenue gains.

Box 11.5: Trolls, troll farms and bot warfare

With **external subversion** campaigns wherein the aim is to manipulate public opinion and information conditions from abroad to achieve some sort of strategically favorable outcome, few methods have received as much attention as has trolling and bot-based disinformation. Though the general impression of **trolls** is that they are Internet personas created for the specific purpose of attacking the credibility and substance of existing content, the reality is that trolls are simply any such persona that plays a part in manipulating the information environment around a particular topic. Trolling can include positive propagandist activities to spread enthusiastic content, often falsified or shaped to present a particular narrative, about a person or cause.

Russia is perhaps the best known employer of trolls as a component part of broad-scoped information warfare efforts to interfere in the social and political processes of states in the former Soviet sphere, in the European Union and in North America. Since at least 2003, Russia has been regularly linked to what are often called **web brigades** made up of an indeterminate number of paid trolls that work to manipulate online discourse on a given topic. During the influence operation campaigns against Ukraine, Germany, the United Kingdom and the United States in the 2010s – campaigns that are, according to public disclosures from Western defense communities, still underway – Russia was linked specifically to **troll farms**, which are specialized locations where a number of paid commentators create disinformation and work to spread it in strategically advantageous ways.

Trolling is not always instigated directly by human operators. Alongside such efforts, **bots** are employed to spread content in much the same way that robocalls might be directed to automatically make calls to unsuspecting names in the phone book. Bots, which are increasingly sophisticated, are fake accounts that engage target audiences via reference to a specific series of instructions. Often the point is simply to spread content as far and as wide as possible. At other times, the idea is to actively convince an audience of the existence of hostility toward or support for a particular cause to encourage individuals to take steps on their own that serve the broader trolling strategy.

Finally, it would not do to fail to mention *corporations* as critical actors in cybercrime episodes. Naturally, corporations often take the role of defender and victim. However, corporations in authoritarian states are sometimes the cause of illegal intrusions authorized or prompted by national governments. Moreover, it is worth considering the arguably criminal role that corporations fill in the aftermath of cyberattacks. Though firms tend not to engage in overt acts of cybercrime (as doing so would incur the wrath of national governments and lead to economic sanctions), the right of a corporation to "hack-back" is less clearly defined in most countries. Does a corporation have the right to "follow" a hacker back to the source and either attempt to delete information stolen or take down a hacker's computer entirely (as discussed previously)? If so, at what point does hacking back cease to be an act of extended self-defense and become a retaliatory criminal act? And what if the hacker is, as is often the case, operating from another country? Who has jurisdiction to decide whether or not the corporation's action is legal? This dynamic constitutes a major debate in the development of national responses to cybercrime regulation and in international efforts to coordinate. Ultimately, this is a particularly pressing issue, as many of the most innovative and productive firms in any country have an international footprint. Such non-state actors face the possibility of operating on two sides of potential interstate cyber conflicts.

For king and country: the non-state proxies of interstate cyber conflict

Again, though economic crime and political "crime" are remarkably different threats faced by the international community, they share similarities in the context of cyberspace in that they both somewhat emerge from distinct marketplaces for intrusion skills and technologies. They also both, with the unique exception of *cyberspies*, emerge from the self-interest of the non-state actors involved. "Cyber spies" is a term broadly used to describe non-state actors that intrude into the networks of foreign states, monitor developments and steal information useful to their state sponsor. Here, we are going to use the term to more specifically refer to direct employees of a state that are engaged in espionage activities. That we might consider these employees non-state actors in a meaningful sense – different, for example, from listing individual soldiers or diplomats as non-state actors – is a unique condition of the way in which countries set up their cyber espionage programs. Beyond even the traditional detachment spies have from their state controllers, at least some countries actively try to distance themselves from the actions of their employees. China, for instance, has reportedly gone so far as to issue false identification information across a number of formats for state-employed cyber spies and devotes resources to maintaining distance (i.e. paying for off-site facilities and equipment) for the purposes of being able to deny liability when espionage activities are detected.

That said, non-state proxies – i.e. non-state actors that act in line with the directions and strategic motivations of state actors – are more commonly at least one step further detached from the apparatus of governments. Proxies generally fall into one of two categories. First, **cyber mercenaries** are for-hire hackers that serve as extended cyber forces of a national military, intelligence community or paramilitary organization.[43] Here, there are perhaps the strongest parallels with economic cybercriminals insofar as the incentive to hack is financial gain for the self. Different from criminals that seek to steal information or money, mercenaries hire out their abilities to states (or other state proxies). Of note, transnational organized criminal networks might situationally be described as mercenaries. This is largely due to the fact that such networks often control botnets and other resources useful to the realization of political objectives. Though even authoritarian states rarely contract criminal elements, there are demonstrable connections between individuals in government and in organized crime networks in countries like Russia that have clearly – as in the case of Estonia in 2007 – led to the employment of criminal hacking resources for state objectives. Corruption and nepotism, in short, often link states to criminals for mercenary purposes. Additionally, groups of mercenaries are often referred to as *cyber militias*. Even where the opportunities for financial gain are minimal, mercenaries often work in broader groups for the purpose of undertaking additional criminal acts that are "protected" by the aura of the state relationship.

Second, **patriot hackers** are malicious non-state cyber actors whose motivations stem from the desire to help their country during a conflict or crisis.[44] Patriot hackers most often aim to disrupt the operations of those perceived to be enemies of the state. A number of patriot hacker collectives are fairly well known today. A Serbian group called the Black Hand (named after the group that assassinated Archduke Franz Ferdinand in 1914 and sparked World War I) vandalized websites in Albania during the Kosovo conflict.[45] A large number of other such collectives are based in Russia and China. Though no official statement of state patronage exists, both governments have been permissive of the actions of such groups in the past. Specifically, the Russian government has appeared broadly permissive of actions taken by patriot hackers during the Kosovo conflict, the 2014–2017 Ukraine conflict and the 2007 Estonian conflict to vandalize the websites of foreign governments and prosecute denial of service attacks against media outlets. And the Chinese government has appeared supportive of nationalistic citizen antagonism since at least 1997, when patriotic hackers protested the Indonesian government's persecution of Chinese citizens.[46]

Political hackers, when not directly employed by the state, are motivated by self-interest in the form of economic gain, self-defined patriotic duty or thrill seeking. But why might states choose to employ non-employee non-state actors in their operation? In reality, states stand to benefit from various elements of such relationships. First, though cyber aggression broadly benefits from secrecy and mechanisms that ensure degrees of anonymity, states are extensively invested in surveillance designed to mitigate foreign cyber threats. Thus, in hiring or encouraging non-state belligerents to act on their behalf, states gain the potential for additional surprise in launching cyber operations. Second, and perhaps most obviously, states can plausibly deny involvement in aggressive cyber operations if the direct belligerents are not officially tied to the government. Third, non-state actors bring unique abilities to states' portfolios of cyber contention. Where a particular country's cyber arsenal is well understood, non-state actors stand to offer unique means for expanding the scope of potential disruption. Fourth, state use of non-state proxies means that resources need not necessarily be diverted to developing a country's own cyber offense portfolio. Tapping existing talent beyond the public sector negates the need to train employees, retain them and provide the infrastructural resources for success. Relatedly, states can often magnify their capabilities by employing non-state actors in a way that is simply not possible with in-house capabilities. Given that patriot hackers and mercenaries alike emerge from diverse market environments – online social spaces (forums, etc.) as well as marketplaces full of collaborators and competitors – the opportunities for rapidly scalable mobilization of resources is immense. States can retain a technically competent force that can act reliably and rapidly at little cost. And finally, the use of non-state proxies takes advantage of the ambiguous context of international law on the use of such belligerents in interstate relations. This ambiguity is a useful shield against the scrutiny of others in the international community.

Naturally, there are a great many drawbacks to the use of non-state proxies for political hacking. Most obviously, states do not directly control their proxies in the way that they might affect control over military forces. At best, non-state proxies function like contractors with diminished oversight. At worst, they are the result of intangible connections between the state and patriotic elements of society (i.e. encouraging rhetoric). This lack of direct control means that states are taking on a number of risks when they employ proxies. Lack of oversight or familiarity with methods means that a state might give the go-ahead for an operation that ends up producing unintended collateral damage. More specifically, if proxies are operating in close proximity to the country employing them – i.e. to help put down protests or to attack targets in a bordering country – it is possible that networks and services related to the initiating state might be affected (e.g. unintended localized Internet outages that affect citizens of the initiating government). And governments might make themselves vulnerable to non-state aggression in the future through current employment, particularly if non-state actors are in possession of information that could implicate state actions that are either illegal or diplomatically problematic.

Finally, states employing proxies for the purposes of political hacking actually do risk legal action. Though the use of non-state actors allows states to take advantage of the current ambiguities of international law, attribution is not entirely outside the realm of possibility, and the potential exists for double-cross by a non-state actor. If a state uses non-state proxies to launch cyberattacks, whether aimed at vandalism, information theft or disruption, being caught could be disastrous on a number of fronts. The Law of Armed Conflict is hazy on what kinds of non-kinetic attack produce a violation of international law, but cyberattacks by one country against another definitely violate the spirit of the treaty. Moreover, the use of non-state actors in conflict essentially constitutes the endorsement of "non-privileged combatants" in warfare. This term denotes a distinction between soldiers employed by a state who are protected from persecution from various forms of aggression in conflict and non-affiliated criminals. Use of non-state proxies could, in short, allow for a charge of supporting terrorism to be levied against a state. And beyond the technicalities of international law, where there are presently few precedents to refer to, high-level attribution of the kind that could convince the international community of a country's complicity in cyberattacks risks escalation to kinetic forms of conflict. Particularly in situations where there are deaths from the employment of cyberattacks – for instance, caused by the inability of emergency services to respond following a power outage – clear aggression could lead to conflict spirals wherein armed forces are mobilized beyond cyberspace.

Making sense of the global cyber ecosystem

Having catalogued various issues, actors and incentives bound up in different cyber threats to national security, the next chapters take up the challenge of making sense of different national and international approaches to cybersecurity issues. Specifically, we consider issues of Internet governance and cyber conflict issues from the perspective of the international community. This includes debates both on the ethics of cyber warfare and on the possibility that norms might be developed to constrain malicious actions via the web. This leads us to a point where we might finally think logically and analytically about the future of cybersecurity and of cyber conflict in a world where geopolitical and technological circumstance are both constantly transforming.

Notes

1 Orman, Hilarie. "The Morris Worm: A Fifteen-Year Perspective." *IEEE Security & Privacy*, Vol. 99, No. 5 (2003), 35–43.

2 Sprinkel, Shannon C. "Global Internet Regulation: The Residual Effects of the ILoveYou Computer Virus and the Draft Convention on Cyber-Crime." *Suffolk Transnat'l Law Review*, Vol. 25 (2001), 491.

3 For extensive treatments of digital activism, see *inter alia* Joyce, Mary C. *Digital Activism Decoded: The New Mechanics of Change*. London: IDEA, 2010; Carty, Victoria. *Social Movements and New Technology*. Boulder, CO: Westview Press, 2015; Earl, Jennifer, and Katrina Kimport. "Movement Societies and Digital Protest: Fan Activism and Other Nonpolitical Protest Online." *Sociological Theory*, Vol. 27, No. 3 (2009), 220–243; Earl, Jennifer, and Katrina Kimport. *Digitally Enabled Social Change: Activism in the Internet Age*. Cambridge, MA: MIT Press, 2011; and Rid, Thomas. *Cyber War Will Not Take Place*. New York: Oxford University Press, 2013.

4 See, for example, Morozov, Evgeny. *The Net Delusion: The Dark Side of Internet Freedom*. London: PublicAffairs, 2012; Lagerkvist, Johan. "The Rise of Online Public Opinion in the People's Republic of China." *China: An International Journal*, Vol. 3, No. 1 (2005), 119–130; Huschle, Brian J. "Cyber Disobedience: When Is Hacktivism Civil Disobedience?." *International Journal of Applied Philosophy*, Vol. 16, No. 1 (2002), 69–83; and Shantz, Jeff, and Jordon Tomblin. *Cyber Disobedience: Representing Online Anarchy*. John Hunt Publishing, 2014.

5 Rid (2013).

6 See Howard, Philip N. *The Digital Origins of Dictatorship and Democracy: Information Technology and Political Islam*. Oxford, UK: Oxford University Press, 2010; Tong, James W. *Revenge of the Forbidden City: The Suppression of the Falungong in China, 1999–2005*. Oxford, UK: Oxford University Press, 2009; and Diamond, Larry. "Liberation Technology." In *In Search of Democracy*, 132–146. Oxford, UK: Routledge, 2015.

7 Denning, Dorothy E. "Activism, Hacktivism, and Cyberterrorism: The Internet as a Tool for Influencing Foreign Policy." *Networks and Netwars: The Future of Terror, Crime, and Militancy*, Vol. 239 (2001), 288.

8 Ibid.

9 Ibid.

10 See, for instance, King, Gary, Jennifer Pan, and Margaret E. Roberts. "How Censorship in China Allows Government Criticism but Silences Collective Expression." *American Political Science Review*, Vol. 107, No. 2 (2013), 326–343.

11 Denning (2001).

12 See Diamond, Larry, and Marc F. Plattner, eds. *Liberation Technology: Social Media and the Struggle for Democracy*. Baltimore, MD: Johns Hopkins University Press, 2012; Christensen, Christian. "Discourses of Technology and Liberation: State Aid to Net Activists in an Era of 'Twitter Revolutions'." *The Communication Review*, Vol. 14, No. 3 (2011), 233–253; and Ziccardi, Giovanni. *Resistance, Liberation Technology and Human Rights in the Digital Age*. Vol. 7. Dordrecht, The Netherlands: Springer Science & Business Media, 2012.

13 See Heydemann, Steven. *Upgrading Authoritarianism in the Arab World*. Saban Center for Middle East Policy at the Brookings Institution, 2007; Cavatorta, Francesco. *Arab Spring: The Awakening of Civil Society—A General Overview*. na, 2012; and Hindman, Matthew. *The Myth of Digital Democracy*. Princeton, NJ: Princeton University Press, 2008.

14 See Hurak, Ihor, and Paul D'Anieri. "The Evolution of Russian Political Tactics in Ukraine." *Problems of Post-Communism*, Vol. 69, No. 2 (2022), 121–132.

15 For much more detailed accounts of cyber disobedience and hacktivism, see *inter alia* Olson, Parmy. *We Are Anonymous*. London: Random House, 2013; Sauter, Molly. *The Coming Swarm: DDOS Actions, Hacktivism, and Civil Disobedience on the Internet*. New York: Bloomsbury Publishing USA, 2014; Beyer, Jessica L. *Expect Us: Online Communities and Political Mobilization*. Oxford, UK: Oxford University Press, 2014; and Brevini, Benedetta, Arne Hintz, and Patrick McCurdy, eds. *Beyond WikiLeaks: Implications for the Future of Communications, Journalism and Society*. Dordrecht, the Netherlands: Springer, 2013.

16 Klein, Adam G. "Vigilante Media: Unveiling Anonymous and the Hacktivist Persona in the Global Press." *Communication Monographs*, Vol. 82, No. 3 (2015), 379–401.

17 For an early overview of the field of study of cyberterrorism, see Conway, Maura. "What Is Cyberterrorism? The Story so Far." *Journal of Information Warfare*, Vol. 2, No. 2 (2003), 33–42. For a more recent account, see Imran Awan, "Debating the Term Cyber-terrorism: Issues and Problems." *Internet Journal of Criminology*, Vol. 2045, No. 6743 (2014), 1–14.

18 For literature on this broad area, see *inter alia* Bjørgo, Tore, ed. *Root Causes of Terrorism: Myths, Reality and Ways Forward*. Oxford, UK: Routledge, 2004; Crenshaw, Martha. "The Causes of Terrorism." *Terrorism Studies: A Reader* (eBook, 2012); and Hoffman, Bruce. *Inside Terrorism*. New York: Columbia University Press, 2006.

19 See Huntington, Samuel P. "The Clash of Civilizations?" *Foreign Affairs*, Vol. (1993), 22–49; and Huntington, Samuel P. *The Clash of Civilizations and the Remaking of World Order*. Mumbai: Penguin Books India, 1997.

20 It is worth noting up front that there is considerable debate as to what "cyberterrorism" looks like. Is it the use of computer network attack techniques to attempt to coerce behavioral changes in a target population, or is it simply any element of the terrorist enterprise that makes use of ICT? For more on this debate, see Dan Verton, *Black Ice: The Invisible Threat of Cyber-Terrorism*. New York: McGrawHill Osborne, 2003; and Pollitt, Mark. "Cyberterrorism-Fact or Fancy?" (2001). www.csgeorgetown.edu/~denning/infosec/pollitt.html.

21 Gabriel Weimann. "The Sum of All Fears?" *Studies in Conflict and Terrorism*, Vol. 129, No. 135 (2005).

22 Denning, Dorothy. "Cyberwarriors: Activists and Terrorists Turn to Cyberspace." *Harvard International Review*, Vol. 23, No. 2 (2001), 70.

23 Stenersen, Anne. "The Internet: A Virtual Training Camp?" *Terrorism and Political Violence*, Vol. 20, No. 2 (2008), 215–233; and Edwards Charlie, and Luke Gribbon. "Pathways to Violent Extremism in the Digital Era." *The RUSI Journal*, Vol. 158, No. 5 (2013), 40–47.

24 Goldschlag, David, Michael Reed, and Paul Syverson. "Onion Routing." *Communications of the ACM*, Vol. 42, No. 2 (1999), 39–41.

25 For the original white paper on the technology, see Nakamoto, Satoshi. "Bitcoin: A Peer-To-Peer Electronic Cash System." https://bitcoin.org/bitcoin.pdf. (2008).
26 Discussed, among other places, in David C. Benson, "Why the Internet is Not Increasing Terrorism." *Security Studies*, Vol. 23, No. 2 (2014), 293–328.
27 See Beilenson, Lawrence W. *Power Through Subversion*. New York: Public Affairs Press, 1972, 5–6.
28 Ibid., p. 56.
29 See Owen IV, John M. *The Clash of Ideas in World Politics: Transnational Networks, States, and Regime Change*. Princeton, NJ: Princeton University Press, 2010.
30 Ibid., pp. 48–52.
31 External subversion has recently again become a focus of some scholars, with interest in the concept revolving around the injection of web-based room for exploitation and disruption across global society. See, among others, Deibert, Ronald J. "Subversion Inc: The Age of Private Espionage." *Journal of Democracy*, Vol. 33, No. 2 (2022), 28–44; Radin, Andrew, Alyssa Demus, and Krystyna Marcinek. "Understanding Russian Subversion." *Perspective*. Santa Monic, CA: RAND Corporation, 2020; Maschmeyer, Lennart, "The Subversive Trilemma: Why Cyber Operations Fall Short of Expectations." *International Security*, Vol. 46, No. 2 (2021), 51–90; and Lee, Melissa M. *Crippling Leviathan: How Foreign Subversion Weakens the State*. Cornell University Press, 2020.
32 See William Rosenau, "Subversion and Insurgency." *RAND Counterinsurgency Study, Paper* 2. Santa Monica, CA: RAND Corporation, 2007.
33 See Rid, Thomas. *Cyber War Will Not Take Place*. Oxford, UK: Oxford University Press, 2013.
34 See Rosenau, "Subversion and Insurgency," p. 6.
35 See Lathem, Niles, "Qaeda Claim: We 'Infiltrated' UAE Government." *New York Post*, February 25, 2006.
36 See, for instance, "Protesters Storm Baghdad's Green Zone Again, Dozens Hurt." *Thompson Reuters*, May 20, 2016.
37 See *inter alia* Inspectors General, *Interagency Assessment of Iraq Police Training*, Washington, DC: U.S. Department of State and U.S. Department of Defense, July 2005; and "Insurgents 'Inside Iraqi Police'." *BBC News*, September 21, 2005.
38 See Pike, Douglas. *Viet Cong: The Organization and Techniques of the National Liberation Front of South Vietnam*. Vol. 7. Cambridge, MA: MIT Press, 1966. Also see Prados, John, "Impatience, Illusion, and Asymmetry: Intelligence in Vietnam." In Marc Jason Gilbert, ed., *Why the North Won the Vietnam War*. New York: Palgrave, 2002; U.S. Information Service, Office of Policy and Research, "The Viet Cong: The United Front Technique." R-13–67, Record 128321, Douglas Pike Collection: Unit 06 – Democratic Republic of Vietnam, April 20, 1967; and U.S. Central Intelligence Agency (CIA), Directorate of Intelligence, "The Vulnerability of Non-Communist Groups in South Vietnam to Political Subversion." record 31052, CIA Collection, May 27, 1966.
39 See Marighella, Carlos. *Minimanual of the Urban Guerrilla*. Seattle, WA: Praetorian Press LLC, 2011; and Molnar, Andrew R., *Undergrounds in Insurgent, Revolutionary, and Resistance Warfare*. Washington, DC: Special Operations Research Office, November 1963.
40 For a good overview of the nexus of cybercrime and terrorism, see Taylor, Robert W., Eric J. Fritsch, and John Liederbach. *Digital Crime and Digital Terrorism*. Upper Saddle River, NJ: Prentice Hall Press, 2014. Also see McGuire, Mike, and Samantha Dowling. "Cyber Crime: A Review of the Evidence." *Summary of Key Findings and Implications. Home Office Research Report*, Vol. 75 (2013).
41 Lewis, James, and Stewart Baker. *The Economic Impact of Cybercrime and Cyber Espionage*. New York: McAfee, 2013.
42 Ibid.
43 See McFate, Sean. *The Modern Mercenary: Private Armies and What They Mean for World Order*. Oxford, UK: Oxford University Press, 2017; Maurer, Tim. *Cyber Mercenaries*. Cambridge, UK: Cambridge University Press, 2018; and Maurer, Tim. " 'Proxies' and Cyberspace." *Journal of Conflict and Security Law*, Vol. 21, No. 3 (2016), 383–403.
44 See *inter alia* Berson, Thomas A., and Dorothy E. Denning. "Cyberwarfare." *IEEE Security & Privacy*, Vol. 9, No. 5 (2011), 13–15; Sigholm, Johan. "Non-State Actors in Cyberspace Operations." *Journal of Military Studies*, Vol. 4, No. 1 (2013), 1–37; and Applegate, Scott D. "Cybermilitias and Political Hackers: Use of Irregular Forces in Cyberwarfare." *IEEE Security & Privacy Magazine*, Vol. 9, No. 5 (2011), 16–22.
45 See Deibert, Ronald. *Black Code: Surveillance, Privacy, and the Dark Side of the Internet*. Toronto, ON: Signal, 2013.

46 Kloet, Jeroen. "Digitisation and Its Asian Discontents: The Internet, Politics and Hacking in China and Indonesia." *First Monday*, Vol. 7, No. 9 (2002).

Further readings

Beyer, Jessica L. *Expect Us: Online Communities and Political Mobilization*. Oxford, UK: Oxford University Press, 2014.

Deibert, Ronald J. "Subversion Inc: The Age of Private Espionage." *Journal of Democracy* 33, no. 2 (2022): 28–44.

Denning, Dorothy E. "Activism, Hacktivism, and Cyberterrorism: The Internet As a Tool for Influencing Foreign Policy." *Networks and Netwars: The Future of Terror, Crime, and Militancy* 239 (2001): 288.

Denning, Dorothy E. "Terror's Web: How the Internet Is Transforming Terrorism." In Y. Jewkes and M. Yar eds., *Handbook of Internet Crime*. Uffculme, UK: Willan Publishing, 2010, 194–213.

Diamond, Larry, and Marc F. Plattner, eds. *Liberation Technology: Social Media and the Struggle for Democracy*. Baltimore, MD: Johns Hopkins University Press, 2012.

Macdonald, Stuart, Lee Jarvis, and Simon M. Lavis. "Cyberterrorism Today? Findings From a Follow-on Survey of Researchers." *Studies in Conflict & Terrorism* (2019): 1–26.

Marsili, Marco. "The War on Cyberterrorism." *Democracy and Security* 15, no. 2 (2019): 172–199.

Maurer, Tim. *Cyber Mercenaries*. Cambridge, UK: Cambridge University Press, 2018.

Olson, Parmy. *We Are Anonymous*. London: Random House, 2013.

Sauter, Molly. *The Coming Swarm: DDOS Actions, Hacktivism, and Civil Disobedience on the Internet*. New York: Bloomsbury Publishing USA, 2014.

Shandler, Ryan, Michael L. Gross, Sophia Backhaus, and Daphna Canetti. "Cyber terrorism and Public Support for Retaliation—a Multi-Country Survey Experiment." *British Journal of Political Science* 52, no. 2 (2022): 850–868.

Shantz, Jeff, and Jordon Tomblin. *Cyber Disobedience: Representing Online Anarchy*. London: John Hunt Publishing, 2014.

Stevens, Tim. "Strategic Cyberterrorism: Problems of Ends, Ways and Means." In *Handbook of Terrorism and Counter Terrorism Post 9/11*. Edward Elgar Publishing, 2019.

Theohary, Catherine A. *Terrorist Use of the Internet: Information Operations in Cyberspace*. Philadelphia, PA: DIANE Publishing, 2011.

12 Norms, ethics and international law for offensive cyber operations

Norms, which are shared expectations of appropriate behavior, exist at various levels and can apply to different actors.[1] In the international arena, these nonbinding shared expectations can, to some degree, constrain and regulate the behavior of international actors and, in that sense, have a structural impact on the international system as a whole. For example, early in the age of nuclear weapons, Lt. Gen. James Gavin expressed the contemporary wisdom when he wrote, "Nuclear weapons will become conventional for several reasons, among them cost, effectiveness against enemy weapons, and ease of handling."[2] However, as the nuclear era advanced, a constraining norm developed that made states reluctant to possess or use nuclear weapons. International security and U.S. national security may be enhanced by the emergence of regulative norms for offensive cyber operations (OCOs), similar to norms that developed in the past for these emerging-technology weapons such as nuclear, chemical and biological weapons. In February 2022 Director of National Intelligence Avril Haines testified that many actors China, Russia, Iran, North Korea and many other lessor actors are pursuing cyber means to achieve high impact and even strategic-level effects.[3] This is likely due in part to the current lack of consensus on constraining international norms, which will be discussed in detail in this chapter. Previously, former Directors of National Intelligence testified that the growing international use of these emerging-technology weapons to achieve strategic objectives was outpacing the development of a shared understanding or norms of behavior and thus increasing the prospects for miscalculations and escalation.[4]

Today, relatively early into the age of cyber conflict, many hold a view regarding the inevitability of significant use of force in cyberspace like that held early in the nuclear era. In expectation that norms will emerge, in May 2011 the Obama administration issued the *International Strategy for Cyberspace*.[5] One pillar of this strategy recognizes the "borderless" international dimension of cyberspace and identifies the need to achieve stability and address cyber threats through the development of international norms. In 2013 Michael Daniel, the former White House cybersecurity coordinator, told computer security practitioners that diplomacy, including fostering international norms and shared expectations, is essential to prevent OCOs against U.S. economic interests. More recently, the Biden administration's Interim National Security Strategic Guidance says the administration will renew the U.S. "commitment to international engagement on cyber issues, working alongside . . . allies and partners to uphold existing and shape new global norms in cyberspace."[6] This chapter discusses how constraining norms for OCOs are actually developing and offers some predictions – based on norm evolution theory for emerging-technology weapons – for how they will develop in the future.[7]

DOI: 10.4324/9781003246398-12

This chapter does so by first introducing key concepts regarding norms and international law, then examining available evidence that offers clues as to the current state of norms, followed by a discussion of current organizational platforms and norm leaders' efforts to reach consensus and codify norms, and finally offering some predictions and conclusions based on norm evolution theory for emerging-technology weapons.

Key concepts

Norms are standards of right and wrong that form a prescription or proscription for behavior.[8] Essentially norms are nonbinding, shared expectations that can be helpful in constraining and regulating the behavior of international actors and, in that sense, have a structural impact on the international system. International norms cover a wide range of issues, from the practice of dueling to human rights. Specific to warfare, multiple regulative norms have emerged regarding particular categories of weapons and modes of warfare, such as weapons of mass destruction, strategic bombing, antipersonnel land mines, leadership assassination and dueling. Norms for weapons and warfare can affect a variety of functions and activities, such as weapon possession and use. While not always successful (with the demise of the constraining strategic bombing norm in World War II being perhaps one of the better examples), some of these norms for warfare have helped restrain the widespread development, proliferation and use of various weapons. Norms are obviously one of many variables impacting state behavior and the international system, with material state interests and power dynamics also playing a major, perhaps dominant role. Norms also provide the foundation for international law, which is the set of rules generally regarded and accepted as binding in relations between states. When international law is based on shared expectations of behavior, its source is considered "peremptory norms" or *jus cogens/ius cogens*. Consistent state practice or customs and eventually codification in agreements or treaties serve as additional sources of international law.

Indicators of early norms in the age of cyber conflict

There are some CNA-style OCOs that provide insight into the emerging customary practice of states and related emerging norms in regard to this most serious type of hostile cyber operation. Consciously or unconsciously, early cyber actors are acting as the early norm leaders as they help establish customary practice for hostile operations in cyberspace. Many small CNA-style operations involve DDoS attacks to degrade access to websites, such as the Code Red attack in 2001, which involved malware that launched a DDoS attack against White House computers.[9] It is believed that approximately 100 million to 150 million botnets are utilized to conduct these frequent DDoS attacks.[10] However, there are limited examples of major OCOs. Eleven are summarized in Table 12.1 for the purposes of evaluating the norms and contemporary state practice of OCOs: the purported attacks on a Siberian gas pipeline in 1982, the DDoS attacks on Estonia in 2007, the Israeli Operation Orchard attacks on Syria in 2007, the attacks on Georgia in 2008, the notorious Stuxnet attack on Iran disclosed in 2010, the Shamoon virus attack on Saudi Aramco in 2012, Izz ad-Din al-Qassam's Operation Ababil attack against financial institutions in 2012, North Korea's attack on Sony Corp. in 2014, the Russian attack on Ukrainian power utilities in 2015, the Russian NotPetya attacks on Ukraine in 2017 and the Russian attack on Ukrainian government websites in early 2022 as a precursor to the Russian invasion of Ukraine. The table identifies the nature and target of the attacks.

Table 12.1 Selected cyber attacks and implications for norm emergence

Attack Name	*Date*	*Target*	*Effect*	*Suspected Sponsor*
Trans-Siberian Gas Pipeline	June 1982	Soviet gas pipeline (civilian target)	Massive explosion	United States
Estonia	April–May 2007	Commercial and governmental web services (civilian target)	Major denial of service	Russia
Syrian Air Defense System as part of Operation Orchard	September 2007	Military air defense system (military target)	Degradation of air defense capabilities allowing kinetic strike	Israel
Georgia	July 2008	Commercial and governmental web services (civilian target)	Major denial of service	Russia
Stuxnet	Late 2009–2010, possibly as early as 2007	Iranian centrifuges (military target)	Physical destruction of Iranian centrifuges	United States
Saudi-Aramco	August 2012	State-owned commercial enterprise (civilian target)	Large-scale destruction of data and attempted physical disruption of oil production	Iran
Operation Ababil	September 2012–March 2013	Large financial institutions (civilian target)	Major denial of service	Iran
Sony	November 2014	Commercial networks and data (civilian target)	Large-scale disruption of access to networks and loss of confidentiality of information due to public disclosures	North Korea
Ukrainian Power Utilities	December 2015	Civilian critical infrastructure (civilian target)	Temporary loss of power for 230,000 civilians, permanent damage to SCADA systems	Russia

These 11 cyberattacks collectively provide some insight into the emergence of international norms through the customary practice of OCOs. There are three main conclusions from the attacks. First, the majority (9 of 11) of the attacks were aimed at civilian targets, showing that a norm constraining targeting to explicitly military targets or objectives has not yet arisen. Second, to the extent attacks did strike exclusively military targets, they were suspected to have been launched by Western nations (the United States and Israel). This seems to indicate there may be competing, and in some cases more permissive norms regarding OCOs depending on which nation is associated with it. This is consistent with the expected competitive environment in the early days of norm emergence. Third, experience with OCOs is very limited at this point. No known deaths or casualties have yet resulted from cyberattacks, and the physical damage caused, while impacting strategically significant items such as Iranian centrifuges or Ukrainian power utilities, has not been particularly widespread, permanent or severe.

Norms and recent examples of total restraint

In addition to actual attacks, decisions *not* to employ OCOs also indicate an emerging customary practice and potential nascent norm. Fear of collateral damage of civilian targets and an inability to discriminately target military objectives have led some OCO plans to be called off. For example, in advance of a physical invasion in 2003, the United States was planning a massive CNA-style cyberattack on Iraq to freeze bank accounts and cripple and disrupt government systems. Despite possessing the ability to carry out such attacks, the Bush administration canceled the plan out of a concern that the effects would not be contained to Iraq but instead would also have a negative effect on the civilian networks of allies across the region and in Europe.[11] In 2011, the United States allegedly considered using cyber weapons to disrupt Libya's air defenses but chose not to, in part due to the limited time available and greater suitability of conventional weapons to achieve the desired effects.[12] The United States again declined to launch kinetic or cyberattacks against the Syrian regime in August and September 2013, in part due to concerns about causing unintentional collateral damage as well as concerns regarding Syria's and Iran's ability to retaliate in cyberspace against U.S. banks and other targets (which Iran did following Stuxnet).[13] And as of late spring of 2022, Russia had not launched major OCOs as part of its ongoing invasion of Ukraine. This restraint offers a glimmer of hope for the emergence of constraining norms for OCOs, although other factors and practical considerations likely played a role in these examples of nonuse. Healey identifies this restraint against engaging in "full-scale strategic OCOs" as a de facto norm.[14] That said, no state has protested any of these CNA-style cyberattacks as a violation of international law, although Georgia did protest the ongoing physical invasion of their country by Russian forces.[15] This further suggests that this level of OCO is permitted by existing norms.

Overall, it appears cyber conflict is becoming more destructive, remaining largely covert with limited public discussion, involving an increasing and continued mix of state and non-state actors, and more U.S., Russian, Chinese and Iranian offensive cyber operations (among others).[16] More destructive and sophisticated cyber weapons are likely in part due to the success and example provided by Stuxnet and the interest in and proliferation of cyber weapons it has spawned along with the absence of constraining norms on developing such weapons. The United States, in spite of its interest in developing constraining cyber norms, has continued to pursue secretive military and intelligence CNA capabilities since the 2000s.[17] Thus, cyber conflict capabilities will play an increasingly decisive role in military conflicts and are becoming deeply integrated into states' doctrine and military capabilities. Many countries have taken steps to incorporate OCO capabilities into their military planning and organizations, and the use of OCOs as a "brute force" weapon is likely to increase.[18]

Organizational platforms and norm leaders' efforts to reach consensus

While the preceding information makes it apparent that few, if any, normative constraints governing OCOs exist, increased attention and discussion have helped spur efforts to reach a consensus on and codify emerging norms for OCOs. Norm evolution theory indicates that norm emergence is more likely to occur when norm entrepreneurs with organizational platforms and key states acting as norm leaders are involved.

Box 12.1: Cyberspace and the law of armed conflict: the *Tallinn Manual*

When selecting material to include in this book, we opted to write this chapter on the nature of constraining norms around conflict and the likelihood that cyber conflict norms might come into existence for a few reasons. More than anything else, we sought here to write about cyber warfare and not simply about what many have called cyberpolitics or just Internet governance. As a result, it seemed more important that we describe international effects pertaining to conflict than simply focus, as many other resources do, on how the networked world is governed and regulated.

Secondarily, we chose to focus on norms here instead of the formal trappings of international law, because it is the case that the applicability of international law to cyber conflict issues quite simply remains a hazy area at this time. There are, at present, only two major cybersecurity agreements between states that seek to constrain particular kinds of behavior. Both focus on criminal activity. Other bilateral behavior where there has been agreement on the relevance of particular legal standards has largely taken the form of informal arrangements, often arranged at leadership summits behind closed doors. In short, while it is certainly true that there are obvious issue areas of legal applicability to discuss, there is not much of a landscape to overview at this time.

That said, it would not do to move through this chapter without at least some notion of the close-in issues of cyber conflict in the eyes of international law. We do that in this and subsequent breakout boxes across the rest of this chapter.

Broadly, international law as it pertains to interstate warfare takes the form of the **law of armed conflict** (LOAC, or law of war), which is constituted of all elements of international law that pertain to the conduct of states during wartime and the role of both combatants and non-combatants. With cyber conflict, the landmark effort to assess the applicability of the LOAC has taken the form of the *Tallinn Manual* (and an updated version). This document is a nonbinding effort by practitioners from NATO countries to assess how certain forms of cyberattack – disruptive, degradative, etc. – should be either permitted or constrained under international law.

Specifically, the *Tallinn Manual* attempts to slot what is known about different forms of cyber operations into the framework provided by **just war theory**. Just war theory is a doctrine on the ethical use of force that serves as the benchmark and progenitor framework for much international law as it appears in the **Geneva and Hague Conventions**, in the wording of the United Nations Charter, etc. The doctrine addresses criteria that should be consulted both prior to going to war (*jus ad bellum*) and during wartime (*jus in bello*). Criteria that must be met for a war to be considered justified include the need for just cause, the exhaustion of other options, a probability of success, and that a decision to wage war be made legitimately. For a war to be fought justly, the use of force should be strategically appropriate, proportionate, not cruel, and not against those who surrender.

Important to note, the *Tallinn Manual* does not necessarily attempt to address all cyber-enabled conflict. As such, its scope is limited and, though one might think from reading Chapter 9 in this book that such a document could not possibly be written with realistic conditions of interstate conflict in mind, is defensible in its remit. In the remaining breakout boxes in this chapter, we address a few specific prominent issues that are proving particularly difficult to grapple with in the international arena.

The two primary intergovernmental bodies and organizational platforms currently being used to promote emerging norms for OCOs are the UN and NATO. Other key multilateral efforts to encourage the development of cyber norms are the London Conference on Cyberspace (and subsequent conferences) and academic cyber norm workshops. Efforts in the UN have initially primarily been led by Russia although the United States and other western countries are now playing a more active role, while efforts in NATO have been led by the United States. In addition to these two main forums focused on OCOs, the EU's Council of Europe Convention on Cybercrime is a regional yet important treaty that came into force in 2004.[19] The convention criminalizes non-state cybercrime and obliges state parties to prevent non-state actors from launching cyberattacks from their territory.[20] Additionally, the UN has organized a series of events under the umbrella of the World Summit on the Information Society, which, like the EU's efforts, includes actions against cybercrime.[21] However, these other efforts are only indirectly focused on the issue of OCOs conducted between nation states.

The UN as an organizational platform for cyber norm emergence

Since the UN Charter entered into force in 1945, international law and norms have been based on Article 2(4), which directs that "all Members shall refrain in their international relations from the threat or use of force against the territorial integrity or political independence of any state, or in any other manner inconsistent with the purposes of the UN."[22] How do OCOs fit into this construct? Within the UN the main focus on OCOs has occurred in the UN General Assembly's First Committee (Disarmament and International Security Committee) as well as various subsidiary organs and specialized agencies, particularly the ITU, UNIDIR and the CTITF working group.[23] Serious focus on OCOs began in 1998 with the Russian resolution in the First Committee, "Developments in the Field of Information and Telecommunications in the Context of Security," to establish cyber arms control similar to other arms control agreements.[24] Richard Clarke, a former senior official who led the U.S. government opposition to the treaty, "viewed the Russian proposal as largely a propaganda tool, as so many of their multilateral arms control initiatives have been for decades."[25] The U.S. position is that the laws of armed conflict (LOAC) apply to state behavior in cyberspace, and so a prohibition on offensive cyber weapons is unnecessary.[26] Interestingly, China, another key actor in cyberspace, has largely been quiet on the Russian proposals and efforts within the UN to develop and codify cyber norms, and has supported them only in recent years.[27] While the Russian proposal in 1998 was adopted by the General Assembly without a vote every year between 1998 and 2004 (a sign of a lack of consensus and weak reception), in 2005 a vote was taken; 14 other nations, including China, signed on as co-sponsors, and the United States was the only country voting against the resolution.[28] The proposal established a group of government experts (GGE) in 2004, which raised the profile of the issue of OCOs but initially failed to achieve consensus on whether the LOAC were sufficient to address the threat. Perhaps due to the cyberattacks in Estonia and Georgia, the Russian proposal reverted back to being adopted without a vote in 2009.[29] Then, in 2010, the United States reversed its opposition and supported adoption, perhaps a tactical move by the Obama administration's "reset" with Russia.[30] The United States also began co-sponsoring the proposal in 2010, along with 35 other nations. Clearly momentum is building.

Box 12.2: Protected entities in cyberspace?

One major issue under constant debate in international conversations about the applicability of the LOAC to cyberspace is the status of protected entities and the practicality of awarding a protected status. Traditionally, there is a category of actors and types of infrastructure that would be considered as off-limits for targeting during a conventional conflict. This would include civilian infrastructure associated with a particular humanitarian prohibition, such as a hospital or a school. The question with regard to cyber conflict is fairly simple: should these targets be off-limits as targets of CNA?

The answer that all states involved in debate over the applicability of international law to cyber conflict is yes. However, the practicalities of the matter are not clear cut. What constitutes an attack on the networks and computers at such institutions? Surely if there is a deleterious effect on the operations of, say, a hospital from a cyberattack, then offensive action can be said to have taken place. However, is it the case that a cyberattack has occurred if a computer on a school or library network has been compromised, not for disruptive purposes but to enable further intrusive activity elsewhere? As we discussed in another chapter, both the early Cuckoo's Egg and Moonlight Maze cases included initial attacks against secondary targets for the purpose of gaining access and then elevating the attack toward other targets. Russian hackers in the late 1990s compromised a library computer in an effort to evade the detection of the FBI, while East German hackers in the 1980s attacked the Berkeley National Labs before jumping off into the **MILNET**. Were these the kinds of attacks covered by international law?

Even beyond the definitional issue, there is the challenge of designating protected entities such that they would not be attacked. In the physical world, a standard approach would be to use an emblem or badge that denotes an entity's protected status. Russian and U.S. interlocutors have suggested that hospitals, schools, etc., might consider using specialized domain registries to do the same thing (i.e. to be "generichospital.un" or "generichospital.protected"). The problem here is that not all network activity related to the business of protected entities occurs on local networks (e.g. inside a hospital or school structure). Would distance learners not enjoy the same protections in an ideal world? What about medical telepresence, where doctors use Skype- or FaceTime-like applications to provide medical services to remote locations? Would those activities not be protected? Again, the answer is obviously yes. But how would their protected status be noted in a meaningful fashion? Furthermore, who bears the costs of monitoring such actions and enabling such protections? These questions are, at present, unresolved.

The draft UN cyber resolution the United States supported in 2010 lacked the important reference to the need to develop definitions of key terms, which would be the first real step in developing a cyber arms control treaty.[31] While momentum in favor of the Russian proposal is clearly building, this change also weakens the tangible effect of the resolution insofar as it could actually lead to a binding agreement for OCOs. In addition to this annual cyber resolution, in 2011 Russia and China offered an additional proposal, "International Code of Conduct for Information Security," which has the "aim of achieving the earliest possible consensus on international norms and rules guiding the behavior of States in the information space."[32] This proposed code, which has 11 main points, seems to tack back against the United States and Western concerns regarding Russian interests in limiting Internet freedom by prohibiting not only CNA-style attacks but also "information warfare" and the free exchange of ideas.[33]

In addition to growing support for the Russian cyber resolution (as amended) in the First Committee, there are other indicators of activity to address OCOs in the UN. Unlike the 2004 GGE, which failed to achieve consensus, more recent GGEs have had far greater success. In 2010 and 2013, the GGEs established by the First Committee cyber resolutions were able to achieve consensus and generate several recommendations.[34] For example, the 2010 GGE consensus report called for a sustained dialogue on "norms for state use of information and communications technologies to reduce risk and protect critical infrastructures" and "confidence-building and risk reduction measures, including discussion of information and communication technologies in conflict."[35] The GGE's 2013 report broke new ground in affirming that "the application of norms derived from existing international law" was relevant to OCOs and "essential to reduce risks to international peace, security and stability." This seems to represent a break toward the long-standing U.S. position that existing international law and agreements regarding the use of force are sufficient to address the new challenge of OCOs. The 2013 GGE report also recommends additional study to promote shared understanding regarding how this existing international law and norms apply to state behavior in cyberspace given the unique characteristics of OCOs, noting that "additional norms could be developed over time." This last component could be a nod toward the Russian position that new and more constraining and specific norms (such as an outright ban or prohibition on first use) could eventually be adopted. Finally, the 2013 GGE offered voluntary confidence-building measures (CBMs) to promote trust, increase predictability, and reduce misperception. These include measures such as exchanging views and information on national strategies, policies and organizations as well as the creation of bilateral and multilateral consultative frameworks such as seminars and workshops.[36] In 2016, the Organization for Security and Cooperation in Europe (OSCE) also agreed to a range of cyber CBMs.[37] This echoes the conclusions of the Security Defense Agenda's 2012 report, which called for cyber CBMs as an alternative to a global treaty or at least as a near-term stopgap measure.[38] ICT4Peace, an international organization that spun off the UN's World Summit on the Information Society activities, has taken a leadership role in developing cyber CBMs and issued a report in October 2013 identifying a process to do so.[39] The 2015 GGE reaffirmed the discussion on CBMs in the 2013 GGE report and made some modest strides toward the U.S. position through the inclusion of a mention that states should respond to requests for assistance and refrain from cyber activity that intentionally damages or impairs CI or CERTs.[40] However, there is no firm definition of CI, which limits the utility of this new language. Later GGEs made no additional major progress. Additionally, the fact that the 2015 GGE report also included language asserting that the fact that an attack originated from a state's territory is insufficient for attributing the attack to the state, may be a step backward in regard to constraining norms that inculcate responsible behavior in cyberspace. The 2017 GGE failed to reach any consensus. In 2021, a major milestone was achieved when the GGE's 2015 effort to develop norms achieved a significant endorsement by all UN member states participating in the Open-Ended Working Group (OEWG), where an agreement was reached on a framework for responsible state behavior in cyberspace. The establishment of the OEWG after numerous GGEs is considered the beginning of the UN's "parallel" or "two-track" approach to cyber norms. The noteworthy distinction of the OEWG is it was open to all member states, rather than just the 15–25 that were part of each GGE. The OEWG also allowed some non-state actors, such as NGOs and academics, to participate as well. Many of the OEWG-endorsed norms focus on protecting critical infrastructure and are considered legally binding, however there is still no accountability mechanism to take action against those violating the framework. In addition to the efforts of the First Committee, the ITU, UNIDIR and CTITF working group have also taken steps to promote the emergence of norms for OCOs. Of these, the ITU, as a treaty-level UN organization, is perhaps the most significant organizational platform for norm emergence. The ITU, partnering with the World Federation of Scientists, initially approached OCOs at the request of various states but has since acted autonomously as a norm entrepreneur in pursuit of its own "cyber peace" agenda seeking to prevent the use of force in cyberspace.[41] The ITU secretary-general submits quarterly cyber threat assessments to the UN secretary-general and maintains a database of experts to be consulted in the event of a major cyberattack.[42] Many of the ITU's cyber efforts are focused more on general cyber hygiene and cybercrime, but it has advocated for norms related to OCOs. At the 2010 World Telecom Development Conference, the ITU secretary-general proposed a "no first attack vow" for OCOs as well as an obligation of states to prevent independent or non-state attacks from originating from their territory.[43] With the World Federation of Scientists, the ITU helped develop various declarations, such as the Quest for Cyberpeace, which advocate for these two norms.[44] Finally, the ITU also maintains a global cybersecurity index to help measure the cybersecurity capabilities of nation states.

UNIDIR has also played a role in fostering cyber norms. Germany has acted as a key norm leader by sponsoring ongoing UNIDIR research titled "Perspectives on Cyber War: Legal Frameworks and Transparency and Confidence Building."[45] Russia previously sponsored much of UNIDIR's OCO-related activities. UNIDIR's effort seeks to raise general awareness of OCOs and generate multilateral discourse through publications and various meetings and conferences. UNIDIR staff also serve the GGE sessions supporting the First Committee's efforts. Additionally, in order to advance transparency that is essential to norm development, the UNIDIR Security and Technology Program has led the development of a series of research papers outlining national capabilities to conduct international cyber operations and related national doctrines. As of 2022, this has included research focused on 15 countries from multiple regions.[46]

NATO as an organizational platform for cyber norm emergence

Following the major cyberattacks on Estonia (a NATO member) in 2007 and Georgia (a then aspiring NATO member) in 2008, NATO began to focus more seriously on the threat of OCOs.[47] In 2008 NATO established the NATO CCD COE, located in Tallinn, Estonia.[48] Its mission is to enhance NATO's cyber defense through research, education and consulting. In 2012 the organization published the *National Cyber Security Framework Manual* to help member nations better develop national policies for cyber defense. In 2016, the organization published *International Cyber Norms: Legal, Policy, and Industry Perspectives* to assist efforts to agree on common norms in the cyber domain.[49] NATO's commitment to addressing OCOs extends beyond this center of excellence. In November 2010, NATO adopted a new strategic concept which recognized that OCOs "can reach a threshold that threatens national and Euro-Atlantic prosperity, security and stability."[50] Additionally, NATO and its partners regularly participate in large-scale cyber defense exercise. In general, NATO, led by the United States, has approached OCOs from a perspective that seeks to apply the existing LOAC to cyberattacks rather than pursue more comprehensive and new restrictions like those proposed by Russia in the UN.

Box 12.3: Should non-state netizens be involved or considered?

As previous chapters have discussed, non-state actors – as well as what we've called "**semi-state**" actors, such as cybersecurity vendors or backbone operators – are immensely important stakeholders in cyberspace. Traditionally, the LOAC is applied only to state actors. Given the relative asymmetry of power in cyberspace, two issues arise.

The first – and, we would argue, lesser – issue pertains to the applicability of legal standards bound up in the LOAC for responsible stakeholders. We return here to the notion of contested sovereignty in the digital age where a number of semi-state stakeholders have unprecedented power to block state cyber conflict actions and an unusual potential claim to legitimacy in doing so if such actions go against public interests. Given this dynamic, should international law apply to such actors and compel them to interfere in state cyber operations if, for instance, they observe a violation of the stipulation that protected entities should not be targeted? In other words, if a state actor is observed (given some high degree of attribution) to attack a hospital in order to create a stepping stone toward a more disruptive attack, should an ISP intervene and block the attack?

Second, how should states treat non-state actors caught hacking state systems? Common Article 3 of the Geneva Convention outlines humanitarian protections to be given to prisoners of war (POWs). Following the events of September 11, 2001, the United States argued that the article did not apply to certain terrorist actors as they had disavowed national allegiances and had engaged in hostile conflict actions. Therefore, they did not have the same rights as POWs. Would hackers acting as a proxy for state authorities fall into the same category? States often use proxy agents to extend their plausible deniability, to take advantage of external resources and to engage a broader political movement. Part and parcel of this kind of activity, however, is a disavowal of the use of proxy hackers. So, if a hacker claimed to be acting on behalf of a state but that state denies the claim, does international law apply?

NATO's most important activity in this effort was the development of the *Tallinn Manual on the International Law Applicable to Cyber Warfare*. The *Tallinn Manual*, which does not reflect official NATO opinion, but rather the personal opinion of the authors (an "international group of experts"), was sponsored by the NATO CCD COE and three organizations acting as observers: NATO, U.S. Cyber Command and the International Committee of the Red Cross.[51] Also noteworthy is an independent yet similar effort by Israel, led by Col. Sharon Afek, which reached similar conclusions regarding the LOAC and OCOs in early 2014.[52] The *Tallinn Manual* represents not only the consensus view of these NATO-affiliated participants but also the main positions of the U.S. government.[53] This is based on a September 2012 speech by a U.S. State Department legal advisor, Harold Koh, who articulated the U.S. positions on international law and cyberspace, which are consistent with the positions articulated in the *Tallinn Manual*.[54] In addition, the 2011 "International Strategy for Cyberspace" specified that the "long-standing international norms guiding state behavior – in times of peace and conflict – also apply in cyberspace."[55] Both the Koh speech and the *Tallinn Manual* go further to flesh out the U.S.-NATO position that the international LOAC are adequate and applicable to OCOs and reject the Russian position that OCOs require new and distinct international norms and agreements.

This argument was made in the past by other U.S. leaders, including Gen. Keith Alexander, commander of U.S. Cyber Command, who testified that military operations in cyberspace "must comply with international law that governs military operations," essentially equating cyber weapons with guns and bombs and implying that the rules that apply to those also apply to OCOs.[56] The interpretation of the LOAC to OCOs can be challenging, and a consensus on application needs to be developed.[57] This lack of clarity is similar to the confusion over definitions and applications of norms for airpower in the early part of the 20th century. The following areas of application are currently being resolved: What constitutes the use of force in cyberspace? What constitutes an armed attack in cyberspace? What constitutes legitimate military objectives in cyberspace? How does the principle of distinction and noncombatant immunity apply? What role does state sovereignty play in cyberspace?[58] Both the United States (as interpreted through the Koh speech) and the group of experts that developed the *Tallinn Manual* attempt to clarify these issues. For example, they identify cyberattacks that "result in death, injury, or significant destruction" as the use of force in cyberspace, and they determined that "whether a cyber use of force qualifies as an armed attack" depends on its "scale and effects" and that cyberattacks must exercise distinction and be aimed at legitimate military objectives, although defining a military objective in cyberspace is particularly complicated.[59] This last point is an area where the United States may stand apart from the authors of the *Tallinn Manual* in supporting a fairly broad definition of military objectives that includes "war-sustaining" objectives in addition to "war-fighting" objectives.[60] Media response to the *Tallinn Manual* has varied. Some have misinterpreted and twisted its conclusions to claim that it justifies "killing hackers."[61] However, it represents a significant step forward in developing the emerging OCO norm tied to the existing warfighting norms codified in the LOAC. Today all major powers except China agree that the LOAC apply to OCOs.[62]

Other multilateral forums acting as organizational platforms for cyber norm emergence

The United Kingdom has been the most active individual norm entrepreneur. Its Foreign and Commonwealth Office hosted the London Conference on Cyberspace in November 2011 to "discuss the vital issues posed for us all by a networked world connected ever more closely together in cyberspace." The conference involved over 700 participants from 60 nations, including then U.K. Prime Minister David Cameron and then U.S. Vice

President Joseph Biden. Among the many cyber topics addressed, discussions included international security and OCOs. The conference chair reported that:

> [a]ll delegates underlined the importance of the principle that governments act proportionately in cyberspace and that states should continue to comply with existing rules of international law and the traditional norms of behavior that govern interstate relations, the use of force and armed conflict, including the settlement by states of their international disputes by peaceful means in such a manner that international peace, security and justice are not endangered.

The participants agreed the next steps should be to focus on further developing "shared understanding" (norms) through efforts such as the UN First Committee's GGE and other organizations. However, the participants did not have the "appetite . . . to expend effort on legally-binding international instruments."[63] Following the conference in London, a follow-up conference was held in Budapest in October 2012, and then in Seoul in October 2013.[64] The conference in Seoul had over 1,000 participants from approximately 90 countries and generated the "Seoul Framework for and Commitment to Open and Secure Cyberspace" as well as plans for the fourth conference held in 2015 in The Hague.[65] The "Seoul Framework" reaffirmed the conclusion of the London conference that existing norms and international law apply to OCOs and that states should prevent non-state actors from launching attacks from their territory. The framework also noted that additional norms could be developed over time.[66]

In addition to this U.K.-initiated global multilateral effort to address cyber issues, including OCOs, regional multilateral efforts have included efforts at the Organization for Security and Cooperation in Europe and the Association of Southeast Asian Nations Regional Forum.[67] These groups began discussions on "cyber confidence-building measures" heading into 2014.[68] CBMs have been used to reduce uncertainty and potential for miscalculation with other weapons. As part of the 1986 Second Review Conference of the BWC, states agreed to implement a number of CBMs in order to "prevent or reduce the occurrence of ambiguities, doubts and suspicions and in order to improve international co-operation in the field of peaceful biological activities."[69] Another example of a regional multilateral effort to cultivate norms for OCOs is a group of government leaders and security experts (80 senior military officials from 23 Asia-Pacific nations), met in Seoul less than a month after the 2013 Seoul Cybersecurity Conference for the Seoul Defense Dialogue, which also addressed OCOs.[70] Chinese professor Jia Qingguo summarized the results of the defense dialogue, saying that clearer definitions of OCOs and rules of engagement were needed and that:

> [t]hose who engage in OCO should make necessary efforts to avoid attacking civilian infrastructures that may harm the civilians . . . [and] major countries should work together to develop an agreement on a code of conduct in cyberspace and a set standards.[71]

There are also numerous bilateral dialogues that involve key cyber actors, such as China, the United States and Russia.[72] This includes activities such as the cyber agreement Russia and the United States signed in June 2013, which established a communication hotline for a cyber crisis. In light of North Korea's increasingly bellicose cyber posture, South Korea and the United States established a Cyber Cooperation Working Group in early 2014 to discuss cooperation on OCO issues.[73]

Predicting norm evolution going forward

OCOs pose a real and growing threat, a threat that is growing faster than the development of constraining international norms, increasing the prospects for miscalculations and escalation.[74] Some scholars think that great powers will inevitably cooperate and establish rules, norms and standards for cyberspace.[75] While it is true that increased competition may create incentives for cooperation on constraining norms, norm evolution theory for emerging-technology weapons leads one to conclude that constraining norms for OCOs will face many challenges and may never successfully emerge.

Some of these challenges were also presented by the advent of the other emerging-technology weapons, in historic cases such as chemical and biological weapons, strategic bombing and nuclear weapons. An analysis of these three historic examples offers valuable lessons that led to the development of norm evolution theory tailored for emerging-technology weapons, which can then be applied to OCOs to better evaluate whether or not the authors' conclusions are well founded. This chapter does exactly that, first by defining emerging-technology weapons and norm evolution theory, then briefly reviewing the current state of international norms for OCOs. Next it illustrates norm evolution theory for emerging-technology weapons – grounded in the three historic case studies – and prospects for current norms between China, Russia and the United States. Third, it presents a refined theory of norm development as a framework to evaluate norm emergence that contradicts the authors' thesis. This argument leads to the conclusion that a constraining international order in cyberspace is far from inevitable.

Emerging-technology weapons and norm evolution theory

Emerging-technology weapons are weapons based on new technology or a novel employment of older technologies to achieve certain effects. Given that technology is constantly advancing, weapons that initially fall into this category will eventually no longer be considered emergent. For example, the gunpowder-based weapons that began to spread in 14th century Europe would clearly be classified as emerging-technology weapons in that century and perhaps in the 15th century, but eventually these weapons were no longer novel and became fairly ubiquitous.[76] Chemical weapons up to the early 20th century could be considered an emerging-technology weapon. Likewise, strategic bombing up to World War II also falls into this category. Nuclear and biological weapons could be considered emerging-technology weapons during World War II and the immediate years that followed. Today cyber weapons used to conduct CNA are emerging-technology weapons. In general, norm evolution theory identifies three major stages in a norm's potential life cycle. These three stages are (1) norm emergence, (2) norm cascade and (3) norm internalization.[77] The primary hypothesis of norm evolution theory for emerging-technology weapons is that a state's self-interest will play a significant role, and a norm's convergence with perceived state self-interest will be important to achieving norm emergence and a state acting as a norm leader. It further states that norms are more likely to emerge when vital actors are involved, specifically key states acting as norm leaders and norm entrepreneurs within organizations. The two primary intergovernmental bodies and organizations currently being used to promote emerging norms for OCOs are, as discussed earlier in this chapter, the UN and NATO. Additionally, there are some other key multilateral efforts to encourage the development of cyber norms, such as the London Conference on Cyberspace and academic cyber norm workshops.

The case for norm evolution theory

What does norm evolution theory for emerging-technology weapons predict regarding the development of constrictive international norms? The three examples of chemical and biological weapons, strategic bombing and nuclear weapons are particularly salient historic case studies when considering norm evolution for OCOs due to a variety of reasons.

Chemical and biological weapons and cyber weapons are both non-conventional weapons that share many of the same special characteristics with significant international security implications. They include challenges of attribution following their use, attractiveness to weaker powers and non-state actors as asymmetric weapons, use as a force multiplier for conventional military operations, questionable deterrence value, target and weapon unpredictability, the potential for major collateral damage or unintended consequences due to "borderless" domains, the multi-use nature of the associated technologies and the frequent use of covert programs to develop such weapons.[78] Due to these characteristics, both of these weapons are also attractive to non-state actors or those seeking anonymity resulting in a lack of clarity regarding the responsible party.

Strategic bombing – particularly with the advent of airpower as an emerging-technology weapon and the early use of airplanes to drop bombs on cities – forced states to grapple with a brand new technology and approach to warfare; as is now the case with OCOs. As with chemical and biological weapons, strategic bombing shares some special characteristics with OCOs. Strategic bombing made civilian populations highly vulnerable, was difficult to defend against and used technology which also had peaceful applications (air travel and transport) – all of which can also be said about OCOs today. The effort to constrain strategic bombing through normative influences was mixed and at times completely unsuccessful, which makes it particularly well suited as an exemplar of the limits of norms and how other factors may impede or reverse norm development.

Finally, nuclear weapons, like airpower before them and perhaps cyber weapons today, presented states with a challenge of a completely new and emerging war fighting technology. Nuclear weapons and cyber weapons, like the other emerging-technology case studies, share many of the same special characteristics with significant international security implications. These include the potential for major collateral damage or unintended consequences (due to fallout, in the case of nuclear weapons) and covert development programs. Because of these common attributes, lessons regarding norm development can be learned and a framework developed that is applicable to predicting the prospects of constraining norms as a tool to address the use of cyber weapons.

Box 12.4: Restricting cyber "weaponry": what would qualify?

A final point worthy of discussion regarding the applicability of international law to cyber conflict revolves around the notion that some weapons of war are inhumane and cruel and therefore subject to a global moratorium on their use. Generally, banned weapons are those that cause unnecessary suffering or widespread and long-term damage to the natural environment. The list is long and its contents would be largely unsurprising to the average global citizen, including various kinds of booby trap, anti-personnel landmines, poisonous gases, dum-dum bullets and various forms of incendiary weapons.

The obvious question here is whether or not certain forms of cyber "weaponry" should be included under international law. The obvious problem is that such weaponry is not uniform. At the heart of the thing, we're talking about code that can be maliciously weaponized and employed to nefarious effect. As such, cyber analogues to the kinds of weapons described here are unclear. One might certainly conceive of a scenario wherein a cyberattack leads to the release of, say, nuclear or otherwise toxic waste that causes massive damage to natural habitats. But the form of the attack, tailored as it would surely be in the details, is not substantially different from what might be used to less inhumane effect elsewhere. If the answer is that certain effects should be off-limits, then how do we define such effects so as to capture all future threat scenarios effectively? If the answer is that certain targets should be off-limits, we have the same issues of protected status that we have previously discussed.

Examining how norm evolution theory, informed by the three historic case studies mentioned, specifically applies to norms for emerging-technology weapons allows for a more informed prediction regarding the prospects of norm emergence for OCOs.[79] When these three case studies are considered, the primary reason for developing constraining norms for emerging-technology weapons is the perception among powerful or relevant states that such norms are in their national self-interest. That is, a direct or indirect alignment of national self-interest with a constraining norm leads to norm emergence. The extent to which it is aligned with key or powerful states' perception of self-interest will determine how rapidly and effectively the norm emerges. The role of national self-interest as the primary ingredient leading to norm emergence also helps explain why, when challenged with violations of a young and not-yet-internalized norm, a state is quick to abandon the norm and pursue its material interest by using the previously constrained emerging-technology weapon, as was seen with both chemical and biological weapons and strategic bombing in World War I and strategic bombing in World War II.

Prospects for OCO norms

The key principle of norm evolution theory for emerging-technology weapons is that norm emergence is more likely to occur when powerful, relevant actors are involved, specifically key states acting as norm leaders and norm entrepreneurs within organizations. There are a variety of intergovernmental bodies and organizations currently being used by a variety of states to promote various emerging norms for OCOs. Through these organizations a variety of actors, motivated by a number of factors and employing a range of mechanisms, have promoted various candidate cyber norms, ranging from a total prohibition on cyber weapons and warfare, to a no first-use policy, or the applicability of the existing LOAC to OCOs. Norm evolution theory would thus seem to interpret this as a sign of progress for norm emergence. However, if one examines these efforts more closely, the prospects are less hopeful.

Powerful states, constraining norms and self-interest

Powerful self-interested state actors will play a significant role in norm emergence. Additionally, the perceived state self-interest will be important for norms to emerge and for a state to become a leader of a particular norm. Successful norm emergence requires states as norm leaders and increasing multipolarity is unlikely to help. After all, there were eight great powers in 1910 and that complicated rather than fueled the convergence of a constraining norm for strategic bombing. Since there is generally less exposure and understanding surrounding cyber weapons as well as different rates of weapon adoption and cyber vulnerability, states will be reluctant to lead on the issue of norms because they may be unable to determine the utility of such weapons relative to their own interests. However, such calculations are essential if important and powerful states are going to become a strong norm leader and help promote the emerging norm. Additionally, specific to China, Russia and the United States – the pre-eminent cyber actors – an analysis of their respective cyber doctrines indicates that there appears to be a perspective that each nation has more to gain from engaging in OCOs than from significantly restricting it or giving it up entirely. National investments in OCO capabilities and the development of doctrine and strategies for OCOs provide insight into state perceptions of self-interest and the expectations for behavior and emerging norms for OCOs. So where do state OCO programs stand today in China, Russia and the United States? The three key states discussed here are the most significant, both due to the breadth and sophistication of their capabilities and activities as well as the likelihood that they are serving as the model for many other nations preparing to operate in cyberspace.

Chinese interest in OCOs

China's early activity and interest in OCOs indicate that it likely does not consider the emergence of constraining norms in its self-interest. It has been largely unconstrained by cyber norms and is preparing to use cyber weapons to cause economic harm, damage critical infrastructure and influence kinetic armed conflict. As such, it is unlikely to be a vocal norm leader. China is best known for its expansive efforts conducting espionage-style cyber operations. For example, in February 2013, the U.S. cybersecurity firm Mandiant released a study detailing extensive and systematic cyberattacks, originating from Chinese military facilities, of at least 141 separate U.S.-affiliated commercial and government targets. In May 2014, the Department of Justice indicted five Chinese military hackers for CNE activity in the United States.[80] These attacks have led the U.S. DoD to classify China as "the world's most active and persistent perpetrators of economic espionage" and point out that they are also "looking at ways to use cyber for offensive operations."[81] It is this latter point that is of most interest to this chapter. China is increasingly developing and fielding advanced capabilities in cyberspace, while its interests in OCOs appear to be asymmetric and strategic. While China and the United States agreed in September 2015 to not knowingly conduct CNE theft of intellectual property for commercial advantage, there is evidence China has not lived up to its end of the bargain.[82] Recent deterioration of the broader U.S.-Sino relationship and President Trump's efforts to address perceived trade imbalances with China further destabilize the situation, with little change in course in the first years of the Biden administration. The U.S. Director of National Intelligence testified in 2022 that the U.S. Intelligence Community assesses "that China presents the broadest, most active and persistent cyber espionage threat to

U.S. Government and private sector networks. China's cyber pursuits and export of related technologies increase the threats of attacks against the U.S. homeland, suppression of U.S. web content that Beijing views as threatening to its control and the expansion of technology-driven authoritarianism globally."[83]

Russian interest in OCOs

Like that of China, Russia's early OCO activity – especially the attacks on Estonia, Georgia and Ukraine – indicates that it is largely unconstrained by restrictive cyber norms and is preparing to use cyber weapons in a wide range of conflicts and against a variety of targets. It likely does not consider the emergence of constraining norms in its self-interest. As such, one would think it unlikely to be a vocal norm leader. However, Russia has been a leading proponent of a total ban on cyber weapons. This is similar to the Soviet Union's efforts early in the nuclear era to demonize U.S. possession of nuclear weapons while simultaneously pursuing such weapons themselves. It helps illustrate how powerful states acting in their own self-interest can inadvertently act as norm leaders despite flouting the candidate norm themselves. However, Russia's confusing support for fully constraining norms for OCOs (based on its behavior in the UN and its proposal for an "International Code of Conduct for Information Security") may be based on its broader definition of OCOs and its interest in using a constraining norm to prevent what it perceives as "propaganda" inside Russia and in its near abroad.[84] But, its position may also be disingenuous, as it was when supporting the Biological Weapons Convention while simultaneously launching a massive, illicit, biological weapons program. To achieve any real convergence among the main cyber actors the authoritarian interest in constraining free speech must be addressed, which could deflate Russian support. Further, Russian doctrine now states that future conflict will entail the "early implementation of measures of information warfare to achieve political objectives without the use of military force, and in the future to generate a favorable reaction of the international community to use military force."[85] Russian interference in the 2016 U.S. presidential election is only one more example of Russia's bellicose nature in cyberspace and elsewhere, as is its repeated use of cyber aggression against Ukraine, to include in the lead up to the 2022 invasion.

U.S. interest in OCOs

While China is perhaps the noisiest and Russia the most secretive when it comes to OCOs, the United States is the most sophisticated. The United States is in the process of dramatically expanding its military organization committed to engaging in OCOs and regularly engages in "offensive cyber operations."[86] However, unlike Russian attacks and Chinese planning, it appears to exercise restraint and avoid targeting non-military targets. This seems to indicate that the United States is acting as a norm leader for at least a certain category of constraining cyber norms, although its general "militarization" of cyberspace may be negating the norm-promoting effects of this restraint. While the United States has recently developed classified rules of engagement for OCOs, it has articulated few, if any, limits on its use of force in cyberspace or response to hostile cyberattacks. For example, the May 2011 *International Strategy for Cyberspace* states that the United States "reserves the right to use all necessary means" to defend itself and its allies and partners, but that it will "exhaust all options before [the use of] military force."[87] Additionally, the former

U.S. Deputy Secretary of Defense, William Lynn, clearly asserted that "the United States reserves the right, under the law of armed conflict, to respond to serious cyberattacks with an appropriate, proportional, and justified military response."[88] Further, U.S. Cyber Command was recently established as a full-fledged combatant command. Ultimately, the U.S. behavior and interest in OCOs indicate that it does not consider the emergence of robust constraining norms in its self-interest.

Secondary factors affecting norm emergence

Norm evolution theory for emerging-technology weapons also recognizes secondary reasons for development.[89] This comprehensive theory of norm evolution for emerging-technology weapons is a framework for predicting the likelihood of norm development for cyber-related weapons and warfare and will be used in the remainder of this chapter to offer additional predictions for cyber norms.

Coherence with existing dominant norms unlikely

Should current trends continue, the outlook for coherence with exiting norms is not favorable when applied to OCOs. First, cyber norms will have difficulty achieving coherence with and grafting onto existing norms. Unfortunately, the success of a norm candidate for emerging-technology weapons also will depend in large part on the ability to achieve coherence by connecting the new weapon type to an existing category and thus beginning the process of grafting the new norm onto existing norms. While cyber weapons and OCOs have some commonalities with certain weapons, particularly unconventional and emerging-technology weapons, overall they are truly unique. In fact, they are so unique as to operate in their own new, man-made domain outside the normal domains of land, sea, air and space. As such, cyber norms lack obvious coherence with many prominent norms, and thus it is difficult for norm entrepreneurs to graft the candidate norms to existing norms. Perhaps the best option for success is the humanitarian norm underlying the existing LOAC, particularly the norm regarding the protection of civilians and minimization of collateral damage.[90] This is precisely what NATO's *Tallinn Manual on the International Law Applicable to Cyber Warfare* attempts to achieve by arguing that the LOAC apply to OCOs.[91] However, despite recent progress in the GGEs and OEWG, the lack of agreement on key terms and accountability mechanisms as well as confusion over the spectrum of hostile cyber operations make coherence and grafting complex and difficult.[92]

Too late to preemptively establish norms for OCOs

Another challenge for norm emergence is that it will be more successful if the candidate norm can be permanently and preemptively established before the weapon exists or is fully capable or widespread. With OCOs, the train has already left the station so to speak. Looking back beyond the past decade, between 2006 and 2013, James Lewis and CSIS identify 16 significant CNA-style cyberattacks.[93] These include major attacks across the globe, occurring in the former Soviet states of Estonia and Georgia to Iran and Saudi Arabia. And since then there have only been additional cyber incidents. The opportunity for permanent preemptive establishment of a norm has long since passed.

Differing perspectives on future capability and threat inflation

There will be challenges arising from both differing perspectives as to future capability and the prospect for threat inflation. While it is true that OCOs have been demonstrated to some degree (e.g. Stuxnet), the hidden and secretive nature of cyberspace make the actors and their intent unclear and thus limit the true demonstrative value of recent cyberattacks. This has the effect of creating competing theories and arguments as to future effectiveness and strategic impact. A case in point, some argue (including former U.S. Secretary of Defense Leon Panetta) that OCOs pose a major threat and warn of a cyber "Pearl Harbor" or "cyber 9/11" moment when CI is attacked. Others have argued that statements such as Panetta's are pure hyperbole and that OCOs pose no such dire threat and that it in fact may not even constitute warfare as properly defined.[94] In the December 2013 edition of *Foreign Affairs*, Thomas Rid argued that not only is cyberattack not a major threat, but it will also in fact "diminish rather than accentuate political violence" by offering states and other actors a new mechanism to engage in aggression below the threshold of war,[95] and Erik Gartzke argued further that OCOs are "unlikely to prove as pivotal in world affairs . . . as many observers seem to believe."[96] However, cybersecurity is a huge and booming business for IT-security firms, with industry market research firms predicting the global cybersecurity market will grow by tens of billions of dollars each year. IT-security expert and privacy activist Bruce Schneier has alleged that these firms benefitting from cyber growth have, along with their government customers, artificially hyped the cyber threat.[97] Some critics have gone so far as to refer to this dynamic as "cyber doom" rhetoric or a "cyber security-industrial complex" similar to the oft-derided "defense-industrial complex."[98] Norm evolution theory applied in this case indicates that these vastly different perceptions as to the impact and role of OCOs in international relations and conflict will impair norm emergence, as was the case early in the 20th century when the role and impact of strategic airpower was highly contested.

Defenseless perception impact

The idea that cyber weapons cannot be defended against will fuel interest in a constraining norm but also limits the effectiveness of reciprocal agreements and can lead to weapon proliferation. As a result, once convention-dependent norms are violated, intense domestic pressure can build for retaliatory violations of the norm. Defenses against cyber weapons are largely viewed as inadequate. A report from the DoD's Defense Science Board reported that the United States "cannot be confident" critical IT systems can be defended from a well-resourced cyber adversary.[99] The nature of cyberspace, with intense secrecy and "zero-day" vulnerabilities makes defense particularly difficult and fuels interest in other strategies to manage the threat, including constraining international norms. This explains the broad range of actors and organizations involved in early norm promotion and is a positive factor for the successful emergence of norms for OCOs. However, the experience of norms for emerging-technology weapons with similar perceptions regarding the weakness of defenses also indicates that while this may fuel interest in cultivating norms, they will be fragile and largely apply to use and not proliferation, as actors will continue to develop and pursue the weapons because they believe they cannot rely on defenses and seek deterrence-in-kind capabilities. Further, if the early norm is violated, given the inability to defend against continued violations, there may be domestic pressure to respond-in-kind, leading to a rapid erosion of the norm. Should early cyber norms be violated, such domestic pressure for an in-kind response could build. In fact, the Iranian attack on Saudi Aramco in August 2012 is largely viewed as one of Iran's responses to Stuxnet.[100] Further, the United States conducted similar retaliatory actions against the Russian Internet Research Agency to temporarily limit its operational capability to interfere in U.S. elections. The challenge of attribution in cyberspace may accentuate this dynamic by making retaliatory responses even easier than with prior emerging--technology weapons.

Unitary dominance and delayed proliferation and adoption

Finally, weapon proliferation and adoption will play a significant role in norm emergence as it will influence state interest in constraining norms. For OCOs, there is not the kind of unitary dominance of a single actor as there was with the U.S. nuclear monopoly early in the age of nuclear – giving the United States significant influence on norm emergence regarding nuclear restraint. Additionally, given the ongoing proliferation of cyber weapons, the multiuse nature of the technology and the relatively low cost of entry, delays in proliferating cyber weapons is unlikely. However, there will likely be varied rates of adoption of cyber weapons, with some nations such as the United States, China, Russia and Israel possessing the most sophisticated cyber warheads.[101] Experience with norm development for emerging-technology weapons indicates that states with powerful cyber weapons are more likely to resist the emergence of any constraining norms. This is especially true with strong bureaucratic actors, such as the NSA in the United States or the Federal Agency of Government Communications and Information in Russia, potentially advocating for permissive norms. While the Russians have been major advocates in the UN for a total prohibition on cyber weapons, their interest may be driven by a perception that the United States is the dominant cyber power or, perhaps more cynically, it could be akin to the Soviet Union's disingenuous early promotion of the constraining biological weapon and nuclear norms while simultaneously pursuing biological and nuclear weapons. Regardless, the varied rates of adoption and development of cyber capabilities indicates that there will be divergent perspectives on constraining norms, making consensus difficult. This helps explain why despite the many actors and organizations involved in developing candidate norms for OCOs, they have not been successful in achieving any broad consensus with accountability measures beyond perhaps the budding consensus regarding the theoretical application of the LOAC.

Ultimately, if current trends continue, norm evolution theory for emerging-technology weapons predicts that the emergence and early development of constraining norms will be challenged and may not occur at all. Key states – especially China, Russia and the United States – are unlikely to perceive the emergence of robust constraining norms in their self-interest. Further, limited options for coherence and grafting, inability to preemptively establish a prohibition, lack of unitary dominance, increased proliferation and adoption of cyber weapons and the lack of powerful self-interested state actors converging on a candidate norm present serious hurdles for norm emergence. However, the connection with the idea that cyber weapons cannot be adequately defended against as well as industry and government hyping of the threat have spurned significant general interest in constraining norms for OCOs – leading to a rise of many actors and organizational platforms. To move past this point and achieve success, a consensus on cyber norms will need to build, and such a consensus does not seem inevitable at this point or in the near future.

Prospects for OCO norm cascade and internalization

While norm evolution theory for emerging-technology weapons predicts low odds for constraining OCO norms, should norms emerge it is worth briefly examining what the theory predicts about achieving a norm cascade and internalization. These latter two phases in the norm life cycle are important if a norm is to have a structural impact on the international system. If a constraining cyber norm emerges and approaches a norm cascade, then a tipping point may actually be more likely. Certain indicators are important to achieving a norm cascade, such as potential technological improvements that mitigate the attribution challenge, the unconventional characterization afforded cyber weapons and the expansive international arms control and disarmament bureaucracy. However, should the norm cascade occur, internalizing it will be less likely largely due to secrecy and the multiuse nature of cyber technologies which pose their own barriers to internalization and blunt international pressure for conformity and private sector support. As a result, norm internalization is likely to be most successful for norms governing usage rather than development, proliferation and disarmament. Table 12.2 summarizes the implications of the various hypotheses of norm evolution theory for emerging-technology weapons when applied to OCOs.

Table 12.2 Norm evolution theory for emerging-technology weapons' implications for norms for OCOs[102]

Primary Hypothesis		*Implications for Cyber Norms*
	Direct or indirect alignment of national self-interest with a constraining norm leads to norm emergence, and the extent to which it is aligned with key or powerful states perception of self-interest will determine how rapidly and effectively the norm emerges	Negative
Secondary Hypotheses for Norm Emergence		**Implications for Cyber Norms**
1	Coherence and grafting with existing norms	Negative
2	Permanently establishing a norm before the weapon exists or is fully capable or widespread	Negative
3	Undemonstrated emerging-technology weapons	Negative
4	Connections with the idea that the weapon can't be defended against	Positive
5	Initial weapon proliferation/adoption	Negative
Secondary Hypotheses for Norm Cascade		**Implications for Cyber Norms**
1	Improvements in technology	Positive
2	Characterizing the weapon type as unconventional or otherwise granting it a special status	Positive
3	Public demonstrations of the weapon type, enabled by real-time media	Negative
4	The international arms control and disarmament bureaucracy and the increasing regulation and legalization of armed conflict	Positive

(*Continued*)

Table 12.2 (Continued)

Primary Hypothesis		*Implications for Cyber Norms*
Secondary Hypotheses for Norm Internalization		**Implications for Cyber Norms**
1	Internalization of aspects of a norm governing usage rather than aspects governing development, proliferation and disarmament	Positive
2	Congruent support and involvement from the public and private sectors	Negative
3	Secrecy and the multiuse nature of the technology	Negative
4	International pressure for conformity, enabled by real-time media coverage of the weapon's use	Negative

Conclusions

While cyber conflict is no longer in its infancy, it is still a relatively new aspect of international security and there are multiple possibilities for how this newer mode of warfare will evolve over the coming decades. However, reasonable conclusions can be drawn regarding the prospects for the emergence of a constraining norm for OCOs based on norm evolution theory for emerging-technology weapons.[103] The theory indicates that there are many hurdles facing development of constraining norms for OCOs and predicts that if current trends continue, meaningful constraining norms for OCOs will have trouble emerging and may not ever reach a norm cascade. This is principally due to the fact that powerful state actors are unlikely to perceive a convergence between a robust constraining norm and their self-interest.

Notes

1 Katzenstein, Peter, Wendt, Alexander, and Jepperson, Ronals, "Norms, Identity, and Culture in National Security." In P. Katzenstein (ed.) *The Culture of National Security: Norms and Identity in World Politics*. New York: Columbia University Press, 1996, 54.
2 Gavin, James. *War and Peace in the Space Age*. New York: Harper Brothers, 265.
3 Haines, Avril. "Statement for the Record: Worldwide Threat Assessment of the U.S. Intelligence Community." March 8, 2022
4 Clapper, "Statement for the Record." 2013, 1–3.
5 *International Strategy for Cyberspace: Prosperity, Security, and Openness in a Networked World*, White-House.gov, May 2011, www.whitehouse.gov/sites/default/files/rss_viewer/international_strategy_for_cyberspace.pdf.
6 The White House. *Interim National Security Strategic Guidance,* 2021
7 Mazanec, Brian. *The Evolution of Cyber War: International Norms for Emerging Technology Weapons*. New York: Potomac Books, 2015.
8 Katzenstein, Wendt, and Jepperson, "Norms, Identity, and Culture." 54.
9 Brown, Gary, and Poellet, Keira, "The Customary International Law of Cyberspace." *Strategic Studies Quarterly* (Fall 2012), 130.
10 Paget, Francois. "How Many Bot-Infected Machines on the Internet?." *McAfee Labs*, January 29, 2007. http://blogs.mcafee.com/mcafee-labs/how-many-bot-infected-machines-are-on-the-internet.
11 John Markoff, and Thom Shanker, "Halted '03 Plan Illustrates U.S. Fear of Cyber Risk." *New York Times*, August 1, 2009, www.nytimes.com/2009/08/02/us/politics/02cyber.html.
12 Nakashima, Ellen. "U.S. Cyberweapons Had Been Considered to Disrupt Gaddafi's Air Defenses." *Washington Post*, October 17, 2011, www.washingtonpost.com/world/national-security/

us-cyber-weapons-had-been-considered-to-disrupt-gaddafis-air-defenses/2011/10/17/gIQAET-pssL_story.html.

13 Menn, Joseph. "Syria, Aided by Iran, Could Strike Back at U.S. in Cyberspace." *Reuters*, August 29, 2013, www.reuters.com/article/2013/08/29/us-syria-crisis-cyberspace-analysis-idUSBRE97S04Z20130829.

14 Healey, Jason. *A Fierce Domain: Conflict in Cyberspace, 1986 to 2012*. Vienna, VA: The Atlantic Council and Cyber Conflict Studies Association, 2013, 23.

15 Brown and Poellet, "The Customary International Law of Cyberspace," 132.

16 Healey, Jason. *A Fierce Domain*, 2013, 86.

17 Healey, Jason. "How Emperor Alexander Militarized American Cyberspace." *Foreign Policy*, November 22, 2013. www.foreignpolicy.com/articles/2013/11/06/how_emperor_alexander_militarized_american_cyberspace?wp_login_redirect=0#sthash.63nueywc.dEQQ8Uxd.dpbs.

18 Lewis, James, and Timlin, Katrina. *Cybersecurity and Cyberwarfare: Preliminary Assessment of National Doctrine and Organization*. Washington, DC: Center for Strategic and International Studies, 2011; Liff, Adam P. "Cyberwar: A New 'Absolute Weapon'? The Proliferation of Cyberwarfare Capabilities and Interstate War." *Journal of Strategic Studies* Vol. 35, No. 3 (2012), 401–428.

19 Council of Europe, Convention on Cybercrime, November 23, 2011, http://conventions.coe.int/Treaty/en/Treaties/Html/185.htm.

20 Carr, Jeffrey. *Inside Cyber Warfare: Mapping the Cyber Underworld*. Sebastapol, CA: O'Reilly, 2009, 63.

21 United Nations, "Basic Information: About WSIS." www.itu.int/wsis/basic/about.html.

22 United Nations, Charter of the United Nations and Statute of the International Court of Justice, Article 2, Paragraph 4, http://treaties.un.org/doc/Publication/CTC/uncharter.pdf.

23 The Second Committee, the Economic and Social Council and other UN organizations have looked at other aspects of cyber security, such as cybercrime and cyber espionage, but these efforts fall outside the scope of this project.

24 Carr, *Inside Cyber Warfare*, 34.

25 Clarke, Richard A., and Knake, Robert. *Cyber War: The Next Threat to National Security and What to Do About It*. New York: HarperCollins, 2010, 220.

26 International Humanitarian Law, based on the concepts of *jus ad bello*, is defined to be the law of war. This means that the laws involved are meant to be active in a situation of an armed conflict or during war

27 Maurer, Tim. "Cyber Norm Emergence at The United Nations: An Analysis of the Activities at the UN Regarding Cyber-Security." *Harvard Kennedy School Belfer Center for Science and International Affairs*, 2011, 20.

28 Maurer, "Cyber Norm Emergence at the United Nations." 22.

29 Maurer, "Cyber Norm Emergence at the United Nations." 22.

30 Maurer, "Cyber Norm Emergence at the United Nations." 6.

31 Maurer, "Cyber Norm Emergence at the United Nations." 21.

32 United Nations General Assembly, A/66/359, Letter dated September 12, 2011 from the Permanent Representatives of China, the Russian Federation, Tajikistan, and Uzbekistan to the United Nations addressed to the Secretary-General, September 14, 2011, https://ccdcoe.org/sites/default/files/documents/UN-110912-CodeOfConduct_0.pdf.

33 Maurer, "Cyber Norm Emergence at the United Nations." 25.

34 Paul Meyer, "Cyber Security Takes the UN Floor," *ICT4Peace Foundation*, November 11, 2013. http://ict4peace.org/?p=3000.

35 United Nations Office for Disarmament Affairs. "Fact Sheet: Developments in the Field of Information and Telecommunications in the Context of International Security." June 2013.

36 United Nations General Assembly, A/68/98, "Report of the Group of Governmental Experts." 2–3, 9.

37 OSCE. *OSCE Participating States, in Landmark Decision, Agree to Expand List of Measures to Reduce Risk of Tensions Arising from Cyber Activities*. March 10, 2016. www.osce.org/cio/226656.

38 Grauman, Brigid. "Cyber-Security: The Vexed Question of Global Rules." *Security Defence Agenda and McAfee* (February 2012), 5.

39 ICT4PEACE. "ICT4PEACE Special Session: Norms and CBMS. Moving towards a More Inclusive Agenda, Summary Report." October 17, 2013, http://mercury.ethz.ch/serviceengine/Files/ISN/172917/ipublicationdocument_singledocument/082e6a53–16ee-4218-b643-4b8fc965bdb0/en/ICT4+Peace+Report+plus+Statement+Seoul+-+Norms+and+CBMs+-+final.pdf

40 United Nations General Assembly, A/70/174. "Group of Governmental Experts on Developments in the Field of Information and Telecommunications in the Context of International Security." July 22, 2015. www.un.org/ga/search/view_doc.asp?symbol=A/70/174.

41 Maurer, "Cyber Norm Emergence at the United Nations." 9–10.
42 Maurer, "Cyber Norm Emergence at the United Nations." 29.
43 Wegener, Henning. "Cyber Peace." *The Quest for Cyber Peace*. Geneva: International Telecommunication Union and World Federation of Scientists, January 2011, 81.
44 Wegener, "Cyber Peace." 81.
45 Maurer, "Cyber Norm Emergence at the United Nations." 28.
46 The countries covered thus far in the UNIDIR research papers are Australia, Brazil, Canada, China, France, Germany, India, the Islamic Republic of Iran, Israel, Japan, the Republic of Korea, the Russian Federation, Saudi Arabia, the United Kingdom of Great Britain and Northern Ireland and the United States.
47 Ackerman, Spencer. "NATO Doesn't Yet Know How to Protect Its Networks." *Wired.com*, February 1, 2012.
48 Klimburg, Alexander. *National Cyber Security Framework Manual*. Brussels: NATO CCD COE Publication, 2012, 4.
49 NATO CCD COE. *International Norms: Legal, Policy, and Industry Perspectives*. Geneva, 2016.
50 NATO, "Active Engagement, Modern Defence: Strategic Concept for the Defence and Security of the Members of the North Atlantic Treaty Organisation, Adopted by Heads of State and Government at the NATO in Lisbon, November 19–20, 2010." Paragraph 12, www.nato.int/strategic-concept/pdf/Strat_Concept_web_en.pdf.
51 Atlantic Council, "Fact Sheet: The Tallinn Manual on the International Law Applicable to Cyber Warfare," March 28, 2013, www.atlanticcouncil.org/events/past-events/tallinn-manual-launch-defines-legal-groundwork-for-cyber-warfare.
52 Cohen, Gili. "Israeli Expert Seeks Ethics Code for Cyber Warfare," *Haaretz*, January 20, 2014, www.haaretz.com/news/diplomacy-defense/1.569450.
53 Schmitt, Michael N. "International Law in Cyberspace: The Koh Speech and Tallinn Manual Juxtaposed." *Harvard International Law Journal*, Vol. 54 (2012), 14.
54 Schmitt, "International Law in Cyberspace." 14.
55 White House, United States, *International Strategy for Cyberspace: Prosperity, Security, and Openness in a Networked World*, May 2011, 9.
56 Keith Alexander, "Responses to Advance Questions for Lieutenant General Keith Alexander, USA, Nominee for Commander, United States Cyber Command." April 15, 2010, http://epic.org/Alexander_04-15-10.pdf.
57 Schmitt, "International Law in Cyberspace." 17.
58 See International Group of Experts, Schmitt, Michael, N., ed. *Tallinn Manual on The International Law Applicable to Cyber Warfare*. Cambridge, UK: Cambridge University Press, 2013, 17–32.
59 Schmitt, "International Law in Cyberspace." 19; International Group of Experts, *Tallinn Manual*, 195.
60 Schmitt, "International Law in Cyberspace." 27.
61 For example, see "NATO Cyberwar Directive Declares Hackers Military Targets." *Russia Today*, March 19, 2013, http://rt.com/usa/nato-publishes-cyberwar-guidelines-502/; Aaron Souppouris, "Killing Hackers Is Justified in Cyber Warfare, Says NATO-Commissioned Report." *Verge*, March 21, 2013, www.theverge.com/2013/3/21/4130740/tallin-manual-on-the-international-law-applicable-to-cyber-warfare.
62 Hurwitz, Roger. "An Augmented Summary of the Harvard, MIT and U. of Toronto Cyber Norms Workshop." May 2012, 12, www.yumpu.com/en/document/view/18791277/cyber-norms-workshop-ecir-mit.
63 UK Foreign and Commonwealth Office, "London Conference on Cyberspace: Chair's Statement." November 2, 2011, www.gov.uk/government/news/london-conference-on-cyberspace-chairs-statement.
64 "2012 Budapest Conference." 2013 Seoul Cyber Conference, www.mofa.go.kr/english/visa/images/res/Cy_Eng.pdf; "Overview." 2013 Seoul Cyber Conference, www.mofa.go.kr/english/visa/images/res/Cy_Eng.pdf.
65 "Citizen Lab Fellows Tim Maurer and Camino Kavanagh on the 2013 Seoul Conference on Cyberspace." Munk School of Global Affairs, University of Toronto, October 28, 2013, https://citizenlab.org/2013/10/citizen-lab-fellows-tim-maurer-camino-kavanagh-2013-seoul-conference-cyberspace/.
66 "2013, Seoul Framework for and Commitment to Open and Secure Cyberspace." 2013 Seoul Cyber Conference, October 18, 2013, www.mofat.go.kr/english/visa/images/res/SeoulFramework.pdf.

67 "Attachment II: Best Practices," 2013 Seoul Cyber Conference, October 18, 2013, www.mofa.go.kr/webmodule/common/download.jsp?
68 "UK Contribution to the Seoul Conference on Cyberspace, Next Steps: Key Activities to Take Forward the London Agenda." 2013 Seoul Cyber Conference, October 18, 2013, 3, www.google.com/url?sa=t&rct=j&q=&esrc=s&frm=1&source=web&cd=1&ved=0CB4QFjAA&url=https%3A%2F%2Fwww.gov.uk%2Fgovernment%2Fuploads%2Fsystem%2Fuploads%2Fattachment_data%2Ffile%2F251569%2FNext_Steps.docx&ei=t5Q5VaqfBPj7sATH6YHADg&usg=AFQjCNEV3lx7mQA6-hgldx_Atg5vk6WEow.
69 United Nations Office for Disarmament Affairs, Biological Weapons Convention, www.un.org/disarmament/WMD/Bio/.
70 Kim Eun-jung, "Seoul Forum Calls for International Rules for Cyber Warfare," Yonhap News Agency, November 12, 2013, http://english.yonhapnews.co.kr/news/2013/11/12/23/0200000000AEN20131112007100315F.html.
71 Eun-Jung, "Seoul Forum Calls for International Rules for Cyber Warfare."
72 "Attachment II: Best Practices."
73 "S. Korea, U.S. Hold Working-Level Talks on Cybersecurity," *Global Post*, February 7, 2014, www.globalpost.com/dispatch/news/kyodo-news-international/140207/s-korea-us-hold-working-level-talks-cybersecurity.
74 James Clapper, "Statement for the Record: Worldwide Threat Assessment of the U.S. Intelligence Community," Senate Select Committee on Intelligence, March 12, 2013, www.dni.gov/files/documents/Intelligence%20Reports/2013%20ATA%20SFR%20for%20SSCI%2012%20Mar%202013.pdf
75 Forsyth, James, Billy Pope. "Structural Causes and Cyber Effects: Why International Order is Inevitable in Cyberspace." *Strategic Studies Quarterly*, Winter (2014), 113–130.
76 Norris, John. *Early Gunpowder Artillery: 1300–1600.* Marlborough, UK: Crowood Press, 2003.
77 Finnemore, Martha, and Sikkink, Kathryn. "International Norm Dynamics and Political Change." *International Organization*, Vol. 52, No. 4 (1998), 887–917.
78 Koblentz, Gregory, and Brian Mazanec. "Viral Warfare: The Security Implications of Cyber and Biological Weapons," *Comparative Strategy*, Vol. 32, No. 5 (2013), 418–434.
79 Mazanec, Brian. *The Evolution of Cyber War: International Norms for Emerging Technology Weapons.* New York: Potomac Books.
80 Wan, William, and Ellen Nakashima. "Report Ties Cyberattacks on U.S. Computers to Chinese Military." *Washington Post*, February 19, 2013. http://articles.washingtonpost.com/2013-02-19/world/37166888_1_chinese-cyber-attacks-extensive-cyber-espionage-chinese-military-unit; U.S. Department of Justice Office of Public Affairs, "U.S. Charges Five Chinese Military Hackers for Cyber Espionage Against U.S. Corporations and a Labor Organization for Commercial Advantage." May 19, 2014. www.justice.gov/opa/pr/us-charges-five-chinese-military-hackers-cyber-espionage-against-us-corporations-and-labor.
81 Mulrine, Anna. "China Is a Lead Cyberattacker of U.S. Military Computers, Pentagon Reports," *Christian Science Monitor*, May 18, 2012. www.csmonitor.com/USA/Military/2012/0518/China-is-a-lead-cyberattacker-of-US-military-computers-Pentagon-reports.
82 Associated Press. "China Already Violating U.S. Cyberagreement, Group Says." October 19, 2015. www.cbsnews.com/news/crowdstrike-china-violating-cyberagreement-us-cyberespionage-ntellectual-property/.
83 Haines, Avril. "Statement for the Record: Worldwide Threat Assessment of the U.S. Intelligence Community." March 8, 2022
84 United Nations General Assembly. *A/66/359: Letter dated 12 September 2011 From the Permanent Representatives of China, the Russian Federation, Tajikistan, and Uzbekistan to the United Nations addressed to the Secretary-General*, September 14, 2011. www.citizenlab.org/cybernorms/letter.pdf.
85 Heickerö, Roland. "Emerging cyber threats and Russian Views on Information Warfare and Information Operations." *Swedish Defence Research Agency* (2010), 27.
86 Healey, Jason. "How Emperor Alexander Militarized American Cyberspace." *Foreign Policy*, November 22, 2013. www.foreignpolicy.com/articles/2013/11/06/how_emperor_alexander_militarized_american_cyberspace?wp_login_redirect=0#sthash.63nueywc.dEQQ8Uxd.dpbs.
87 United States. *International Strategy for Cyberspace: Prosperity, Security, and Openness in a Networked World.* Washington, DC, May 2011, 14; Nakashima, Ellen. "In Cyberwarfare, Rules of Engagement Still Hard to Define." *The Washington Post*, March 10, 2013. www.washingtonpost.com/world/national-security/in-cyberwarfare-rules-of-engagement-still-hard-to-define/2013/03/10/0442507c-88da-11e2–9d71-f0feafdd1394_story.html.

88 Lynn, William J. III, "The Pentagon's Cyberstrategy, One Year Later." *Foreign Affairs*, September 28, 2011.
89 Such as: coherence and grafting with existing norms; permanently establishing a norm before the weapon exists or is fully capable or widespread; threat inflation regarding the possible effects of the weapon often by the private sector via industry and lobbying groups; the idea a weapon cannot be defended against will fuel interest in a norm; unitary dominance of a single actor with the particular weapon-type gives that actor significant influence in norm emergence for that weapon-type; and delays in proliferation (often due to technological barriers) can create added time for a constraining norm to emerge.
90 Finnemore, Martha. "Cultivating International Cyber Norms." *America's Cyber Future: Security and Prosperity in the Information Age* (Center for a New American Security, June 2011), 99.
91 International Group of Experts, *Tallinn Manual*, 2013.
92 Brown, Gary and Keira Poellet. "The Customary International Law of Cyberspace." 2012, 141.
93 Based on the author's analysis of Lewis, James. "Significant Cyber Events since 2006." *Center for Strategic and International Studies*, July 11, 2013. http://csis.org/publication/cyber-events-2006.
94 Rid, Thomas. *Cyber War Will Not Take Place*. Oxford, UK: Oxford University Press, 2012.
95 Rid, Thomas. "Cyberwar and Peace: Hacking Can Reduce Real-World Violence." *Foreign Affairs*, December 2013. www.foreignaffairs.com/articles/140160/thomas-rid/cyberwar-and-peace.
96 Gartzke, Erik. "The Myth of Cyberwar: Bringing War in Cyberspace Back Down to Earth." *International Security*, Vol. 38, No. 2 (2013), 42.
97 Schneier, Bruce. "Threat of 'Cyberwar' Has Been Hugely Hyped." *CNN*, July 7, 2010. www.cnn.com/2010/OPINION/07/07/schneier.cyberwar.hyped/.
98 Brito, Jerry, and Tate Watkins. "Loving the Cyber Bomb? The Dangers of Threat Inflation in Cybersecurity Policy." *Mercatus Center George Mason University*, April 26, 2011. http://mercatus.org/publication/loving-cyber-bomb-dangers-threat-inflation-cybersecurity-policy; *Russia Today*, "Is Cyberwar Hype Fuelling a Cybersecurity-Industrial Complex?." February 16, 2012. http://rt.com/usa/security-us-cyber-threat-529/.
99 United States Department of Defense. *Defense Science Board Task Force Report: Resilient Military Systems and the Advanced Cyber Threat* (January 2013), 1. www.acq.osd.mil/dsb/reports/ResilientMilitarySystems.CyberThreat.pdf.
100 Perlroth, Nicole. "In Cyberattack on Saudi Firm, U.S. Sees Iran Firing Back." *The New York Times*. October 23, 2012. www.nytimes.com/2012/10/24/business/global/cyberattack-on-saudi-oil-firm-disquiets-us.html.
101 As evidenced by the examination of customary state practice of cyber warfare reviewed earlier in this chapter.
102 Mazanec, Brian. *The Evolution of Cyber War: International Norms for Emerging Technology Weapons*. New York: Potomac Books, 2015.
103 Forsyth, James, Billy Pope. "Structural Causes and Cyber Effects: Why International Order is Inevitable in Cyberspace." *Strategic Studies Quarterly* (Winter 2014), 123.

Further readings

Dinniss, Heather Harrison. *Cyber Warfare and the Laws of War*. Vol. 92. Cambridge, UK: Cambridge University Press, 2012.

Kello, Lucas. *The Virtual Weapon and International Order*. New Haven, CT: Yale University Press, 2017.

Lupovici, Amir. "The Dog That Did Not Bark, the Dog That Did Bark, and the Dog That Should Have Barked: A Methodology for Cyber Deterrence Research." *International Studies Review* 23, no. 4 (2021): 1672–1698.

Mazanec, Brian M. *The Evolution of Cyber War: International Norms for Emerging-Technology Weapons*. Lincoln, NE: University of Nebraska Press, 2015.

Philbrick, Ian Prasad, and Andrew McCoy, eds. *Georgetown Journal of International Affairs: International Engagement on Cyber VI*. Vol. 17. No. 3. Washington, DC: Georgetown University Press, 2017.

Schmitt, Michael N., ed. *Tallinn Manual on the International Law Applicable to Cyber Warfare*. Cambridge, UK: Cambridge University Press, 2013.

Schmitt, Michael N., ed. *Tallinn Manual 2.0 on the International Law Applicable to Cyber Operations*. Cambridge, UK: Cambridge University Press, 2017.

Soesanto, Stefan, and Max Smeets. "Cyber Deterrence: The Past, Present, and Future." In *NL ARMS Netherlands Annual Review of Military Studies 2020*, pp. 385–400. TMC Asser Press, The Hague, 2021.

Tor, Uri. "'Cumulative Deterrence'as a New Paradigm for Cyber Deterrence." *Journal of Strategic Studies* 40, no. 1–2 (2017): 92–117.

United Nations, General Assembly, *Open-Ended Working Group on Developments in the Field of Information and Telecommunications in the Context of International Security, Final Substantive Report*, A/AC.290/2021/CRP.2 (March 2021), available from https://front.un-arm.org/wp-content/uploads/2021/03/Final-report-A-AC.290–2021-CRP.2.pdf.

13 Evolution

How the logics of cyber conflict might change

A common perspective on the many cybersecurity and broader cyber conflict issues that we have touched upon in this volume is that insecurity and strategic behavior emerge from certain rules of the digital environment that mankind has constructed and iterated upon for more than half a century. This "logic of the domain"[1] point of view, in fact, is quite arguably the dominating position among many practitioners, policymakers and scholars of global cyber conflict. The underlying technologies of the Internet dictate the dynamics of engagement for actors attempting to use cyberspace for sociopolitical or economic purposes, the argument goes, including the speed of interactions, the inherent challenges of attribution, the logical barriers to building more secure core technologies and more. Real-world patterns of governance, commerce and social use of information technologies may determine the value in using the web in a manipulative fashion, of course, but the rules of such usage are – in broad terms – anchored to architectural facts and infrastructural decisions determined decades ago.

A growing number of experts have pointed out that this perspective on the shape of global cybercrime, conflict and competition is technologically deterministic, a reductionist perspective that sees technology affecting culture and institutions but not so much the other way around.[2] This case is remarkably easy to make. Even just acknowledging that there are distinct ways in which different nations have chosen to build cyber capacity and employ it for strategic gain – as we have discussed in past chapters' discussions of diverse Western, Russian, Chinese, Iranian and other approaches to cyber conflict – is sufficient reason to think that geopolitical context, individual cognition and both national and institutional cultures are all important determinants of cyber conflict. And yet, we must be careful to give technological evolution and revolution their due as measurable elements of sociopolitical change. The significance of the technological tie-in remains relevant regardless of which particular factors one might find as having more power for explaining global cyber conflict trends than the rest. American efforts to build web surveillance tools at scale in the post-Cold War era may have been geopolitically motivated, but the shape of such efforts was determined by the global distribution of physical Internet infrastructures.[3] Likewise, research on decision-making in cyber conflict has demonstrated not only that personal and organizational context drives perception of the value of information about cybersecurity events but also that stakeholders believe that there *are* technologically determined rules of the domain, even if articulation of such rules varies across actors.[4]

Given this, it seems reasonable for us here to assume that certain macro features of Internet architecture and the functionality of key web technologies will continue to be, at the very least, key intervening factors that help delineate the macro shape of global cyber conflict. While technology may not be the sole driver of particular digital behaviors and

DOI: 10.4324/9781003246398-13

outcomes, it is not impossible to imagine new science and innovation that comes to dramatically reshape the logics of engagement that have dominated the first several decades of cyber engagement.

The coming metamorphosis of cybersecurity and conflict

Scholars have grappled with the question of what constitutes "cyber power" for nearly three decades.[5] What does it mean for a state to be cyber powerful? Given that the Internet is a General-Purpose Technology (GPT),[6] what specific operational capabilities equate to a strong strategic capacity to secure national interests and favorable geopolitical outcomes via cyberspace?

As any reader of this book will likely recognize, of course, the answers to these questions have clearly changed over time. Just as recent years have seen various intensifications in patterns of global cyber conflict and crime, so too have they witnessed an immense diversification in the types of digital threats that face countries. This diversification has been determined by numerous factors. These include evolving understanding of norms of acceptable behavior in cyberspace, changing geo-strategic conditions for various actors and the mainstreaming of discourse about cybersecurity as a national security concern across a growing number of national institutions and stakeholders. But changes in those technologies most closely linked to stores of social, political, military and economic value have clearly also fed these trends. The socialization of web services and platforms brought about by the privatization of network infrastructure in the mid-1990s – i.e. the transition toward what would become Web 2.0 – massively increased the value proposition of weaponizing disinformation, manipulating consumer platforms and, generally, hacking to create sociopolitical interference rather than infrastructure disruption or espionage. Countries like Russia, arguably, became more cyber powerful than they had been through this period not because they cultivated better hackers over time or built better cyber institutions but because they fit their capabilities to conditions.[7] The relative power gain involved was premised, in other words, on underlying socio-technological transformation and the foresight to take advantage of it. Thus, the bottom line is that cyber power – as with behaviors and outcomes in cyberspace more generally – is *clearly* a function of sociopolitical, economic and geo-strategic conditions that are shaped by broad underlying technological fundamentals.

This point underscores a core proposition that this book makes, which we discuss in this chapter and expand on in the next, final chapter. In short,

> A sufficiently dynamic information technology (or set of information technologies) might change the logic of global cyber conflict not only because essential techniques, capacities and infrastructure might be evolve, but also because said technologies will have transformative effects on the distribution of social, political, economic and security value in global society.

Herein lies a significant takeaway from this book that we hope the preceding chapters have effectively contextualized and conveyed, namely that technology is *not* a neutral gamechanger of human behavior and human conflict.[8] Technology is never a determined outcome. Rather, it is a feature of society that is shaped by prevailing and parochial narratives about the world in which we live. More importantly, it is shaped in the details by the functional and moral decisions of those individuals that are empowered to leverage resources and sociopolitical potential within societal networks.

The fact that technology is never a pre-determined artifact that interacts with human society – but instead a feature of society that is shaped by humans and shapes them in turn – means that value, power and the drivers of human interactions enabled by technology are **metamorphic**. The same technologies developed and employed in one context do not necessarily equate to similar outcomes elsewhere, something that becomes truer the more dynamic a given innovation is (i.e. a technology with exceedingly static qualities like nuclear weapons versus GPTs like electricity, medicine or the Internet). This reality can be seen in various places in the chapters of this book, for instance in differing national styles of cyber conflict behavior or in the way that vulnerabilities emerging from new social media platforms have clearly been more meaningful for democracies than for less open national societies.

The opposite is also true, of course. Technological usage can and does fundamentally shift toward new mediums, logics and techniques even as the objectives of that usage – coercion, military deception or espionage, for instance – do not. Herein is the promise of future advances in information technologies that could support human efforts to spy, disrupt, destroy or subvert in much the same fashion that web technologies have, but via different means and methods. In the remainder of chapter, we speak of one likely candidate for such an evolution in the logics of cyber engagement – artificial intelligence (AI) – and outline how the technologies that constitute AI could sit at the heart of a metamorphosis of global cyber conflict.

The elephant in the room: artificial intelligence and cyber conflict

Artificial intelligence is a catchall term that encompasses an array of technological and scientific advances emerging from a huge variety of disciplines, from mathematics and computer science/engineering to information systems, neuroscience and physics.[910] In broad terms, it seems fair to say the few other technological developments – arguably no other developments – have captured the attention of both scientists and global publics in the same way that AI has. *Narrow* AI, meaning intelligent systems designed to accomplish singular tasks more effectively or efficiently than a human or traditional computer might, has been a constant feature of everyday life for most people in developed countries for years at this stage. Commercial services like Amazon's marketplace or core technology products like Apple's mobile device operating system, iOS, leverage AI for a range of purposes likely familiar to most users. News services like Associated Press have relied on entirely AI-generated short-reports for years. Added to this, media reporting (some, ironically, machine-generated) is increasingly saturated by news of advances in robotics, machine learning and more than portend AI systems capable of quite radically transforming elements of human social, political or economic experiences.

In the context of national security and economics, of course, AI has resultantly become the focus of an incredible amount of hope and hubris about the world of tomorrow. Tech luminaries like Steve Wozniak and Elon Musk have raised concerns about getting the design of such systems right sooner than later.[11] At the same time, national leaders like Emmanuel Macron and Vladimir Putin have expounded on their belief that whichever country manages to harness AI at scale before the rest will accrue immense power in international relations. AI sits at the heart, in short, of so much promise and concern about the future of global security specifically because its applications are prospectively so truly different from anything developed in human history to this point. Naturally, some fear that this will change the nature of core features of human society, including warfighting, and in doing so will thrust humanity into an era of unprecedented uncertainty and tumult. We do not make an argument along these lines here, but it is worth noting that such voices are not as alarmist as some might argue. After all, revolutionary information technologies from the printing press to the telegraph have overturned and reshaped society in just such a fashion in centuries past. What AI is almost certainly destined to do, however, is ensure that at least some forms of human interaction will occur via reference to new systems and new socio-technical reference points forged in the crucible of choices made in the present. For cyber conflict, this means new potential logics of interaction even as the macro context of human interactions for purposes of economic gain, espionage, coercion and more remains relatively static.

A brief history of the artificial intelligence field

Artificial intelligence is nowhere near as recent a discipline as so much punditry or casual media coverage might lead one to believe. Even as a diverse set of consumer applications, AI has made its presence felt in the world economy for at least three decades. The first game-playing computer program with something that could defensibly be called a form of machine learning was developed in the 1950s. The first industrial robot began work for General Motors in 1961 and the first general-purpose self-reasoning robot – called "Shakey" – was developed just five years later. In the years in which mankind landed on the moon, engaged in a nuclear arms race and fought numerous wars as proxies for the global Cold War, scientists built the first conversant self-learning AI interfaces, the first self-driving car and even the first computer programs capable of cooperating with one another.

Intellectuals of all stripes have been interested – even obsessed – with the idea of non-human intelligence capable of acting *without the input of a human* for centuries. In both Greek and Roman mythos, automated constructs appear as the golem guardians and eldritch machinations of strange forces. In what some might recognize as the first predictions of the plot of films like *I, Robot* or *The Matrix* today, prominent thinkers from antiquity to the Renaissance regularly referenced the idea that automated entities might one day become a subservient working class for all humankind. And many such thinkers acted on these designs, creating "artificial animals" that could move in the manner of their biological inspirations by means of intricate gears, switches and levers.[12] In the era of the Industrial Revolution, writers of science fiction expanded so many of the notions that ancient and Renaissance thinkers had about automated intelligence with the abandon of imagination unburdened by a need to do anything other than entertain. And while some ideas were inevitably the stuff of wild fantasy, others turned out to be highly prescient, with men like Isaac Asimov and H. G. Wells articulating issues that now form core areas of real-world discussions about AI safety, equity and ethics.

The modern field of artificial intelligence – which itself is an immensely diverse discipline spanning core subject areas like machine learning, natural language processing, robotics, Big Data and signal recognition – began after World War II. As so many conflicts do, the war against the Axis powers had created a twofold groundswell of new scientific and technological potential in the West. Wartime innovation opened new frontiers at the same time as the almost-total mobilization of national populations produced work forces with specialized skills and new perspectives on industry. Here, in the decade following the declarations of victory in both Europe and Asia, the roots of today's AI revolution were laid.

The common narrative about modern AI development is that it vacillates between boom and bust. In general, the existence of what many call the "summers" and "winters" of AI since about 1950 can be explained by the impact that breakthroughs in engineering first enables and then holds back scientific discovery.[13] Over the last 70 years, advances in technology – particularly computer-related technologies like the microchip – have regularly enabled explosive growth in areas like natural language processing and machine reasoning that have generally been marooned in the depths of mathematical proofing, unable due to one limitation or another from testing novel theories and approaches. As novel applications proliferate at such times, so too does public interest, intellectual buy-in and funding from both public and private sources. As is often found in the human experience, ingenuity often outpaces material capability. With AI, the shortcomings of the technologies of the day have often forced scientists to slow down until a particular problem or set of problems can be addressed. The cascading effects of this downtrend then simply worsen the challenges facing research communities, with investment and public curiosity drying up in the wake of uncertainty about whether another technology revolution is on the horizon or not.

The first boomtime of AI started in the 1950s, specifically emerging from the work of several intellectuals whose postwar research – including field-defining volumes like Norbert Wiener's *Cybernetics*[14] – had laid the groundwork for thinking about intelligence in informational terms. For minds like Wiener, von Neumann, Shannon and McCarthy, the label of an object – biological or mechanical – as a "thinking" thing should be applied where that object could be said to approximate human cognition along a number of lines.[15] In a now-famous workshop at Dartmouth College in 1956, academics agreed that these dimensions linked to the overall idea of a possible "thinking" machine included the use of structured language to communication, a general ability to solve problems as a human might, a capability to learn and self-improve, and a capacity for conceptual abstraction.[16] For the next 18 years, these and other thinkers built task-specific machines that demonstrated the viability of machine intelligence along these lines, with narrow robotic and computer developments geared toward mimicking the actions of mathematicians, military planners and game players.

In 1974, the first boomtime of AI ended, largely due to an inability to apply much of what had been built beyond narrow within-machine "worlds" built by researchers to demonstrate a concept.[17] The problem was easy to understand. Limited programs could be given small amounts to data and could thereafter be made to imitate humans in uncanny fashion. When trained against larger, more general-purpose datasets, programs froze up or offered bizarre outputs. And these results were produced at immense cost of time and resources. The narrow AI applications of the 1960s were intensive in terms of the computing power required to make them work. The demand from funders for continued progress beyond these narrow applications – combined with failures to advance in areas like image recognition that naturally required more computer power from the outset – amounted to an obstacle that scientists simply could not overcome. Until computers could be improved by an order of magnitude, AI research could not proceed.

The first AI "winter" stretched through the 1970s. In the mid-1980s, however, interest in the field resurged as new thinking about how to build intelligent programs convinced major funders to dip their toes back in the water, even though the technical problems of the past decade persisted. Specifically, a revolution of perspective among scientists about how to train general-purpose software led to several task-based programs that wowed both specialist and public audiences by the end of the decade. Working on the principle that intelligence might come from a broad-but-shallow knowledge of topics rather than a deep-but-singular perspective, programs like Deep Thought – a chess automata that famously defeated Bent Larsen, a grandmaster – demonstrated a real ability for machines to match humans' problem solving in real-world settings.[18] Governments around the world jumped in, including the Japanese government which famously funded the Fifth-General Computer System Project (FGCSP) as a partnership with private industry to leap the national economy ahead of competitors.[19] The South Koreans, Americans, British, French and Canadians all mimicked this approach in short order, funding research where it could be found and using all manner of immigration enticement to amass intellectual communities linked to government programming.

Unfortunately for FGCSP and other programs like it, too little progress was made in the first few years and economic downturn saw funding drop off almost before it had arrived. But all was not lost. The organizational efforts surrounding such programs produced committed academic communities across several areas of AI research, with academics institutionally motivated – even without major external funding – to believe that another technological revolution would soon arrive to raise the societal valuation of their efforts.

This patience was rewarding. A new host of focus areas promised, to the delight of many, to help overcome the classification problems that plagued initial eras of narrow-task AI systems. Naïve Bayes, neural networks, decision trees and more helped resolve enduring issues around the manipulation of data and recognition of complex objects, from physical signals in the environment to semantically meaningful symbols and statements. In the mid-1990s, computers appeared finally ready to support new developments in AI applications as well. With processing power improving exponentially over successive hardware generations built in increasingly short windows, the opportunities for expanding on the limited successes of the 1980s were suddenly immense. Deep Thought was made to look obsolete by Deep Blue, a program that beat grandmaster and world champion Gary Kasparov in 1997. From this promising start, a third boomtime of AI development was clearly underway.[20]

Some scholars, like Henry Kautz, have categorized the third "summer" of AI as not really starting until 2011–2012. To classify the 2000s as a winter, however, misrepresents the degree to which previous bust moments almost derailed promising AI research. Much progress was made and many innovative applications were rolled out between roughly 1996 and 2011. What makes 2011–2012 a kind of "year zero" of the current boomtime for many AI researchers is not so much a condition of neglect in the 1990s and 2000s, but rather something revolutionary that took place in 2012 – the development of so-called "deep" neural networks. This development, the "deep" part of which refers to a neural network possessed of multiple layers that can each transform data into slightly more abstracted and composite representations of the real world than the previous, has led to the explosive development of the field of deep learning and deep reasoning. Rapid advances in natural language processing and visual data recognition using these techniques have led to uncanny intelligent agents across a number of settings in the span of just one decade. Moreover, AI researchers have begun to build AI systems that train themselves and even have the capacity – if not the permission, yet – to build new AI systems. As a result of this one breakthrough, AI research has advanced more in ten years than in the preceding 60. Indeed, AI research has so rapidly advanced on all fronts since 2012 that many voices have called into question the notion of a future bust cycle, stating that humanity has finally realized machine intelligence that can so completely mimic human behavior while at the same time being so unique in its approach to "thinking" that a slowdown seems unlikely in the extreme.[21]

Artificial intelligence and cyber conflict dynamics

Recent work on cyber conflict has highlighted AI as a prospective game-changer when it comes to the planning and execution of cyber operations, as well as for efforts to defend against and deter cyber aggression.[22] However, as empirical references remain limited and largely hypothetical – at least in non-classified settings – so too does the opportunity for extrapolating from technical developments to likely operational and strategic effects. At time of writing, it would not be unreasonable to characterize most assessments of AI's potential impact on cybersecurity as the move toward more of the same, just "faster, smarter, bigger, better." Specifically, experts appear to fear that AI will extend the lifecycle of cyber threats and create an expanded footprint for malicious code that will be more difficult for defenders to model, detect and mitigate. Likewise, machine learning stands to make sophisticated malicious cyber action a more accessible reality for the median security actor, a dynamic that further contributes to a less rigid toolkit of tricks upon which attackers must rely and defenders must consider.[23]

It wouldn't be unreasonable for us to consider potential transformations as occurring across the three oft-cited levels of analysis used by national security practitioners to delineate distinct elements of the battlespace of international relations – the (1) tactical, (2) operational and (3) strategic levels of security operation. At the tactical level, which is where direct engagement between fighting parties is planned and executed, the opportunity to increasingly design software that is highly adaptive beyond the requirement for human input to the process portends substantial challenges for cyber defense.[24] Three categories of such adaptability bear particular mention. First, AI-augmented malware implies an emergent capability for technique selection.[25] An autonomous ability to assess environments and select the optimal method of intrusion is worrisome not so much because it differs dramatically from existing capabilities among sophisticated threat actors; rather, but because a more accessible version of the capacity implies a potential rise in the average sophistication of cyber threats. Second, and relatedly, advanced malware might prove itself capable of tactical adaptation beyond simple method selection.[26] This means that malware may increasingly possess, without human input, an ability to abandon one tactic in favor of another based on analysis of environment contours and defenses. And finally, AI may lend itself not only to a technical ability to select tactical orientation but also to an ever-increasing capacity for value calculations vis-à-vis strategic objectives.[27] An AI-enabled ability to "use incoming data obtained via infection of machines to probabilistically judge where and when further infection is likely to lead to some value return" offers a tactical solution for the constraint of necessary operational resource commitment that often plagues offensive cyber operations (OCO) planning. In other words, cyber operations are often more expensive than is generally thought by non-experts and can be immensely resource- and time-intensive from an intelligence planning perspective. Well, if code can be developed that acts as a human operator might in sensitive situations in a trustworthy fashion then some of that cost may be defrayed, making cyber operations an even more appealing capability than at present.[28]

Operationally, machine learning and Big Data analytics most clearly imply an expanding capability for intelligence agencies, security institutions and criminal enterprise alike to develop extremely high-fidelity comprehension of the attack surface of targets, whether those be individuals, facilities, institutions or even national and sub-national units.[29] Simply put, the more information made available to competent cybersecurity stakeholders, the more capable the infrastructure and tools developed for malicious cyber activity are likely to be.[30] For defenders, this operational advantage suggests heightened insecurity on two particular fronts. First, it implies a general diversification in the tools available for procurement by either criminal or political security actors in the artifact and cyber threat services development marketplaces.[31] Second, and more directly, it implies an ever-increasing probability of compromise via lateral access to critical systems (i.e. access gained via indirect targeting of associated institutions, users or architecture).[32]

It is at the strategic level where the impact of variegated AI technologies and advances on the contours of global cyber conflict remain the most difficult to forecast. On the one hand, everything discussed earlier implies an increased tempo of engagement between capable cyber actors in world affairs, as well as an expansion of the landscape of those that might qualify as "capable" and a pressing need to rely on machine solutions for challenges whose scale and scope might increasingly be shaped more by algorithmic input than human design.[33] On the other hand, such a read of the future of AI and cyber conflict (i.e. the "bigger, faster, smarter" paradigm) ignores the reality that AI incorporated across all facets of national security and society writ large portends new rationales for cyber engagement.[34]

Most simply, as AI systems themselves become synonymous with critical societal and security functions, they become targets of extreme value. Specifically, a malicious actor might either target AI systems with (1) *input attacks* that present intentionally misleading observations to a machine learning algorithm so as to guide its calculations or engage in (2) *poisoning* activities that actually corrupt a learner's dataset, algorithm or models via the provision of manipulated data.[35] In doing so, they might subvert the functionality of systems both bound up in the workings of the fifth domain and beyond it. And while there are, of course, a great many ways to undertake such interference, cyberspace presents as arguably the most direct and critical avenue via which to do so. The strategic implication of AI is thus not only an intensification of current trends but also that the subversive character of cyberspace is set to take on new dimensionality as cyber-enabled artificial intelligence attacks (CAIA) present as an appealing method for interference, espionage, crime and sabotage even beyond the scope of domain-specific activities.[36]

AI hacks us, we hack each other, we all hack AI

The previous section describes artificial intelligence in the context of cyber conflict in fairly simple terms, that is, as a technology that will upgrade current methods of computation sufficient to supercharge certain features of engagement in cyberspace. But to understand the transformative potential of AI for cybersecurity, we need to think more broadly than just the technical variation in software and modeling that will functionally make cyber operations enabled by AI bigger, faster, smarter and better. Specifically, we need to consider the impacts that AI will have on society writ large and then bring these functional considerations back in to that context.

As we've talked about in this book, hacking is an exploitation of some system that subverts the rules or norms of that system. This often hurts the system but is not always something that is explicitly forbidden; much hacking is simply something not anticipated or intended by system designers. Of course, when we think about hacking in the modern context, it is hard not to think of computer or other networked devices. But hacking is an activity that can be generalized to human systems broadly. Systems of economic activity, systems of government, systems of democratic or other governance, systems of social behavior – these are all hackable constructs.

Information technologies, particularly highly dynamic GPTs, tend to hack society in fundamental ways. By this, we mean that the presence of new potential to achieve unprecedented physical, informational or other outcomes leads to changes in the way that society works under the hood. The development of efficient printing presses in Europe in the 1500s, for instance, subverted the "normal" function of social, political and economic systems, simultaneously producing two centuries of religious warfare on the continent *and* unprecedented economic growth leading into the Industrial Revolution.[37] This did not occur only because more people could access stored knowledge; rather, these radical transformations to European (and, thus, global) society happened because the printing press and subsequent iterations of the technology motivated the development of regional and national identities, led to common ways of thinking about time and morality, incentivized the standardization of education systems, and encouraged governments to find ways of mobilizing labor forces sufficient to both sustain economic growth and enable the projection of national power.

Artificial intelligence will hack society. It is already doing so and will increasingly do so in a number of different ways. Over the past two decades, for instance, machine learning developments have produced so many minor consumer conveniences and service features that scientists have now regularly measured changes in the propensity for the average citizen to rely on computer-generated results as fact and to trust constructs that are clearly artificial even under circumstances where the track record of the systems involved is shaky at best. In the 2010s, an immense volume of social media activity around major social or political issues on platforms like Twitter was shown to have been driven by fake bot accounts. Numerous controversies, for instance linked to the World Cup in Brazil in 2014 or the Brexit referendum in 2016, were demonstrably shown to originate either entirely or in large part from the rebroadcasting activities of such accounts. And yet, surveys of social media users in the intervening time have shown that individuals with low tolerance for perceived "fake" activity on such platforms remain highly resilient to changing their own behavior or updating beliefs when given evidence that their own engagement has been influenced by bot machinations. AI is, in other words, already shifting the psychology of those using technology to engage in social, political and economic discourse in meaningful ways. With social media, algorithms are, among other things, altering the underlying resilience of some individuals toward questionably factual information. Elsewhere, similar kinds of cognitive shift are likely shifting the basis of how global publics perceive their relationship to the polity and to each other. AI is, in short, already hacking our brains.

It's hard to predict what else about contemporary society AI might hack beyond simply suggesting that change is likely to be broad-scoped. This book has described the massive and often counterintuitive ways in which web technologies have rewired the function of global society to produce the world we know today. The same thesis and narrative will almost certainly be found in books describing the advent and large-scale adoption of AI several decades from now. But what does such transformation mean for cyber conflict?

From the last section, we know generally how AI will enable our own intentional hacking of the technology – the software and the machines and information it controls – that surrounds us. What must be considered here, however, is that those activities will take place in the context of a society hacked and transformed in an ongoing fashion by AI.

And, of course, part of our hacking won't just be the standard fare of cyber conflict today – espionage, industrial sabotage, etc. We'll target AI itself, poisoning data or manipulating input conditions for AI models often via the medium of cyberspace.

In Chapter 5, we spoke of the various common attributes of cyber conflict often bandied about by practitioners, researchers and policymakers concerned with global cybersecurity challenges. While fathoming out a changed logic of cyber conflict in the age of pervasive AI usage is beyond the scope of this book, it's relatively easy to see how these assumptions might be weakened or invalidated by the new technology. The broad-scoped challenge of attribution in cyberspace, for instance, may be much less relevant to national security planners and cybersecurity investigators in the context of cyber artificial intelligence attacks. If CAIA is always a distinct possible outcome of malicious activity encountered in cyberspace, then possible attribution of an adversary action – technical or political – must always be treated with some skepticism. Was attribution an intended outcome of a given episode? What is the basis of attribution and can that set of foundational assumptions be trusted given the possibility of adversary manipulation of the inputs to defensive AI systems in the past?

At the same time, several of the core attributes of cyber conflict described in Chapter 5 may remain relevant but not descriptive of the same operational dynamics of digital security at work today. The unpredictability of cyber instruments and the possibility of unintended consequences stemming from their use refers generally to how code inserted into complex virtual environments can interact (or fail to interact) with encountered systems in a precisely predictable fashion. Standard attacks may produce no results. Limited intrusions may trigger substantial, unforeseen disruption to a target system or a connected system. In the age of AI, this characteristic of unpredictability about cyber effects remains. However, the source of unintended outcomes likely shifts in one of several ways. On the one hand, automated defense mechanisms, even when well understood in another environment by attackers, may react unpredictably on a case-by-case basis. This is something more than just adverse reactions by programmed code to the intrusion of malware. Intelligent agents are already being designed to operate dynamically as a human might, thus opening the possibility that unintended consequences result from a wider range of intrusive activity. And this likelihood is not something that might be reduced via training and research either, as so much AI operation intrinsically involves "black box" results in which the actual calculations and methods of reaching a particular outcome are not visible to technology users. Another possibility, of course, is that AI systems attacked via cyber operation are themselves referents for yet other AI models being employed across, say, a particular industry or sector of the national security establishment. Here, the idea of unintended consequences takes on new meaning as the implications of even minimally successful compromise of AI targeted via CAIA could be a legacy ensconced in the flaws observed and learned by yet other models.

The human factor, path dependencies and network effects, oh my!

A sufficiently dynamic information technology (or set of information technologies) might change the logic of global cyber conflict not only because essential techniques, capacities and infrastructure might be evolve but also because said technologies will have transformative effects on the distribution of social, political, economic and security value in global society. This axiom guides the thinking of this book's authors and of others who have given thought to the future evolution of cyber conflict. In this chapter, we have described in broad and general terms how this might work with artificial intelligence, perhaps the most scrutinized and popularized forthcoming technological revolution of our day. Simply put, AI stands poised to hack society. Within that context over the next years and decades, humans will both continue to hack one another and, by dint of said societal transformation, begin to use cyber means and infrastructure to directly attack and hack AI itself.

Baked into this chapter's discussion of evolutionary logics of cyber conflict are the core themes introduced early in *Understanding Cyber Warfare's* discussion of digital insecurity in the world today. Just as flaws and shortcomings in human design and use of new information technologies produced the recurrent cybersecurity problems (and the universe of associated digital security challenges) we see at substantial scale in the world today, so too is human inputs to AI development already leading to suboptimal outcomes that could affect future security dynamics. From criminal justice AI models that utilize data infused with decades of racial, social and other biases to systems designed to summarize large quantities of data being favored by decision-makers for aesthetic (rather than functional) reasons, this era of early experimentation with AI for more general purposes is replete with examples of human flaws codified in software. If left to persist in one form or another, these flaws could dictate paradigms of security behavior for years to come. After all, just as early decisions to ignore security management in certain network technologies were made so much more impactful when entire national industries adopted flawed network architectures, so too might early choices made about AI – to make certain standard products less accurate and more robust, for example, thus making false positive results more likely as a rule[38] – produce security imperatives at scale going forward. And in all of this, of course, are the effects of network externalities, the phenomena by which the value and function of key systems is determined by the demands of commercial, consumer and sociopolitical interests.

All this said, "getting it right from the start"[39] is a tall order for any technology, particularly those that are extremely dynamic in terms of the underlying science and engineering, the prospective applications and the societal use cases. With artificial intelligence, a range of issues constitute widespread and ongoing discussions on how to make AI "safe" for human use. Three such issues bear particular mention. First, the issue of **specification** plagues much development of AI systems.[40] Specification essentially refers to task of conveying to a particular machine learning model exactly what the designers would like it to do. Given that machine learning essentially works by the user feeding *both* data and a desired outcome to an algorithm so that a better algorithm might be produced for dedicated task work, the ease of specification varies wildly across tasks. Specification of simple jobs like identifying specific types of images as either fitting a mold or not can be reasonably straightforward. But where the task assigned to a system is more complicated, it can be difficult for the designers and users to mathematically articulate intention. Particularly as machine learner algorithms will infer the parameters of their task from observations in either simulation or past data, the AI system may end up seeing something either slightly or radically differently than the humans in the loop, thus opening the door for potentially massively divergent outputs. Perhaps the best metaphor to describe this problem to the average citizen is that of a genie. You find a lamp and make three wishes sounds simple enough. But as so much storytelling illustrates, the wording of your demand matters. If you phrase a wish too subjectively or even too descriptively, you might get something wildly different than you intended.

Second, AI systems need to be interpretable.[41] What does this mean? Quite simply, humans need to be able to *consistently* predict the outputs of a given algorithm. If experience with a particular AI system finds oneself unclear about why a given result was reached in even a small number of cases, that system immediately presents as less reliable than will be desired – even required – for certain applications. An AI model used to help predict cyber activity and intentionality related to critical infrastructure protection must, in an ideal setup where strategic actions are linked to tactical analyses of intrusion activity, accurately notate anomalous behaviors as distinct from expected engagements. Not doing so could lead to offensive responses not reflective of the original intention of a given adversary, with the result that a simple inability to trust AI findings under certain circumstances could begin a major foreign policy incident.

Finally, though relatedly, AI systems must be explainable to a reasonably high standard. It is not enough simply to be able to predict what a given model will output; rather, humans in the loop must have some substantial degree of understanding as to how and why an algorithm reached that conclusion. This task is nowhere near as simple as it sounds. AI researchers and users commonly observe the "black box" problem[42] wherein the inner workings of an AI model tasked to develop proficiency in a particular task or subject matter area are opaque at the point where the system is in full operation. Indeed, no lecture on AI usability and potential is complete without a series of examples of such systems reaching unexpected outcomes even as a reasonable track record of accurate classification is being observed. Programs set to classify pictures of canids as either being dogs or being wolves will suddenly and unexpectedly classify chihuahuas or pugs as the latter. Algorithms tasked with finding efficient ways to build bridge structures given a set amount of resources will, in unpredictable circumstances, take the unusual step of constructing a tall building and then toppling it over the gap to be bridged. What dictates these odd variations from what otherwise appears to be an accurate ability to undertake the task set? In particular, given that such deviations aren't something you would typically expect from a human, what is it about the cognition of this particular machine that leads to unpredictable outcomes?

Each of these issues clearly exacerbates the various challenges that might be faced via the application of AI systems in global cyber conflict and more broadly in national security establishments. And "getting it right now" is far from a simple task. Indeed, the sheer volume of possible pathways forward with regard to the design, implementation and norm-setting challenges around AI make the exact transformation of cyber conflict by AI virtually impossible to predict with any degree of granularity. All we can say with certainty is that such transformations are coming and they are far from just a function of advancing technological realities. They will emerge from human decision-making, reified and codified in standard practice, and intensified by the overlapping influence of our social, political and economic interests.

From evolution to revolution

Artificial intelligence is far from the only technology – a set of technologies – to portend a dramatic evolution of the core logics of global cyber conflict. Fifth-generation (5G) wireless technology standards and subsequent developments in signals transmission are another area wherein underlying technological advances and the subsequent human-centric development of new infrastructures could have a similar effect.

In the next chapter, we extend this discussion of evolution in global cyber conflict realities to consider technologies that could do more than simply iterate in transformative fashion. Rather, there are technologies – some extant in early stages of development and others entirely theoretical – that could have as substantial an impact on how humans communicate, store and interact with data and cognitively interface with the world around us as did packet switching and the various related breakthroughs in the 20th century that have given us the digitally infused world we have today.

Notes

1 A term used in various cyber conflict works, such as Whyte, Christopher. "Problems of Poison: New Paradigms and "Agreed" Competition in the Era of AI-Enabled Cyber Operations." In *2020 12th International Conference on Cyber Conflict (CyCon)*. Vol. 1300, pp. 215–232. IEEE, 2020.
2 Such as those in Stevens, Tim. "Global Cybersecurity: New Directions in Theory and Methods." *Politics and Governance* 6, no. 2 (2018): 1–4.
3 Kaplan, Fred. *Dark Territory: The Secret History of Cyber War*. Simon and Schuster, 2016.
4 Schneider, Jacquelyn, Benjamin Schechter, and Rachael Shaffer. "A Lot of Cyber Fizzle But Not A Lot of Bang: Evidence about the Use of Cyber Operations from Wargames." *Journal of Global Security Studies* 7, no. 2 (2022): ogac005; Gomez, Miguel Alberto, and Christopher Whyte. "Cyber uncertainties: Observations from cross-national war games." In *Cyber Security Politics*, pp. 111–127. Routledge, 2022.
5 For notable entries in this literature, see Nye Jr, Joseph S. *Cyber power*. Cambridge MA: Harvard University Belfer Center for Science and International Affairs, 2010; Willett, Marcus. "Assessing Cyber Power." *Survival* 61, no. 1 (2019): 85–90; and Dunn Cavelty, Myriam. "Europe's Cyber-Power." *European Politics and Society* 19, no. 3 (2018): 304–320.
6 The best description of which is in Naughton, John. "The Evolution of the Internet: From Military Experiment to General Purpose Technology." *Journal of Cyber Policy* 1, no. 1 (2016): 5–28.
7 An argument forwarded by Gannon, J. Andrés, Erik Gartzke, Jon R. Lindsay, and Peter Schram. *The Shadow of Deterrence: Why Capable Actors Engage in Conflict Short of War*. Working Paper, 2021.
8 This idea has received much attention in the study of social movements in the digital age and the communications studies literature, but is often hard fought elsewhere in the social sciences wherever scholars focus on web technologies in economic, political or other societal context. See Curran, James, Natalie Fenton, and Des Freedman. *Misunderstanding the Internet*. Routledge, 2016.
9 Jensen, Benjamin M., Christopher Whyte, and Scott Cuomo. "Algorithms at war: the promise, peril, and limits of artificial intelligence." *International Studies Review* 22, no. 3 (2020): 526–550.
10 For some of the best historical reading on AI, see Nilsson, Nils J. *The Quest for Artificial Intelligence: A History of Ideas and Achievements*. New York: Cambridge University Press, 2010; and Simon, Herbert. "Artificial Intelligence: An Empirical Science." *Artificial Intelligence* 77, no. 2 (1995): 95–127, https://pdfs.semanticscholar.org/.
11 See, among others, Hawking, Stephen, et al., "Transcendence Looks at the Implications of Artificial Intelligence – but Are We Taking AI Seriously Enough?." *The Independent*, May 1, 2014, www.independent.co.uk/; and Max Tegmark, *Life 3.0: Being Human in the Age of Artificial Intelligence*. New York: Knopf, 2017.
12 Nilsson, Nils J. *The quest for artificial intelligence*. Cambridge University Press, 2009.
13 Floridi, Luciano. "What the near future of artificial intelligence could be." In *The 2019 Yearbook of the Digital Ethics Lab*, pp. 127–142. Springer, Cham, 2020.
14 Wiener, Norbert. *Cybernetics or Control and Communication in the Animal and the Machine*. MIT press, 2019.
15 Bostrom, N. Superintelligence. "Paths, dangers, strategies." (2014).
16 Findler, Nicholas V. "The debt of artificial intelligence to John von Neumann." *Artificial Intelligence Review* 2, no. 4 (1988): 311–312.
17 McCorduck, Pamela, and Cli Cfe. *Machines Who Think: A Personal Inquiry Into the History and Prospects of Artificial Intelligence*. CRC Press, 2004.
18 McCorduck 2004.
19 Copeland, Jack. *Artificial Intelligence: A Philosophical Introduction*. John Wiley & Sons, 1993.
20 McCorduck 2004.

21 Floridi, Luciano. "AI and its new winter: From myths to realities." *Philosophy & Technology* 33, no. 1 (2020): 1–3.
22 See, for instance, Whyte, Christopher. "Poison, Persistence, and Cascade Effects." *Strategic Studies Quarterly* 14, no. 4 (2020): 18–46; Burton, Joe, and Simona R. Soare. "Understanding the strategic implications of the weaponization of artificial intelligence." In *2019 11th International Conference on Cyber Conflict (CyCon)*. Vol. 900, pp. 1–17. IEEE, 2019; and Davis, Zachary. "Artificial intelligence on the battlefield." *Prism* 8, no. 2 (2019): 114–131.
23 Sharikov, Pavel. "Artificial intelligence, cyberattack, and nuclear weapons—A dangerous combination." *Bulletin of the Atomic Scientists* 74, no. 6 (2018): 368–373.
24 Whyte, Christopher. "Poison, persistence, and cascade effects." *Strategic Studies Quarterly* 14, no. 4 (2020): 18–46.
25 Ibid., 24.
26 Ibid., 24–25.
27 Ibid., 24–25.
28 Whyte, Christopher, and Brian Mazanec. *Understanding Cyber Warfare: Politics, Policy and Strategy*. Routledge, 2018, Chapter 5.
29 Leenen, Louise, and Thomas Meyer. "Artificial intelligence and big data analytics in support of cyber defense." In *Research Anthology on Artificial Intelligence Applications in Security*, pp. 1738–1753. IGI Global, 2021. Also see Jahankhani, Hamid, Stefan Kendzierskyj, Nishan Chelvachandran, and Jaime Ibarra, eds. *Cyber Defence in the Age of AI, Smart Societies and Augmented Humanity*. Springer Nature, 2020.
30 Whyte (2020), 25–26.
31 Kantarcioglu, Murat, and Fahad Shaon. "Securing big data in the age of AI." In *2019 First IEEE International Conference on Trust, Privacy and Security in Intelligent Systems and Applications (TPS-ISA)*, pp. 218–220. IEEE, 2019
32 Radanliev, Petar, David De Roure, Rob Walton, Max Van Kleek, Rafael Mantilla Montalvo, La'Treall Maddox, Omar Santos, Peter Burnap, and Eirini Anthi. "Artificial intelligence and machine learning in dynamic cyber risk analytics at the edge." *SN Applied Sciences* 2, no. 11 (2020): 1–8.
33 Whyte (2020).
34 Horowitz, Michael C. "Artificial intelligence, international competition, and the balance of power (May 2018)." *Texas National Security Review* (2018).
35 Comiter, Marcus. "Attacking artificial intelligence." *Belfer Center Paper* (2019): 2019–08. Also see Insua, David Rios, Roi Naveiro, Victor Gallego, and Jason Poulos. "Adversarial machine learning: Perspectives from adversarial risk analysis." *arXiv Preprint arXiv:2003.03546* (2020); and Puthal, Deepak, and Saraju P. Mohanty. "Cybersecurity Issues in AI." *IEEE Consumer Electronics Magazine* (2021).
36 Whyte (2020), 26.
37 Most famously described in Anderson, B. *Imagined Communities. Reflections on the Origin and Spread of Nationalism*. London: Verso, 1983. 240 p.
38 Hoffman, Wyatt. "Making AI Work for Cyber Defense." (2021).
39 For an overview of expert opinion on AI, see Müller, Vincent C. and Nick Bostrom, "Future Progress in Artificial Intelligence: A Survey of Expert Opinion, In Vincent C. Müller ed. *Funda-Mental Issues of Artificial Intelligence*Synthese Library; Berlin: Springer, 2016, 555–72, www.nickbostrom.com/.
40 Rudner, Tim G.J., and Helen Toner. "Key Concepts in AI Safety: Specification in Machine Learning." (2021).
41 Rudner, Tim G.J., and Helen Toner. "Key Concepts in AI Safety: Interpretability in Machine Learning." (2021).
42 Rai, Arun. "Explainable AI: From black box to glass box." *Journal of the Academy of Marketing Science* 48, no. 1 (2020): 137–141.

Further readings

Buchanan, Ben and Andrew Imbrie. *The New Fire: War, Peace, and Democracy in the Age of AI*. The MIT Press, 2022.

Corera, Gordon. *Cyberspies*. Simon and Schuster, 2016.

Domingos, Pedro. *The Master Algorithm: How the Quest for the Ultimate Learning Machine Will Remake Our World*. Basic Books, 2015.

Kania, Elsa B. "Artificial intelligence in China's revolution in military affairs." *Journal of Strategic Studies* 44, no. 4 (2021): 515–542.

The National Security Commission on Artificial Intelligence. *The Final Report*. www.nscai.gov/2021-final-report/

U.S. Government Accountability Office. *GAO-21–519SP, Artificial Intelligence: An Accountability Framework for Federal Agencies and Other Entities*. June 2021. www.gao.gov/products/gao-21-519sp

Whyte, Christopher. "Poison, Persistence, and Cascade Effects." *Strategic Studies Quarterly* 14, no. 4 (2020): 18–46.

14 Revolution

How the nature of cyber conflict might change

As we discussed in the previous chapter, artificial intelligence and other technologies herald a dramatic evolution of the core logics of cyber conflict. In this chapter, we flip the consideration of major technological evolutions and revolutions on its head and look specifically to those advances that might strike at the very heart of the Internet's inherent ability to lock people into its use – the viability of its logical layer. In this volume, we review the history and technical functionality of the Internet that was introduced in Chapter 2, emphasizing the Internet's inherent insecurity because of early design decisions and the path dependency borne of entire industries growing up around its use. Today, some potential innovation areas threaten prevailing cybersecurity practices based on the syntactic design of the Internet or promise to end reliance on the Internet's current protocol-based architecture in favor of alternative means of connectivity. Foremost among these are quantum computing and blockchain and other forms of distributed ledger technologies, but other trends like non-traditional data storage are also relevant. In this concluding chapter, we describe such advances as both a counter to the existing architecture of the Internet and a promising opportunity for bettering the state of global digital security. Whether such promise can be realized, however, depends on having learned lessons from more than half a century of network developments and – arguably more importantly – on quickly building from ongoing debates to effectively reconcile the function of the digital with the function of the democratic. This chapter examines this issue by again reviewing the current inherently insecure nature of the Internet, followed by an examination of these new, potentially revolutionary, technologies. We then conclude with some final thoughts on the future of cyber conflict at this key juncture.

As we discussed in the last chapter, our book underscores a core proposition, essentially that a sufficiently dynamic set of information technologies might change the logic of global cyber conflict not only because essential techniques, capacities and infrastructure might be evolve but also because said technologies will have transformative effects on the distribution of social, political, economic and security value in global society. This chapter examines some of the technologies that could go beyond an iterative transformation. Namely, those that could have as substantial impact on how humans communicate, store and interact with data, and cognitively interface with the world around us. Before examining these potentially revolutionary technologies, let us first look at the existent logic of the Internet, build on packet switching and the various related breakthroughs in the 20th century that have given us the digitally infused world we have today.

DOI: 10.4324/9781003246398-14

The inherent insecurity of today's Internet

As was discussed in Chapter 2, security concerns associated with today's Internet can be traced all the way back to the "handshake protocol," whereby designers sought compatibility across different systems and technologies and did not contemplate or incorporate security considerations. Compatibility and convenience were king, and the Internet was originally intended to link research computers, increase remote access, and harness collective computing power. For example, the Advanced Research Projects Agency (ARPA), early in the development of the Internet, tried to tie together its three different networks so they would function as a single system known as ARPANET. As was previously discussed, ARPANET was not set up nor augmented at various developmental inflection points with security management in mind. Though security considerations were brought up even during the initial development of ARPANET, most famously by Willis Ware and his colleagues, they were ultimately side-lined by the argument that it was more important to get the technology right before saddling it with regulatory requirements.[1] As a result, the Internet and related global network architecture that undergirds the modern world is largely based on trust and relies on approaches such as the aforementioned handshake protocol as well as the Transport Control Protocol, Internet Protocol (IP) and HyperText Transfer Protocol. Compounding these logical insecurities is the human element, which adds yet another level of insecurity, along with increasing localization of the physical infrastructure of the Internet, allowing more authoritarian governments increased opportunity to further undermine privacy and security considerations. While there are potential changes looming to public key and digital signature encryption methods as well as a transition from IPv4 to IPv6 to increase the number of IP addresses – as we discussed in Chapter 2 – none of these portend a fundamental revolution in the security of the Internet. However, there are some new, potentially revolutionary, technologies that could transform the Internet's current inherent insecurity into something different altogether. These technologies are blockchain and distributed ledger technology (commonly referred to as crypto), non-traditional data storage and quantum computing. Each of these will be examined in more detail in the sections that follow.

Transforming security: blockchain, distributed ledger technology and new information storage mediums

Blockchain, a form of distributed ledger technologies (DLT), are a secure way of conducting and recording transfers of digital assets without a central authority or registry. DLT is "distributed" because multiple individuals or participants in a computer network share and synchronize copies of the ledger. New transactions are added in a manner that is cryptographically secured, permanent and visible to all participants in near real time. DLT and blockchain may present a revolutionary paradigm shift for cybersecurity, insofar as it offers a decentralized way of conducting transactions without a need for a central actor. This is potentially of great interest as the Internet undergoes a transition to more of a Balkanized infrastructure with various data localization requirements. Despite these headwinds to international collaboration, DLT could offer a path to improving the architecture of the Internet, for example by enabling device-to-device encryption across the Internet of Things. DLT could also be used to harden the Domain Name System (DNS), which as we discussed in Chapter 2 is a kind of a public directory that links domain names to their IP addresses. Malicious cyber actors have exploited the insecurity of the Internet by accessing the DNS and attacking these links thus crashing sites. With the immutability and decentralized nature of DLT and blockchain, it could potentially be used to store the DNS with enhanced security. However, the potential advantage of DLT could also be its weakness, as a lack of centralized governance and regulatory frameworks could impede its adoption. It also is very reliant on private keys for encryption, which cannot be recovered once lost and could lead to permanent data lost in some scenarios. Despite these challenges, private and state actors are moving toward adoption of DLT to enhance security, for example, Barclays in London has filed a patent to use DLT to enhance the security of fund transfers in traditional banking.

In addition to the logical changes that DLT may bring to the Internet and cyber conflict, a change in the physical nature of data storage may also result in revolutionary changes. For example, alternative data storage technologies – such as synthetic DNA and etched glass – are being developed due to expected future inadequacies of current data storage media with the ongoing explosion of big data. Current data storage media, such as tape, DVDs, and hard drives, meet the current global data storage need of around 97 trillion gigabytes; however, data storage demands are expected to double by 2025. Additionally, the plastic and magnetic materials used in current storage technologies degrade over time, often requiring frequent replacement and the silicon needed to produce more storage using contemporary means is limited. Synthetic DNA and etched glass offer potential alternatives that address these issues; however the technology is not yet mature nor commercially available.[2] Synthetic DNA storage simply uses an artificial DNA strand created in a lab to store data, which is then read by DNA sequencing equipment. The potential capacity is enormous – nature is clearly highly efficient – with synthetic DNA potentially storing over 11 million, million gigabytes in a cubic inch of material.[3] This far exceeds contemporary data storage methods. With etched glass data storage, quartz glass is used along with a very precise laser, similar to those uses in Lasik corrective vision surgery. The laser can make etchings representing binary code, containing five unique attributes (the location on a three-point axis, size and orientation), allowing for greater data storage. But, like synthetic DNA storage, this technology is not yet mature, and it also suffers from issues of very slow write speeds that may limit its broad utility.[4] Regardless, both of these technologies could transform the data storage landscape of cyberspace in the coming decades, bringing revolutionary changes to the nature of cyber conflict through eliminating some existing vulnerabilities and perhaps creating new ones.

The looming quantum computing revolution

Another potential change that could shift the entire framework and paradigm of cyber conflict that we have discussed throughout this volume is a breakthrough in quantum computing technology which gives one state a massive advantage in cyberspace, allowing it to attack previously well-defended targets and penetrate advanced cryptology. Gret Tallant, a manager at Lockheed Martin who has worked on quantum computing, has said that "computationally, quantum computing is the equivalent of the Wright Brothers at Kitty Hawk; it has the potential to be a turning point in our history."[5] As of early 2022, Google has the most powerful quantum computer, a 72-quantum-bit (qubit) processor that is equivalent to a 272-conventional-bit processor and National Security Agency (NSA) allegedly has a billion dollar research program focused on these computing machines, so the technology is not simply speculative science fiction.[6] Quantum computing is commercially available today, although may applications remain conceptual.

However, recent stumbles in keeping pace with Moore's Law may indicate that the pace of technological breakthroughs may be slowing.[7]

Quantum technology is an advanced form of computing based on the physics and engineering concept of quantum mechanics, the set of rules that govern the universe on a subatomic scale. The atomic clock used in GPS satellites and Internet synchronization protocols are currently utilizing the concepts of this technology to bring stability in the form of only a one second loss over a 100-million-year period.[8] The positive applications of this technology are plentiful. Quicker and more powerful computing could eventually allow for a way to solve profound technical and scientific problems that escape the ability of current supercomputers. However, the potential for serious risks to national security and cyber conflict also arise out of this technological advancement. For example, a quantum computer could in theory be able to break any encryption used by computers today. This poses a significant problem for communication security. Quantum communications could enable adversaries to develop secure communications that personnel would not be able to intercept or decrypt. Quantum computing may also allow adversaries to decrypt information, which could enable them to target an adversary or strategic competitor's personnel and military operations.

So what exactly is a quantum computer? It is a device based on Quantum Information Science (QIS) that takes advantage of counterintuitive properties that apply at the smallest scale of physics. The first of these quantum properties is a connection between particles known as "entanglement." Yet another quantum property is known as "superposition," which allows a particle, while unobserved, to be in all possible observable states simultaneously. There are other properties based on QIS that are used along with superposition and entanglement, either in isolation or in combination, within a quantum computer to sense, communicate and compute.[9] As an example, this could entail quantum bits, known as qubits, using superposition and entanglement to process data in more efficient ways.[10] Leveraging these quantum properties does not necessarily allow a quantum computer to process qubits at a faster rate than a regular bit of data in a conventional computer, however because each qubit can hold data in multiple states (rather than traditional 1s and 0s) they can hold so much more information and processing fewer qubits can reach the same output faster. For example, a conventional byte (eight bits) can store a single number between 0 and 256. However, a quantum byte (eight qubits) can represent 256 numbers all at the same time. Quantum computers with sufficient qubits will be able to quickly solve problems that would take a conventional computer decades or centuries. Of note, quantum computers will not fully replace conventional computers, but will supplement them and offer new capabilities in certain areas while simultaneously being incapable of some tasks currently conducted by conventional computers.

QIS also offers some promise beyond computing, such as for sensors, which are outside of the scope of this book. As with BLT and some of the new information storage means discussed earlier in this chapter, quantum computing is still in the relatively early stages of development and the technology does have some noteworthy limitations. Some have compared quantum computers today to the classical computers of the past, when somewhat fragile vacuum tubes filling an entire room or floor of a building were used. Today, quantum computers require very advanced hardware and an incredibly cold environment that is near absolute zero −273.15 Celsius due to the fact that its processers involve subatomic particles.

Despite these challenges, quantum computing presents the potential for a fundamental revolution in cybersecurity and cyber conflict. At the most basic level, the computing power that quantum computing may eventually bring to bear would be able to factor complex prime numbers, rendering encryption of any kind (public or private key) of limited use. The implications of this are enormous, rendering formerly classified or sensitive information and communications visible to whatever state or entity develops the necessary quantum computing capability.[11] This has led to a number of states to pursue efforts to develop quantum-proof encryption, often known as "post-quantum" encryption. For example, the U.S. National Security Agency (NSA) reported in its 2021 NSA Cybersecurity Year in Review that it was making a number of "Cryptographic Modernization Technology" advancements in response to the advances in QIS and adversary threats.[12] NSA's report noted, among other things, that they had partnered with the U.S. National Reconnaissance Office to certify the smallest post-quantum unit ever produced for space applications, enabling cube and small satellites to provide certified, crypto-secured support for the warfighter in a highly contested space environment. The report also noted that NSA provided the first set of updated U.S. cryptographic devices to protect U.S. National Security Systems (classified networks) from potential adversarial quantum computing attacks. The United States is certainly not the only state actor that recognizes the potential national security implications of quantum computing. For example, in 2017 the People's Republic of China announced a Jinan Project, a QIS effort intended to develop "the world's first unhackable computer network."[13] Many other private and state actors are undertaking similar initiatives.

Given the state of the QIS field and the fact that a cryptographically relevant quantum computer has not yet been developed, it is easy to assume that the cybersecurity risks and the revolutionary change this will bring to encryption in cyberspace is at some to-be-determined point in the future. However, this is not the case and to some extent the looming quantum revolution in cyberspace will be retroactive. Given the relative low cost of data storage (setting aside the long-term need for new storage options, such as the aforementioned synthetic DNA storage), many state actors are likely vacuuming up encrypted data that they are unable to read with the hope that at some point in the future they will be able to break through the encryption with a quantum computer. Therefore, the U.S. NSA and others have pressed for near term migration to interim post-quantum approaches, such as "quantum key distribution" which uses quantum principles to distribute traditional mathematically based encryption algorithms.[14] However, even with interim post-quantum encryption, it is still likely that some state actors will have significant archives of adversary encrypted information that will be abruptly rendered accessible at some point in the quantum future. The national security implications of this are hard to quantify or predict, but could be enormous. Given the often classified nature of Offensive Cyber Operations (OCO) and other forms of cyber conflict discussed in this book, this means that much of the unacknowledged cyber activity that has occurred over the past few decades may ultimately be acknowledged or disclosed. This could have a major impact on international norms and expectations regarding cyber behavior, among other potential impacts.

Revolution and beyond

The technological advances discussed this this chapter may result in a revolution that changes the nature of the Internet's inherent insecurity. New forms of post-quantum encryption or DLT may introduce new forms of security that eliminates some of the technological pathways for malicious cyber behavior. Conversely, the advent of quantum computing, especially if its emergence is asymmetric and gives a single state actor a monopoly on cryptographically relevant quantum computing, could be majorly destabilizing to strategic stability and the cyber balance of power. When and how these technology revolutions will play out is uncertain, as is the potential interplay between the evolutionary and revolutionary technologies in question in the two final chapters of this book. For example, might we see a form of artificial intelligence turbo-charged by quantum computing technology and massive new data storage technologies grounded on synthetic DNA? How will the advent of an Internet of Things enabled by ubiquitous connectivity and underwritten by fifth-generation (5G) – and soon sixth generation (6G) – wireless technology intersect with computing that is consistently being enhanced by superposition and other quantum phenomena? Only time will tell. What will not change is that these evolutions that we discussed in the previous chapter and revolutions discussed in this chapter will be grounded in human decision-making, reified and codified in standard practice, and intensified by the overlapping influence of our social, political and economic interests. In short, the digital world of the future will be shaped and molded by human beings and our social and cultural institutions, with digital insecurities very much an uncanny reflection our geopolitics, our economic systems and behaviors and those inherent sociopolitical insecurities that define modern human society.

Notes

1 For perhaps the best historical narrative treatment of the development of the Internet and its lack of security features, see Kaplan, Fred. *Dark Territory: The Secret History of Cyber War*. New York: Simon & Schuster, 2016. Also see Ware, Willis H. “Security and Privacy in Computer Systems.” In *Proceedings of the April 18–20, 1967, Spring Joint Computer Conference, ACM*, 1967, pp. 279–282.

2 Panda, D., Molla K.A., Baig M.J, Swain A, Behera D, Dash M. “DNA as a digital information storage device: hope or hype?.” *3 Biotech*. 2018;8(5):239. doi:10.1007/s13205-018-1246-7.

3 Potomac Institute for Policy Studies. *The Future of DNA Data Storage*. Arlington, VA: 2018.

4 Zhang, J. et al. “Eternal 5D Data Storage by Ultrafast Laser Writing in Glass.” Optical Society of America (2013).

5 Lockheed Martin. “Speedy Qubits Lead the Quantum Evolution.” October 28, 2013. www.lockheedmartin.com/us/news/features/2013/quantum.html?goback=%2Egde_4682795_member_5812625891389366275#%21.

6 Ibid.; Rich, Steven. “NSA seeks to build quantum computer that could crack most types of encryption.” *The Washington Post*, January 2, 2014. www.washingtonpost.com/world/national-security/nsa-seeks-to-build-quantum-computer-that-could-crack-most-types-of-encryption/2014/01/02/8fff297e-7195–11e3–8def-a33011492df2_story.html.

7 Clark, Don. “Intel Explains Rare Moore’s Law Stumble.” *The Wall Street Journal*, November 21, 2013. http://blogs.wsj.com/digits/2013/11/21/intel-explains-rare-moores-law-stumble/.

8 National Science and Technology Council. Advancing Quantum Information Science: National Challenges and Opportunities. Washington, DC, July 22, 2016.

9 Library of Congress. Congressional Research Service. Quantum Information Science: Applications, Global Research and Development, and Policy Considerations. R45409. Washington, DC: Updated November 1, 2019.

10 A general-purpose quantum computer may need over 100,000 qubits.

11 The necessary quantum computing capability for this purpose is often referred to as “cryptographically relevant quantum computing” or CRQC.

12 U.S. National Security Agency. *2021 NSA Cybersecurity Year in Review*. https://media.defense.gov/2022/Feb/03/2002932462/-1/-1/0/2021_NSA_CYBERSECURITY_YEAR_IN_REVIEW.PDF

13 Biercuk, Michael, and Richard Fontaine. “The Leap Into Quantum Technology: A Primer For National Security Professionals.” *War on the Rocks*. November 17, 2017.

14 U.S. National Security Agency. *Quantum Computing and Post-Quantum Cryptography: Frequently Asked Questions*. August 2021. https://media.defense.gov/2021/Aug/04/2002821837/-1/-1/1/Quantum_FAQs_20210804.PDF.

Further readings

John Naughton, "The evolution of the Internet: From Military Experiment to General Purpose Technology." (2016 Journal of Cyber Policy, 1:1, 5–28, DOI: 10.1080/23738871.2016.1157619), p. 10.

David Clark, Thomas Berson, and Herbert S. Lin, "At the Nexus of Cybersecurity and Public Policy: Some Basic Concepts and Issues." (2014, National Academies Press, Washington, D.C.), p. 18.

U.S. Government Accountability Office. GAO-19–704SP, *Science & Tech Spotlight: Blockchain & Distributed Ledger Technologies*. September 16, 2019.

U.S. Government Accountability Office. GAO-20–527SP, *Science & Tech Spotlight: Quantum Technologies*. May 28, 2020.

U.S. National Security Agency. *Quantum Computing and Post-Quantum Cryptography: Frequently Asked Questions*. August 2021. https://media.defense.gov/2021/Aug/04/2002821837/-1/-1/1/Quantum_FAQs_20210804.PDF

Glossary

Advanced Persistent Threat	Any set of hacking processes that involve a dedicated, often sophisticated adversary and their tools.
ARPANET	An early network system, funded by the U.S. Department of Defense, that was the first to utilize TCP/IP packet-switching procedures.
Asymmetric encryption	The practice of using two "keys," one privately held by an interlocutor and another publicly available to all, to securely encrypt and decrypt communications.
Asymmetric warfare	Warfare that occurs between actors possessed of remarkably different levels of capability.
Attribution	The responsibility of an actor for a particular action.
Authenticity	Making sure that the person(s) sending a message are who they say they are.
Autonomous systems	A collection of connected routing systems that ascribe to a single approach to routing policy online.
Availability	The notion that any information system is not entirely functional unless it is able to ensure that authorized users can always access it, even when it is under attack.
Balkanized Internet	A scenario wherein Internet access potential, policy and service offerings are wildly variable across different parts of the world, determined by sociopolitical or commercial interests.
Black hat hacker	A hacker whose intentions are malicious.
Blockchain	The technology underlying cryptocurrencies, based on the concept of a distributed ledger of records that is held in common across a community of peers who validate all transactions to ensure high Byzantine fault tolerance.
Bomba	The name given to an electromagnetic device, first developed by Polish engineers and then further updated by Alan Turing, to mimic the workings of the Enigma machines.

Brute force attack	An information attack wherein the attacker attempts to guess every single possible combination of variables that make up a "key."
Byzantine fault tolerance	The development of a system wherein a Byzantine fault, meaning a fault that is impossible to distinguish from normal conditions, is unlikely to occur.
Byzantine General's Problem	A game theoretical scenario wherein multiple generals, who must cooperate and communicate accurately in order to successfully invade a city, must figure out how to overcome treacherous messaging options so that they might act with certainty.
CIA Triad	The name given to the three pillars of information assurance – confidentiality, integrity and availability.
Ciphertext	The garbled output text of a message involved in cryptographic communications.
Circuit-switching	A method of communications that dedicates specific bandwidth to the transmission of entire information packages.
Classical realism	An IR school of thought that holds that states are the primary actors in international affairs and that their actions are motivated by the fact that human nature is intrinsically concerned with gaining power.
Client	Any device that accesses online services made available via a server.
Cloud Service Provider	Any organization that offers access to cloud-based services and storage.
Coercion	The practice of using force or threat of force to either prompt an opponent to change their behavior (compellence) or maintain their current behavior (deterrence).
Compulsory power	Power that manifests in the active relationships of actors in international affairs.
Computer	Traditionally a human worker whose job it was to "compute," now a name given to any device that mimics human intelligence by virtue of its reprogrammable design.
Computer Emergency Response Team	Any expert group that handles computer security events for a nation, region, sector, state or other regulatory subdivision.
Confidentiality	Making sure that information is not viewed by those who are not authorized to do so as it is transmitted.

Constructivism	An IR school of thought that holds that the fabric of international politics is best understood as social above anything else, wherein socially constructed perceptions of identity and interests shape intentions and actions.
Critical infrastructure	The name given to describe national industrial sectors that function as core assets for the workings of national economies and society.
Cryptocurrencies	The name given to blockchain-based platforms that offer some kind of value services to decentralized communities.
Cryptography	The study and practice of ensuring secrecy and security in communications/information transmission.
Cyber blockade	A type of large-scale offensive cyber operation that aims to deny service to a given region or country for a period of time.
Cyber conflict	A term used to describe all manifestations of cyber conflict, including and beyond cyberwar.
Cybercrime	Criminal activities that are enabled by the Internet
Cyber hygiene	The development and maintenance of routines that ensure good computer security awareness and practices.
Cyber mercenary	A non-state actor that hacks as part of a contract arrangement.
Cyber spammers	A category of cybercriminals that make money from spam content.
Cyberspace	The name given to the domain in which humans operate when they interact via the Internet.
Cyberterrorism	A term used to describe either the use of computer network attack for terroristic coercion or the use of the Internet to upgrade all parts of the terrorist enterprise.
Cyberwar	Generally, any scenario wherein two states (or otherwise powerful near-state actors) fight one another entirely via cyber means.
Dark Web	The name given to those parts of the networked world that exist on darknet overlay networks.
darknet site	A location on an overlay network, access to which requires specialized software.
Deep Web	The name given to those parts of the Internet that exist behind a restricted access lockout (i.e. something like a paywall or email service that requires user credentials to access).
Denial of service	The disruption of legitimate network access, most often at the application level, via flooding target applications and programs with requests.

Department of Defense	The federal agency in the United States responsible for the governance and maintenance of American military and intelligence policy and operations.
Department of Homeland Security	The federal agency in the United States responsible for public security matters, including civilian government and civilian-facing cybersecurity issues.
Digital activism	Social activism upgraded via the use of Internet technologies.
Digital signature encryption	Utilizing the public and private keys of asymmetric encryption in reverse to as to append an un-replicable "signature" to a message.
Distributed denial of service	The disruption of legitimate network access at the network level by flooding a computer with connection requests from numerous sources.
Domain Name System	A hierarchical naming system for Internet-connected devices and services that functions in a decentralized manner.
Electronic warfare	Warfare that involves the use of electromagnetic weaponry or occurs within that spectrum.
Enigma	The name given to a machine developed in Germany in the interwar period that would later become the primary means of encrypting German military communications.
External subversion	The name given to subversive campaigns and efforts resulting from the influence and direct interference of foreign governments.
Federal Bureau of Investigation	The federal agency in the United States responsible for domestic security, intelligence and counterterrorism operations.
Fifth domain of warfare	The term applied to cyberspace by the U.S. military to signify its shape as a piece of terrain, alongside the traditional domains of land, space, sea and air, wherein warfighting can occur.
Firewall	A type of defensive program that sits on information exchange points and lets traffic in (or not) via reference to a specific set of rules.
General Communications Headquarters	The signals intelligence and information assurance organization operating under the auspices of the government of the United Kingdom.
Geneva and Hague Conventions	Treaties and protocols that have established rules of conduct and international law on the humanitarian conduct of warfighting.
Geospatial intelligence	Intelligence collated from sensory data and demographic information about the interaction of humans and terrain.

Grey hat hacker	A hacker who undertakes computer network attacks for idealistic reasons.
Grey zone conflict	Conflict that occurs between war and peace, but most often between actors that could otherwise fight on the battlefield with some degree of parity of capabilities.
Hacker	A generic term used to describe those with specialized skills needed to construct malicious code and employ it.
Hacktivism	The use of malicious code and basic web vulnerabilities to hack in support of a social or political cause.
Handshake protocol	A protocol type that dictates pre-transmission granting of permission for communication by two or more nodes.
Hash function	A unique series of letters and numbers that results from any mathematical function that is easy to compute in one direction but not the other.
Human intelligence	Intelligence gathered by inter-human contact and connections.
HyperText Transfer Protocol	A protocol that allows for seamless multimedia data communication.
Imagery intelligence	Intelligence gathered via the use of image capture techniques, including cameras, infrared sensors and satellites.
Information collection	Any operation aimed at stealing information from adversary systems.
Information disinformation	Any operation that involved doctoring or otherwise manipulating information content to achieve psychological effects.
Information disruption	Any operation that aims to prevent enemies from achieving full information on their strategic environment.
Information exfiltration	The theft of data resulting from a computer network intrusion.
Information layer of the Internet	The name given to all content accessible via the Internet.
Information protection	Any operation designed to prevent negative consequences from military deception, psychological or propaganda operations.
Information transportation	Any operation designed to shape how adversaries communicate with one another.
Information warfare	Warfare that emerges from the manipulation and control of information systems for purposes of sowing disruption and misdirection among opponents.

Institutional power	Power that manifests in the rules of international interactions.
Integrity	Maintaining the accuracy of information within a message as it is transmitted.
Internal security measures	Actions taken to prevent negative consequences from military deception, psychological or propaganda operations.
Internal subversion	The name given to subversive campaigns and efforts waged by dissidents operating inside a given state.
International Relations	The name given to the study of world politics as a distinct sub-field of study within Political Science.
International Telecommunication Union	An organization originally developed to help provide multilateral oversight for telephone communications.
Internet	The global system of networks and network-connected computers that utilize packet-switching technologies for communications.
Internet Architecture Board	A deliberative and organizing forum that provides outreach functions for the IETF under the auspices of the ISOC.
Internet backbone	The routing infrastructure of the Internet, constituted of the cables, satellites and servers needed to maintain Internet functionality.
Internet Corporation for Assigned Names and Numbers	A nonprofit organization, originally set up under the auspices of the U.S. Department of Commerce, responsible for maintaining databases of Internet names and addresses.
Internet Engineering Steering Group	A group that is responsible for the day-to-day management of the IETF.
Internet Engineering Task Force	A nonprofit wing of the Internet Society that offers advice and sets standards on Internet routing technologies and protocols.
Internet Protocol	The primary data transmission protocol for relaying data across network boundaries.
Internet Service Provider	Any organization that offers services for accessing the Internet.
Internet Society	A nonprofit organization set up to provide leadership on Internet technology issues.
Intrusion Detection System	A type of defensive program that monitors data traffic within a network for anomalous behavior, which is then flagged to an administrator.
IP spoofing	The creation of data packets with false Internet Protocol address information for the purposes of spoofing target systems into redirecting responses to malicious sources.

Just war theory	An ethical doctrine dictating the appropriate use of force and the declaration of warfare.
Kernel	A piece of software that sits at the heart of a computer's operating system, often intended to be made up of only the most critical and secure system programs.
Kill Chain	A model that linearly tracks the stages involved in a successful cyber intrusion.
Law of Armed Conflict	All parts of international public law that deal with conflict, with major elements including the provisions of the Geneva and Hague Conventions and those articulations of just right to wage war included in the United Nations charter.
Liberalism	An IR school of thought that holds that states are inclined toward cooperation in the long run, either because democratic and liberal systems impel such judgements (classical liberalism) or because repeated interactions even in an anarchic world prompt cooperative inclinations to develop over time (neoliberalism).
Liberation technologies	The term often used to describe Internet technologies in the context of the affordances provided to non-state activists.
Logical layer of the Internet	The name given to all routing features of networked systems involved in Internet functionality.
Malware	"Malicious software"
Measurement and signature intelligence	Intelligence gathered from non-optical sensors, such as seismographs.
Message Authentication Code	A small piece of information appended to a message that is used to verify the identity of the sender.
Metamorphism	Code that is added to viruses that allow the program to entirely recreate itself to achieve the same functions but to technically appear unrelated to previous versions.
Military deception	Operations that are undertaken to mislead the military planning and observation efforts of foreign militaries.
MILNET	The part of the ARPANET that was specifically dedicated to American military usage.
Multi-stakerism	The notion that global governance issues should be debated and decided upon by both state and relevant non-state stakeholders.
Multilateralism	The notion that global governance issues should primarily be the remit of states.

National Aeronautics and Space Administration	The federal agency in the United States responsible for the administration of space exploration and transportation programming.
National Security Agency	The organization with the United States intelligence community responsible for the signals intelligence mission of the defense establishment.
Neorealism	An IR school of thought that holds that states are the primary actors in international affairs and that their actions are motivated by a concern about the power and intentions of other states.
Non-repudiation	The notion that one party cannot deny having taken an action with regard to communications.
North Atlantic Treaty Organization	The transatlantic alliance formed during the Cold War to combat the threat of the Soviet Union.
Offense-defense balance	The idea that technological potential manifests over different eras in inclinations toward either offensive behavior or defensive behavior, depending on the value of specific military technologies for either.
Offensive cyber operations	Any offensive actions enabled via the use of cyber instruments.
One-time pad	A type of symmetrical cryptographic device that uses two unique and completely randomly-selected algorithms – built into a physical disc – to allow secure communications.
Open source intelligence	Intelligence gathered and collated from publicly-available information sources.
Packet sniffing	The intercept and analysis of packet traffic across networks so as to glean information about would-be targets and/or attackers.
Packet switching	A method of communications wherein information is deconstructed to be packaged inside datagrams, constituted of a header and payload, before being sent across networks and reconstructed on a destination computer.
Patriot hacker	A non-state actor that hacks on behalf of a state entity for nationalistic or civic motivations.
Payload	A term used to describe malware that is functionally linked to a major objective and delivered via the use of other malware or computer network attacks/exploitation techniques.
Phishing	A large-scale technique for gathering private information by spamming groups of people with misleading content.
Physical layer of the Internet	The name given to all physical infrastructural elements involved in Internet functionality.

Plaintext	The normal input text of a message involved in cryptographic communications.
Political warfare	The use of any political means to compel changes in the behavior of opponents, most often via the targeting of foreign populations or information systems.
Polymorphism	Code that is added to viruses to allow the program to rewrite itself in ways that might help in avoiding detection.
Port	Functionally, a communications endpoint that is linked to a specific information transmission process or protocol.
Productive power	Power that manifests in the processes that shape all other forms of power.
Propaganda	The use of false or misleading information to manipulate the preferences of foreign leaders and populations.
Protocol	In communications, a set of rules governing how information systems should communicate with one another.
Protocol stack	A group of protocols that allow the seamless transmission of information from computer hardware to another computer via a network connection.
Proxy	Any actor who presents as the primary belligerent but is in actual fact operating on behalf of another.
Psychological operations	Operations that are undertaken to communicate selectively with target populations or foreign leaders for purposes of misleading them.
Public key encryption	Another name for asymmetric encryption
Response actions	Procedures that dictate how actors go about responding to different intrusion scenarios and conflict incidents.
Root server	The set of high-level name servers that answer queries for domain records among country-level domains.
Script kiddy	A term used to denote those who employ malicious code but lack the specialized skills needed to construct it.
Security dilemma	The notion that security actions by one actors will prompt balancing mobilization by other actors, regardless of prior intention toward conflict.
Semi-state actor	Any entity that plays a significant role in governing and controlling the networked world that is not a state, such as an ISP or major cybersecurity vendor.
Server	Any device, program or piece of computer hardware that offers a specific functional service to clients.

Signals intelligence	Intelligence gathered from the communications of enemies and from commercial information transmission infrastructure.
Social engineering	The manipulation of individuals intended to force them to divulge private information related to accessing information systems.
Spearphishing	A more targeted technique for gathering private information than phishing that makes use of specific information on an individual to craft a more believable hook.
Stability-instability paradox	A situation that results from the possession of nuclear weapons where the threshold for nuclear retaliation is well-known, thus allowing for low-intensity conflict to occur with frequency in the absence of major warfighting.
Structural power	Power that manifests in the norms of interaction and the inclinations of actors toward certain courses of action.
Subversion	The undermining or detachment of the loyalties of significant political and social groups within a target state and their transference, under ideal conditions, to the symbols and institutions of the aggressor.
Surface web	The name given to those parts of the Internet fully discoverable by web crawlers and, resultantly, searchable by individuals with only an Internet connected device.
Symmetric encryption	The practice of using a "key," shared between interlocutors, that can encrypt and decrypt communications to ensure secret information transmission.
SYN flood	A common type of denial of service attack wherein attackers take advantage of common handshake protocol procedures to force a target to devote too much memory to anticipated incoming connections.
The cloud	Virtual servers and storage systems that simulate the function of dedicate hardware and software run locally.
The Onion Browser	The most well-known software that allows access to darknet sites via anonymization of data traffic.
Tragedy of the Commons	A situation wherein a number of individuals have access to a shared resource and where rational individual action unfortunately incentivizes predatory behavior with regard to that resource's use.
Transport Control Protocol	An element of the Internet protocol suite that ensures effective transmission of data from applications to Internet communications, and vice versa.

Trojan horse	A type of malware that masquerades as a legitimate program so as to trick users into downloading it.
User layer of the Internet	The name given to all human users of the Internet.
Vicious employees	A term used to describe insider threats.
Virus	A type of malware that replicates itself via reference to some human action and then takes steps to avoid detection so as to deliver a payload held within.
Waterhole attack	A type of attack that occurs from online locations compromised because they are commonly visited.
White hat hacker	A hacker who aids governments and commercial vendors by finding security loopholes in their systems/products.
World Summit on the Information Society	A two-part world summit on information systems in 21st century society meant to address issues of Internet standards, governance and access.
World Wide Web	The name given to all parts of the Internet underwritten by the HyperText Transfer Protocol.
Worm	A type of malware that replicates itself without reference to human actions and can be used to deliver a payload.
Zombie computer	A computer that has been compromised and is controlled most often in aid of DDoS attacks.

Index

Note: Locators in *italic* indicate figures, in **bold** tables and in ***bold-italic*** boxes.